# CARBON COMPOSITES

*Composites with Nanotubes, Nanomaterials, and Graphene Oxide*

# CARBON COMPOSITES

## *Composites with Nanotubes, Nanomaterials, and Graphene Oxide*

*Edited by*

**Eduardo A. Castro, PhD**

**Ann Rose Abraham, PhD**

**A. K. Haghi, PhD**

AAP | APPLE ACADEMIC PRESS

First edition published 2023

**Apple Academic Press Inc.**
1265 Goldenrod Circle, NE,
Palm Bay, FL 32905 USA
760 Laurentian Drive, Unit 19,
Burlington, ON L7N 0A4, CANADA

**CRC Press**
6000 Broken Sound Parkway NW,
Suite 300, Boca Raton, FL 33487-2742 USA
4 Park Square, Milton Park,
Abingdon, Oxon, OX14 4RN UK



**Library and Archives Canada Cataloguing in Publication**

Title: Carbon composites : composites with nanotubes, nanomaterials, and graphene oxide / edited by Eduardo A. Castro, PhD, Ann Rose Abraham, PhD, A.K. Haghi, PhD.
Names: Castro, E. A. (Eduardo Alberto), 1944- editor. | Abraham, Ann Rose, editor. | Haghi, A. K., editor.
Description: First edition. | Includes bibliographical references and index.
Identifiers: Canadiana (print) 20230145132 | Canadiana (ebook) 20230145175 | ISBN 9781774912508 (softcover) | ISBN 9781774912492 (hardcover) | ISBN 9781003331285 (ebook)
Subjects: LCSH: Carbon composites. | LCSH: Graphene. | LCSH: Carbon nanotubes. | LCSH: Nanostructured materials.
Classification: LCC TA418.9.C6 C37 2023 | DDC 620.1/18—dc23

**Library of Congress Cataloging-in-Publication Data**

CIP data on file with US Library of Congress

ISBN: 978-1-77491-249-2 (hbk)
ISBN: 978-1-77491-250-8 (pbk)
ISBN: 978-1-00333-128-5 (ebk)

# About the Editors

**Eduardo A. Castro, PhD**

*Superior Researcher, Argentina National Research Council, Argentina*

Eduardo A. Castro, PhD, is a full professor in theoretical chemistry at the Universidad Nacional de La Plata and a career investigator with the Consejo Nacional de Investigaciones Cientificas y Tecnicas, both based in Buenos Aires, Argentina. He is the author of nearly 1,000 academic papers on theoretical chemistry and other topics and has published several books. He serves on the editorial advisory boards of several chemistry journals and is often an invited speaker at international conferences in South America and elsewhere.

**Ann Rose Abraham, PhD**

*Assistant Professor, Sacred Heart College (Autonomous), Thevara, Kochi, Ernakulam, Kerala, India*

Ann Rose Abraham, PhD, is currently an Assistant Professor at Department of Physics, Sacred Heart College (Autonomous), Thevara, Kochi, Kerala, India. She has expertise in the field of condensed matter physics, nanomagnetism, multiferroics and polymeric nanocomposites, etc. She has research experience at various reputed national institutes, including Bose Institute, Kolkata, India; SAHA Institute of Nuclear Physics, Kolkata, India; UGC-DAE CSR Centre, Kolkata, India; and she has collaborated with various international laboratories. She is the recipient of a Young Researcher Award in the area of physics and Best Paper Awards–2020, 2021. She served as Assistant Professor and Examiner at the Department of Basic Sciences, Amal Jyothi College of Engineering, under APJ Abdul Kalam Technological University, Kerala, India. Dr. Abraham is a frequent speaker at national and international conferences. She has a good number of publications to her credit in many peer-reviewed high impact journals of international repute. She has authored many book chapters and edited more than 10 books with Taylor and Francis, Elsevier, etc. Dr. Abraham received her MSc, MPhil, and PhD degrees in Physics from the School of Pure and Applied Physics, Mahatma Gandhi University, Kerala, India. Her PhD thesis was on the title "Development of Hybrid Mutliferroic Materials for Tailored Applications.

**A. K. Haghi, PhD**

*Professor Emeritus of Engineering Sciences, Former Editor-in-Chief, International Journal of Chemoinformatics and Chemical Engineering Member, Canadian Research and Development Center of Sciences and Culture*

A. K. Haghi, PhD, has published over 250 academic research-oriented books as well as over 1,000 research papers published in various journals and conference proceedings. He has received several grants, consulted for several major corporations, and is a frequent speaker to national and international audiences. He is the Founder and former Editor-in-Chief of the *International Journal of Chemoinformatics and Chemical Engineering*, published by IGI Global (USA), as well as *Polymers Research Journal*, published by Nova Science Publishers (USA). Professor Haghi has acted as an editorial board member of many international journals. He has served as a member of the Canadian Research and Development Center of Sciences and Cultures (CRDCSC) and the Research Chemistry Center, Coimbra, Portugal. Dr. Haghi holds a BSc in urban and environmental engineering from the University of North Carolina (USA), an MSc in mechanical engineering from North Carolina A&T State University (USA), and an MSc in applied mechanics, acoustics and materials from the Université de Technologie de Compiègne (France), and a PhD in engineering sciences from Université de Franche-Comté (France).

# Contents

# Contributors

**H. Akhina**
International and Inter-University Center for Nanoscience and Nanotechnology, Mahatma Gandhi University, Kottayam, Kerala – 686560, India

**T. R. Anilkumar**
Inter-University Center for Evolutionary and Integrative Biology, University of Kerala, Karyavattom, Trivandrum, Kerala, India, E-mail: anilkumartr97@gmail.com

**Priya Banerjee**
Department of Environmental Studies, CDOE Rabindra Bharati University, Kolkata – 700091, West Bengal, India, E-mail: prya_bnrje@yahoo.com

**Sandipan Bhattacharya**
Department of Chemical Engineering, Jadavpur University, Kolkata – 700032, West Bengal, India

**P. Biji**
Nano-Sensor Laboratory, PSG Institute of Advanced Studies, Coimbatore, Tamil Nadu, India

**Tanmoy Chakraborty**
Department of Chemistry and Biochemistry, School of Basic Sciences and Research, Sharda University, Greater Noida – 201310, Uttar Pradesh, India,
E-mails: tanmoychem@gmail.com; tanmoy.chakraborty@sharda.ac.in

**Papita Das**
Department of Chemical Engineering, School of Advanced Studies in Industrial Pollution Control Engineering, Jadavpur University, Kolkata – 700032, West Bengal, India

**Shayeri Das**
Department of Mechatronics Engineering, Manipal University Jaipur, Dehmi Kalan – 303007, Rajasthan, India; Department of Electrical Engineering, Ideal Institute of Engineering, Kalyani, Nadia, West Bengal – 741235, India

**Bony K. John**
School of Chemical Sciences, Mahatma Gandhi University, Priyadarsini Hills P.O., Kottayam – 686560, Kerala, India

**Neenamol John**
School of Chemical Sciences, Mahatma Gandhi University, Priyadarsini Hills P.O., Kottayam – 686560, Kerala, India

**V. I. Kodolov**
BRHEC of Chemical Physics and Mesoscopics, Udmurt Federal Research Center, Ural Branch of the Russian Academy of Sciences, Izhevsk, Russia;
M.T. Kalashnikov Izhevsk State Technical University, Izhevsk, Russia,
E-mail: vikodolov@mail.ru

**V. V. Kodolova-Chukhontseva**
Basic Research – High Educational Center of Chemical Physics and Mesoscopics, Udmurt Federal Research Center, RAS, Izhevsk, Russia; St. Petersburg Institute of Macromolecular Compounds, Russian Academy of Sciences; Peter Great St. Petersburg Polytechnic University, St. Petersburg, Russia, St. Petersburg, Russia , E-mail: vera_kodolova@mail.ru

**A. M. Lipanov**
Keldysh Institute of Applied Mathematics, Russian Academy of Sciences, Moscow, Russia

**Beena Mathew**
School of Chemical Sciences, Mahatma Gandhi University, Priyadarsini Hills P.O., Kottayam – 686560, Kerala, India, E-mail: beenamathew@mgu.ac.in

**Jincy Mathew**
School of Chemical Sciences, Mahatma Gandhi University, Priyadarsini Hills P.O., Kottayam – 686560, Kerala, India

**M. Ya. Mel'nikov**
M.V. Lomonosov Moscow State University, Moscow, Russia

**R. V. Mustakimov**
Basic Research – High Educational Center of Chemical Physics and Mesoscopics, UFRC, RAS, Izhevsk, Russia

**Keerthi G. Nair**
Department of Science and Humanities, Federal Institute of Science and Technology, Angamaly, Ernakulam, Kerala, India; Nano-Sensor Laboratory, PSG Institute of Advanced Studies, Coimbatore, Tamil Nadu, India, E-mail: drkeerthignair@gmail.com

**Vinod Kumar Nathan**
School of Chemical and Biotechnology, Sastra University, Thanjavur, Tami Nadu – 613401, India

**S. Neelima**
School of Pure and Applied Physics, Mahatma Gandhi University, Kottayam, Kerala, India

**Tressia Alias Princy Paulose**
Post-Graduate and Research Department of Chemistry, Bishop Moore College, Mavelikara, Alappuzha, Kerala – 690110, India, E-mail: tressiapaulose@gmail.com

**Aparna Raj**
School of Pure and Applied Physics, Mahatma Gandhi University, Kottayam, Kerala, India

**D. Rithesh Raj**
Department of Electronics and Communications, School of Electrical and Electronics Engineering, Sastra University, Thanjavur, Tami Nadu – 613401, India, E-mail: ritheshraj@ece.sastra.edu

**Prabhat Ranjan**
Department of Mechatronics Engineering, Manipal University Jaipur, Dehmi Kalan – 303007, Rajasthan, India, E-mails: prabhat23887@gmail.com; Prabhat.ranjan@jaipur.manipal.edu

**I. N. Shabanova**
Udmurt Federal Research Center, Ural Branch of the Russian Academy of Sciences, Izhevsk, Russia

**N. S. Terebova**
Udmurt Federal Research Center, Ural Branch of the Russian Academy of Sciences, Izhevsk, Russia

**Riju K. Thomas**
Bharata Mata College, Thrikkakara, Ernakulam, Kerala, India, E-mail: rijukthomas@gmail.com

**L. Vidya**
School of Pure and Applied Physics, Mahatma Gandhi University, Kottayam, Kerala, India

**I. S. Vidyalakshmi**
SB College, Changanacherry, Kottayam, Kerala, India

# Abbreviations

| | |
|---|---|
| $(Br)_n$-SWCNT | brominated SWCNTs |
| 4-NP | 4-nitrophenol |
| AAO | anodic aluminum oxide |
| AC | activated carbon |
| Ach | acetylcholine |
| AcPO | acetophenone |
| AFP | alpha-fetoprotein |
| AG-Fe-Ce | alginate-GO-iron-cerium |
| AIBN | 2,2'-azobisisobutyronitrile |
| AmβCD | mono-(6-amino-6-deoxy)-β-cyclodextrin |
| AN | acid number |
| ANN | artificial neural network |
| ANOVA | analysis of variance |
| AOP | advanced oxidation procedure |
| APPh | ammonium polyphosphate |
| AQ | anthraquinone |
| AR | anthrone |
| BBD | butanediamine |
| B-CNTs | B-doped CNTs |
| BPA | bisphenol A |
| CBZ | carbamazepine |
| CCD | central composite design |
| CDGO | β-cyclodextrin GO |
| CdS DETA | CdS-diethylenetriamine |
| CDs | carbon dots |
| CEA | carcinoembryonic antigen |
| CeO2NSs | ceria mesoporous nanospheres |
| CIP | ciprofloxacin |
| CK-MB | creatine kinase MB |
| cMWCNTs | carboxylated MWC-NTs |
| CNFs | carbon nanofibers |
| CNTs | carbon nanotubes |
| COD | chemical oxygen demand |

| | |
|---|---|
| cSWCNTs | carboxylated SWCNTs |
| cTnI | cardiac troponin I |
| Cu | copper |
| CuO | copper oxide |
| CV | cyclic voltammetry |
| CVD | chemical vapor deposition |
| CWAO | catalytic wet air oxidation |
| CyH | cyclohexane |
| DCF | diclofenac |
| DFT | density functional theory |
| DHP | dihexadecyl phosphate |
| DI | deionized |
| DMAP | 4-dimethylaminopyridine |
| DMF | N,N-dimethylformamide |
| DMP | 2,4-dimethyl phenol |
| DNP | 2,4-dinitrophenol |
| DRG | dorsal root ganglia |
| dsDNA | double-stranded DNA |
| DWCNTs | double-walled carbon nanotubes |
| EB | ethyl benzene |
| ECs | electrochemical capacitors |
| EDC | 1-ethyl-3-(3-dimethylaminopropyl) carbodiimide |
| EDLCs | electrochemical double-layer capacitors |
| EED | ethylenediamine |
| EFC | enzymatic biofuel cell |
| EIS | electrochemical impedance spectroscopy |
| EOC | emerging organic contaminants |
| EP | emerging pollutants |
| EPD | electrophoretic deposition |
| EPR | electron paramagnetic resonance |
| FA | folic acid |
| FAK | focal adhesion kinase |
| Fe | iron |
| FeAlPO | FeIII-substituted aluminum phosphate |
| fGO | functionalized GO |
| FPL | flupentixol |
| FT | Fischer-Tropsh |
| FTIR | Fourier transform infra-red |

| | |
|---|---|
| Ga | gallium |
| Gd | gadolinium |
| GHGs | greenhouse gases |
| GO | graphene oxide |
| GQDs | graphene quantum dots |
| $H_2$ | hydrogen |
| $H_2O$ | water |
| $H_2O_2$ | hydrogen peroxide |
| $H_2S$ | hydrogen sulfide |
| HAS | higher alcohol synthesis |
| HCC | hepatocellular carcinoma |
| HD | hydration–dehydration |
| HMF | hydroxymethyl furfural |
| HMImBr | 1-hydroxyethyl-3-methyl imidazolium bromide |
| hPAN | hydrolyzed polyacrylonitrile |
| HRTEM | high-resolution transmission electron microscope |
| ILs | ionic liquids |
| ITO | indium-tin oxide |
| LiCL | lithium chloride |
| LIG | laser-induced graphene |
| Mb | myoglobin |
| MCNT-$TiO_2$ | magnetic carbon nanotube-TiO2 |
| Met/C NC | metal/carbon nanocomposites |
| MFC | microbial fuel cell |
| miRNAs | microRNAs |
| MO | methyl orange |
| MPc | metal phthalocyanine |
| MPor | metalloporphyrin |
| MSP | mesoporous silica particles |
| MWCNTs | multi-walled carbon nanotubes |
| MWCO | molecular weight cut-offs |
| NAD | non-absorbent |
| NaOH | sodium hydroxide |
| NC | nanocomposite |
| NCNTs | nitrogen containing CNTs |
| NHPI | N-hydroxyphthalimide |
| NHS | N-hydroxysuccinimide |
| NIR-II | near-infrared II |

| | |
|---|---|
| NSAID | non-steroidal anti-inflammatory drug |
| N-$TiO_2$ | nitrogen-doped TiO2 |
| nZVI | nano scale zero valent iron |
| $O_2$ | oxygen |
| $O_3$ | ozone |
| ODH | oxidative dehydrogenation |
| OPD | o-phenylenediamine |
| ox-GNRs | oxidized graphene nanoribbons |
| P3HT | poly(3-hexylthiophene) |
| P3OT | poly(3-octylthiophene) |
| PAH | poly(allylamine)hydrochloride |
| PAN | polyacrylonitrile |
| PANI/G/CNTs | polyaniline/graphene/carbon nanotubes |
| PANI-SWCNT | polyaniline/single-walled carbon nanotube |
| PB | Prussian blue |
| PCBM | [6,6]-phenyl-C61-butyric acid methyl ester |
| Pd | palladium |
| PDA | polydopamine |
| PDDA | poly(dimethyldiallylammonium chloride) |
| PdHx | palladium hydride |
| PEI | polyethyleneimine |
| PEPA | polyethylene polyamine |
| PES | polyethersulfone |
| PLCL | poly(L-lactic acid-co-caprolactone) |
| PLLA | poly(L-lactic acid) |
| PPCP | pharmaceuticals and personal care products |
| PPD | p-phenylenediamine |
| ppm | parts per million |
| PQ | phenanthraquinone |
| PSA | prostate-specific antigen |
| PSF | polysulfone |
| PSO | particle swarm optimization |
| Pt | platinum |
| PVA | polyvinyl alcohol |
| PVP | polyvinylpyrrolidone |
| PWP | pure water permeability |
| RB | rose bengal |
| RBM | radial breathing mode |

| | |
|---|---|
| RCT | randomized controlled trials |
| RNA | ribonucleic acid |
| RSM | response surface methodology |
| RT | room temperature |
| S | sulfur |
| SGM | scanning gate microscopy |
| SiO2 | silica |
| SOFs | side glowing optical fibers |
| SRI | serotonin uptake inhibitor |
| STP | sewage treatment plant |
| SWCNTs | single-walled carbon nanotubes |
| TAM | tamoxifen |
| TC | tetracycline |
| TEM | transmission electron microscopy |
| TEVGs | tissue engineering vascular grafts |
| THF | tetrahydrofuran |
| $TiO_2$ | titanium dioxide |
| TMDs | transition dichalcogenides |
| TNF-$\alpha$ | tumor necrosis factor-alpha |
| TOC | total organic carbon |
| TOF | turnover frequency |
| TON | turnover number |
| TPO | temperature-programed oxidation |
| WAO | wet air oxidation |
| WGA | wheat-germ agglutinin |
| WHO | World Health Organization |
| WWTPs | wastewater treatment plants |
| XPS | X-ray photoelectron spectroscopy |
| XRD | X-ray diffraction |
| ZnO | zinc oxide |

# Preface

Carbon composites have a unique place in nanotechnology and nanoscience owing to their exceptional properties. They find diverse applications in areas such as super strong composite materials, smart sensors, energy storage and conversion, and supercapacitors. Unlike other available books on composites, this new title focuses on materials rather than mechanics. This, in fact, will reflect an effective role in the R&D of composite materials.

This volume is a result of the research work of international experts in the applied research and development on carbon composite and its derivatives. The book also reviews and presents modern experimental methods to obtain chemically modified carbon composites. It also reviews the recent methods on modifications of carbon-based composite materials along with their potential applications in the real world and clearly interprets the interesting physical effects obtained for the first time for carbon materials under consideration.

This book covers new methodologies and modern strategies adopted in the carbon materials research area, including:

- synthesis, characterization, and functionalization of carbon nanotubes (CNTs) and graphene.
- surface modification of graphene.
- carbon-based nanostructured materials.
- applications of carbon nanomaterials for energy sectors.
- detailed studies on reinforced carbon nanotube composites with metallic matrix and non-metallic matrix.
- detailed studies on current state-of-the-art carbon nanotubes composited, which are reinforced with metal matrix, and those composites that are reinforced using non-metallic technology from their synthesis to their myriad potential end-use applications.
- methods for quantification and improved control of carbon nanotube distributions.
- recent research and design trends for carbon nanomaterials-based sensors for a variety of applications.

- the application of nanostructured carbon materials as adsorbents for water purification.
- research progress reported in environmental monitoring and healthcare.

The book is useful for postgraduate students and senior scientists, and engineers in industrial sectors. It also appeals to research students involved in materials science and nanotechnology.

## CHAPTER 1

# Carbon Nanotubes as Metal-Free Catalysts

TRESSIA ALIAS PRINCY PAULOSE[1*] and H. AKHINA[2]

*[1]Post-Graduate and Research Department of Chemistry, Bishop Moore College, Mavelikara, Alappuzha, Kerala – 690110, India*

*[2]International and Inter-University Center for Nanoscience and Nanotechnology, Mahatma Gandhi University, Kottayam, Kerala – 686560, India*

*[*]Corresponding author. E-mail: tressiapaulose@gmail.com*

## ABSTRACT

Most industrial catalytic processes employ metals, including precious metals or metal oxides, as catalysts. Many of these catalytic processes are energy-consuming, and the catalysts are mostly expensive, toxic, and neither very selective nor recyclable. Metal-free catalysis has been a hot area of research for green, environment-friendly chemistry, a new mantra for a sustainable future. Based on literature reports, carbon nanotubes (CNTs) have shown high potential as metal-free, recyclable, and economic catalysts with good activity in various catalytic reactions at relatively milder reaction conditions contributing to a greener environment. Their extraordinary properties, such as large specific surface area, chemical inertness, excellent thermal/oxidative stability, etc., have contributed to their wide application in catalysis. The functionalization of CNTs further enhances their catalytic activity. In this chapter, the contributions of CNTs as metal-free heterogeneous catalysts are addressed. Preparation and

Carbon Composites: Composites with Nanotubes, Nanomaterials, and Graphene Oxide.
Eduardo A. Castro, Ann Rose Abraham, & A. K. Haghi (Eds.)

morphological studies of functionalized CNTs are also discussed in recent studies.

## 1.1 INTRODUCTION

Ever since Iijima [1, 2] discovered carbon nanotubes (CNTs) in 1991, their synthesis, characterization, and application have been a hot research topic in many laboratories across the world. Due to their unique structural, mechanical, and electronic properties, CNTs have a wide range of potential applications in medicine, electronics, optic devices, catalysis, energy storage materials, gas storage materials, composites, sensors, etc., making them an important material in science and technology. CNTs can be divided essentially into single-wall carbon nanotubes (SWCNTs) and multi-wall carbon nanotubes (MWCNTs). Ideally, SWCNTs are made of graphene sheets rolled up into cylinders, thus providing a non-planar sp. [2] hybridized nanostructure, which may be closed at the ends with semi-fullerene caps. MWCNTs can be considered as concentric SWCNTs with increasing diameter coaxially disposed. The number of walls present can vary from two (double-walled carbon nanotubes (DWCNTs)) to several tens, so that the external diameter can reach 100 nm. The concentric walls are regularly spaced by 0.34 nm. CNTs exhibit conducting/semi-conducting properties, depending on the number of layers and on the helicity introduced by the rolling up of each graphitic sheet.

CNTs have been widely used in catalysis, both as catalytic supports and catalysts. In many catalytic reactions, they have proven to be better catalytic supports than conventional supports (such as active carbon, graphite, and $Al_2O_3$). The applications of CNTs as such, as heterogeneous catalysts, have also attracted much interest due to their unique porous structure with nano diameters, lengths ranging from nm to mm, rich surface chemistry, and interesting chemical properties. The high surface area of CNTs facilitates the adsorption of reactants/products, and the specific mesoporous structure decreases the mass-transfer limitations of the reactants/ products. Surface functionalization further enhances their catalytic potential by creating surface centers that attract reactant molecules to the surface of the CNTs. Depending on the size and geometry of the reactant molecules, the reaction can take place inside or on the CNTs. When the reactants are confined inside the nanodimensions of the tube, reactions such as dimerization, coupling, etc., are favored. CNTs have good thermal and mechanical stability and are resistant to acidic/basic mediums, which makes them suitable for severe reaction conditions. CNTs have been used as such as metal-free heterogeneous catalysts in literature for several chemical transformations, some of which are discussed in further sections.

## 1.2 PREPARATION AND MORPHOLOGICAL STUDIES OF CARBON NANOTUBES (CNTS)

CNTs are fascinating nanostructured materials possessing applicability in various fields owing to their unique properties. Catalysis by CNT-based materials has attained great popularity, and continues to be a very productive research field. Catalytic reactions are sensitive to the catalyst's structure, morphology, and size of the catalyst. CNTs are chemically stable in most media and have a high surface area. Apart from that, CNTs can be chemically modified and functionalized to improve their catalytic potentials. Thus, one can tailor the properties of the CNTs according to the requirement. Besides arc discharge, a number of laboratories and industrial synthetic methods, including laser ablation, chemical vapor deposition (CVD), flame synthesis, pyrolysis, electrolysis, electron or ion beam irradiation, template methods, and solar approaches, are employed to manufacture CNTs. The CNTs thus produced has a native hydrophobic nature. Since the surface chemistry of CNTs influences their performance as catalysts, certain types of modifications are sometimes needed to make it hydrophilic and, thus, the desired catalyst. Both the microstructures and macrostructures of CNTs can be well-designed by the careful manipulation of some of the parameters during its processing. In this section, various methods of synthesis, modification, and morphological studies of CNT-based catalysts are discussed.

Samples of commercial MWCNTs and CNTs functionalized by oxidation both in the liquid and gas phases have proven to be efficient catalysts. Lapkin et al. [3] used MWCNTs as such as pristine catalysts without functionalizing the CNTs for the ozonation of atrazine in aqueous solutions. Kang et al. used MWCNTs directly as catalysts for the hydroxylation of aromatic hydrocarbons at low temperatures (50–70°C) and obtained high selectivity without the assistance of any solvent or additive [4]. Pereira et al. [5] studied the influence of ball-milling on the texture and surface chemistry of MWCNTs. Under varied milling conditions, commercial CNTs were ball milled in Retsch MM200 equipment. Periods of up to 360 min at constant vibration frequency (15 vibrations/s) and vibration frequencies from 10 to 20 vibrations/s for 30 min were used for the preparation of the ball-milled samples. Figure 1.1 shows the TEM images of the pristine as well as the ball-milled MWCNTs. Ball-milled samples show evidence of CNT damage due to the ball-milling treatment, which gradually leads to shortened CNTs as the ball-milling time increases. Ball-milled CNTs proved to be an effective ozonation catalyst. Ball milling is an effective and simple method to increase the surface area of commercial CNTs without significant changes in their structural properties.

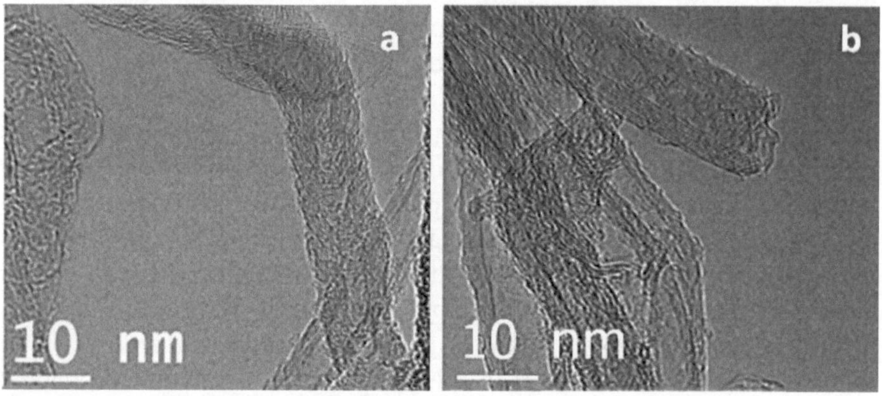


**FIGURE 1.1** TEM images of (a) pristine; and (b) ball-milled MWCNT [5].

*Source*: Reprinted with permission from Ref. [5]. © 2015 Elsevier.

A commercial MWCNTs sample (Nanocyl 3100, purity 95%) having an average diameter of 9.5 nm and an average length of 1.5 μm was used as the starting material (sample MWCNT-Orig.). A set of modified MWNCT catalysts with different levels of acidity and basicity were prepared, details of which are given in Table 1.1, and used as an ozonation catalyst [6, 7]. From the study, it was concluded that oxidation with nitric acid produced highly acidic materials with a large amount of oxygen-containing surface groups and the subsequent thermal treatments selectively removed these groups introduced by nitric acid oxidation. The acidic character of the samples was found to decrease with an increase in thermal treatment temperature. Oxidation in the gas phase introduced mainly basic, neutral, and weakly acidic groups on the surface of MWCNTs.

**TABLE 1.1** Various MWCNT-based Catalysts [6]

| Sample | Starting Material | Treatment |
|---|---|---|
| MWCNT-Orig | Nanocyl 3100 | None |
| MWCNT-$HNO_3$ | MWCNT-Orig | Oxidation with $HNO_3$ 7 M at 130°C for 3 h |
| MWCNT-$HNO_3$_$N_2$_400 | MWCNT-$HNO_3$ | Thermal treatment under $N_2$ flow at 400°C for 1 h |
| MWCNT-$HNO_3$_$N_2$_600 | MWCNT-$HNO_3$ | Thermal treatment under $N_2$ flow at 600°C for 1 h |
| MWCNT-$HNO_3$_$N_2$_900 | MWCNT-$HNO_3$ | Thermal treatment under $N_2$ flow at 900°C for 1 h |
| MWCNT-$O_2$ | MWCNT-Orig | Oxidation with 5% $O_2$ in $N_2$ flow at 500°C for 3 h |
| MWCNT-$H_2O_2$ | MWCNT-Orig | Oxidation with $H_2O_2$ (30%) for 20 h at room temperature |

*Source*: Reprinted with permission from Ref. [5]. © 2015 Elsevier.

Yang et al. [8] used MWCNTs as a catalyst in the wet air oxidation (WAO) of phenol. They prepared MWNCTs by CVD using a $Fe/Al_2O_3$ catalyst. The inside diameter of the MWCNTs was 3–10 nm, the outside diameter was 6–20 nm, and the length/diameter ratio was in the range of 100–1,000. Raw MWCNTs contained some amorphous carbon and some residual catalyst particles with purity higher than 99.5%. Raw MWCNTs were immersed in a 37% HCl solution and sonicated for 20 min, and then the mixture was continued overnight to remove metal catalyst particles. MWCNTs that had been treated with HCl were split into two groups. One part was washed several times with deionized water, dried overnight in the air at 80°C, and then crushed, leading to a powder (A-MWCNTs). The other part was immersed and dispersed in a 67% $HNO_3$ – 98% $H_2SO_4$ (1:3, volume) solution and sonicated for 20 min. The suspension was further refluxed at 50°C for 4, 12, and 16 h. This acid treatment of MWCNTs the surface functionalization of MWCNTs by introducing carboxylic acid and quinone groups. The materials were washed several times with deionized water, dried at 80°C overnight, and crushed (B-MWCNTs/4 h, B-MWCNTs/12 h, and B-MWCNTs/16 h). BET, SEM, TEM, FT-IR, and Raman spectroscopy were used to evaluate the surface areas, morphologies, and surface functional groups of the MWCNTs. Croston et al. [9] employed functionalized MWCNTs as catalysts in the oxidation of aniline. The MWCNTs are functionalized by C=O groups on their surface via refluxing with $HNO_3$ followed by filtration and washed with running distilled water for 2 h. The obtained sample was refluxed for 6 hours in

50 mL of distilled water. The CNTs were filtered out and dried for 24 hours at 100°C.

Figueiredo et al. applied different chemical and thermal treatments to MWCNTs in order to produce materials with different textural and chemical properties, as shown in Figure 1.2, for application as catalysts in the CWAO of oxalic acid [10]. To improve the yield of nitrogen-containing groups on the surface, such as amide and amine groups, original MWCNTs (CNT-O) were first treated with nitric acid at boiling temperature for 3 hours. The oxidized material was rinsed with distilled water multiple times until it attained a neutral pH. The recovered solid was dried at 110°C for 24 h in an oven (sample CNT-N). A treatment with urea solution (1M) was carried out in a stainless-steel high-pressure batch reactor at 200°C for 2 h under a nitrogen atmosphere. The material was washed several times and dried in the oven (sample CNT-NU). Then, a gas-phase thermal treatment was applied to this sample by heating at 10°C $min^{-1}$ until 600°C under a $N_2$ flow and kept at 600°C for 60 min (sample CNT-NUT). The original (CNT-O) and modified MWCNTs (CNT-N, CNT-NU, and CNT-NUT) were investigated as catalysts in the CWAO process, using oxalic acid as a model compound at 140°C and 40 bar of total pressure. Treatment with nitric acid increased the amount of oxygen-containing surface groups (carboxylic acids, phenols, and some anhydrides, lactones, and carbonyl-quinones) on the MWCNTs. A slight increase of the H and N contents is also observed in CNT-N, possibly due to the incorporation of acid and nitrogen groups during the treatment. The urea-treated material (CNT-NU) exhibited an additional increase in the nitrogen content and a decrease in the oxygen content, suggesting the loss of some carboxylic acid groups by the incorporation of nitrogen functional groups on the carbon material.

Sulfonated MWCNTs (MWCNTs–$SO_3H$) has been identified as a useful solid acid catalyst in organic transformations. Typical synthetic routes of sulphonate-functionalized CNTs include acid-assisted thermal decomposition, electrochemical modification, and chemical reduction methods [11, 12]. Wang et al. [13] synthesized sulfate functionalized MWCNT by using chlorosulphonic acid as a sulfonating agent instead of sulfuric acid. In a typical process, MWCNTs were immersed in 37% HCl and sonicated for 20 minutes, and then the mixture was kept overnight to remove residual metal catalyst particles. Then the acid-treated CNTs were immersed and dispersed in a 67% $HNO_3$ + 98% $H_2SO_4$ solution (1: 3 by volume), sonicated and refluxed at 60°C for 2 h and then filtrated and

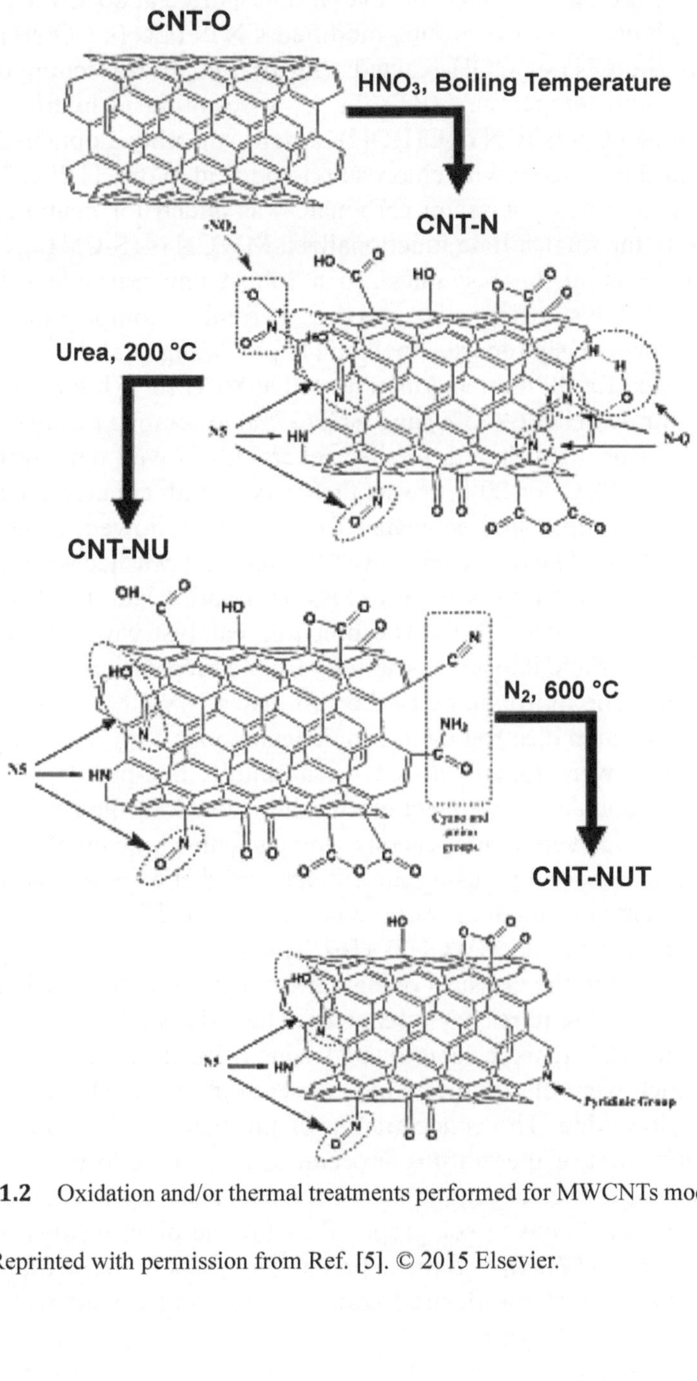


**FIGURE 1.2** Oxidation and/or thermal treatments performed for MWCNTs modification [10].

*Source*: Reprinted with permission from Ref. [5]. © 2015 Elsevier.

washed with deionized water for several times, dried at 80°C for 12 h, and crushed, giving carboxyl groups modified CNTs (CNTs-COOH). In the next step, the CNTs-COOH reacted with $LiAlH_4$ in dry tetrahydrofuran (THF) at room temperature (RT) for 12 h, resulting in hydroxyl group functionalized CNTs (CNTs-$CH_2OH$). Finally, the above obtained CNTs-$CH_2OH$ further reacted with chlorosulphonic acid in dry THF at 70°C for 6 h, and anhydrous potassium carbonate was added for neutralizing the HCl, giving the final sulfate functionalized MWCNTs (S-CNTs).

Alamdari et al. [14] synthesized a robust and reusable sulfonated MWCNTs catalyst using acid assisted thermal decomposition method based on a previously reported method [15] by Wang et al. MWCNTs was first sonicated for 30 min, and then heated at 80°C for 4 h in a mixture of 1:1 concentrated $HNO_3$ (65%) and HCl (37%) to remove residual catalyst particles. It was then filtered, washed several times with deionized water, and dried at 120°C for 20 h. It was then mixed with concentrated $H_2SO_4$ (98%) in a dry flask and sonicated for 30 min, followed by heating at 300°C for 20 h. The suspension was filtered and washed several times with deionized water to remove excess acid and dried at 120°C for 24 h to obtain sulfonated CNTs. The prepared catalyst was reused several times without efficient loss of its activity for the preparation of bisphenolic antioxidants. The morphology of the sulfonated MWCNTs was analyzed using transmission electron microscopic techniques.

MWCNTs were functionalized via a simple, one-pot deprotonation-carbometallation and subsequent electrophilic attack of the bromotriethyl-amine to produce very homogeneous samples with a high number of easily accessible amino groups using an efficient synthetic route based on the direct covalent grafting of the desired basic functional groups onto existing structural defects of the MWCNTs [16]. This procedure is faster and more sophisticated than the classical oxidation–amination route, which involves several steps and requires harsh reagents. The C-H bonds located near the defects in MWCNTs were deprotonated using n-BuLi and replaced by C-Li bonds, which were then subjected to an electrophilic attack using diethyl-aminoethylbromide. The concentration of the functional groups, without any optimization of the grafting procedure, was found to be 1 $mmolg^{-1}$, which is close to the concentration of Bronsted acid sites in zeolites.

Alkaline CNTs have been prepared by the incipient wetness impregnation method [17]. In a typical preparation, CNTs were dipped into $Na_2CO_3$ solution of the desired concentration and mixed well for ca.

30 min. The mixture was placed in the air at ambient temperature for 24 h, and then dried at 120°C for 12 h to obtain the alkaline CNTs. The alkaline CNTs were used for the direct oxidation of $H_2S$ into sulfur at 30°C.

Park et al. [18] developed a variety of MWCNTs grafted with imidazolium-based ionic liquids (CNT-ILs) and utilized them as highly efficient heterogeneous catalysts for the production of cyclic carbonates via cycloaddition reactions of epoxides and $CO_2$. CNT-ILs showed dramatically increased catalytic reactivity towards cycloaddition processes as compared to traditional heterogeneous catalysts that use porous silica and polymer supports. Pristine MWCNTs were oxidized to introduce COOH groups on their surfaces prior to the grafting of the IL on the MWCNTs. The oxidized MWCNTs were dispersed in anhydrous Dimethyl formamide (DMF). 4-Dimethylaminopyridine (DMAP), 1-(3-dimethylaminopropyl)-3-ethylcarbodiimide (EDC), and 1-hydroxyethyl-3-methyl imidazolium bromide (HMImBr) were added to the obtained CNT sol. The esterification reaction was performed at 60C for 24 h. After that, the mixture was filtered and rinsed in deionized water. The resultant HMImBr-grafted MWCNTs (CNT-HMImBr) were collected and dried under vacuum at 60°C for 12 h. A similar procedure was used for the synthesis of a number of CNT-IL heterogeneous catalysts. The schematic for the synthesis of MWCNTs grafted with imidazolium-based ILs is shown in Scheme 1.1. Figure 1.3 represents the TEM image of the CNT-HMImBr heterogeneous catalyst.

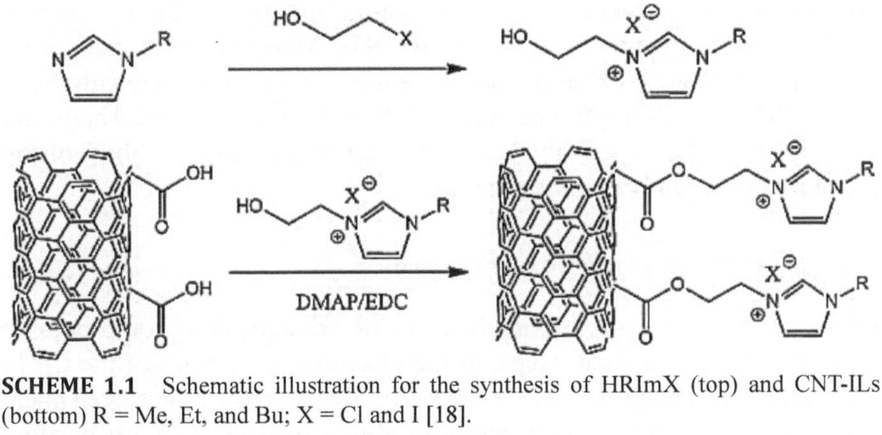


**SCHEME 1.1** Schematic illustration for the synthesis of HRImX (top) and CNT-ILs (bottom) R = Me, Et, and Bu; X = Cl and I [18].

*Source*: Reprinted with permission from Ref. [18]. © 2012 Elsevier.

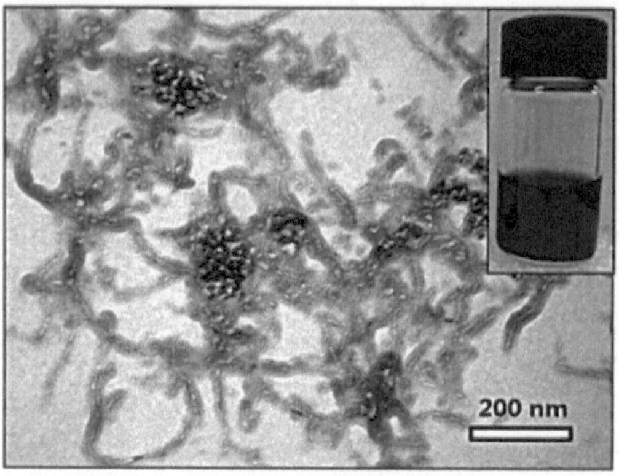


**FIGURE 1.3** TEM image of CNT-HMImBr heterogeneous catalyst. Inset is the typical photograph of CNT-HMImBr sol in AGE [18].

*Source*: Reprinted with permission from Ref. [18]. © 2012 Elsevier.

## 1.3 CARBON NANOTUBES (CNTS) AS CATALYSTS

Although the major use of CNTs in catalysis is as support, they have widely been employed as catalysts in various chemical transformations. Due to their chemical/thermal stability, electronic conductivity, and capability to functionalize with chemically or electrochemically active species, CNTs are promising materials for application in catalysis. Although in most catalytic applications, CNTs are modified by introducing surface functional groups via different oxidative/thermal treatments, in some catalytic studies, CNTs have been used directly (pristine) as catalysts. It is worth noting that residual metallic particles coming from the production process can be found in the inner cavity of MWCNTs. Metal catalysts such as Fe, Ni, and Co that are used in the production of CNTs can remain entrapped in the CNTs, which might interfere with the catalytic reaction. Therefore, a purification step is usually included to eliminate the possibility of the metal particles catalyzing the reactions.

### *1.3.1 PRISTINE CNTS AS CATALYSTS*

Pristine CNTs have been considered to be advantageous for processes requiring electron transfer steps, as the electronic properties of the CNTs are preserved and can be exploited entirely. Hence, efforts have been made towards carrying out catalytic reactions with as-produced or just purified

CNTs (p-CNTs) as catalysts, although the number of reports with pristine CNTs is much smaller when compared to modified CNTs. One of the most critical problems associated with the use of p-CNTs is their high level of aggregation through extended π-π stacking, which makes liquid-phase manipulation difficult. Moreover, for many reactions, the introduction of organic functional groups on the CNTs plays a more direct role in the catalysis, often behaving as the active sites. Despite this, p-CNTs have been used to catalyze several reactions.

#### *1.3.1.1 OZONATION*

Ozonation is an important environmental remediation process in the treatment of water contaminated with organic pollutants, pharmaceutical residues, and pesticides. Due to the relatively low solubility and stability of ozone in water, heterogeneous catalytic ozonation has emerged as a powerful and greener technology for the removal of several refractory pollutants from drinking and wastewater by the transformation of ozone into more reactive species and/or adsorption and reaction of the pollutants on the surface of the catalyst. Activated carbon (AC) has received considerable attention in this regard. However, the performance of AC in ozonation is affected by operating conditions such as catalyst dosage, ozone gas concentration, and pH value. Because it affects the decomposition of ozone, the surface characteristics of the catalyst, and the dissociation of organic pollutants in an aqueous solution, the pH value has a considerable impact on the removal of organic pollutants in catalytic ozonation.

Oxalic acid is a common oxidation product in organic product degradation. Commercial MWCNTs have been used directly, without any purification or functionalization, as a catalyst for the ozonation of oxalic acid in an aqueous solution [19]. The degradation efficiency of oxalic acid increased with the MWCNTs catalyst dosage (50–200 $mgL^{-1}$) and reaction temperature (283–313 K). The degradation efficiency of oxalic acid reached the maximum value at an initial oxalic acid concentration of 1.0 mM, and the degradation efficiency decreased with either increase or a decrease of this initial concentration. The presence of ter-butanol reduced MWCNT activity on oxalic acid removal, and the inhibition of ter-butanol on oxalic acid removal was more significant at initial pH 6.1 than at initial pH 3.0, implying that the contribution of bulk reactions to oxalic acid removal was boosted at a high initial pH value. The oxalic acid removal increased with the initial pH value in the range of 1.0–3.0, but decreased with further increases of the initial pH value in the range of 3.0–6.1.

Short and open-ended MWCNTs obtained via ball-milling of MWCNTs have been used in the catalytic ozonation of oxalic acid [5]. Ball-milling can convert MWCNTs to curved nanotubes, nanoparticles, short and open-tipped nanotubes, and also into amorphous as well as disrupted tubular structures when prolonged ball-milling is applied. The ball-milling time and intensity imparted significant textural and morphological properties to the MWCNTs, such as increased surface area and decreased particle size of agglomerates, without any functionalization of the surface. These modifications were found to have a significant influence on the catalytic ozonation efficiency, which is favored by MWCNTs with high specific surface areas and smaller lengths.

#### *1.3.1.2 HYDROXYLATION*

MWCNTs have been successfully used as a catalyst for the hydroxylation of aromatic hydrocarbons such as benzene, halobenzenes, toluene, and nitrobenzene to phenol, halophenols, cresols, and nitrophenols respectively [4]. The catalytic hydroxylation reactions carried out in a simple one-pot process using $H_2O_2$ as the oxidant, at low temperatures (50–70°C) exhibited a high selectivity of 98%, without the use of any solvent or additive. MWCNTs proved to be highly active, highly selective, and recyclable heterogeneous catalysts in the hydroxylation reactions.

#### *1.3.1.3 OXIDATIVE DEHYDROGENATION (ODH)*

Commercial MWCNTs have displayed remarkable catalytic activity and stability in the oxidative dehydrogenation (ODH) of 1-butene to butadiene at atmospheric pressure and 400°C in a continuous flow fixed-bed quartz reactor [20]. The catalytic reaction was conducted with an oxygen/1-butene ratio of 1 for 50 h, after which the oxygen/1-butene ratio was increased to 2 and maintained for the next 20 h. The performance of CNTs, along with the reaction time, is displayed in Figure 1.4. The by-products are mainly CO and $CO_2$, and only a trace amount of propene was detected. After a 20-h induction period, CNTs stably catalyzed the ODH of 1-butene. The conversion of 1-butene and the butadiene yield stayed at around 45% and

29%, respectively. A decrease in $CO_2$ selectivity during the induction period was related to the activation of CNTs. After the oxygen/1-butene ratio was increased from 1 to 2, a higher conversion of 72% and yield of 43% was obtained, while selectivity remained almost constant at about 60%. The constant selectivity implies that an excess of oxygen on the surface of CNT has no effect on the selective sites.

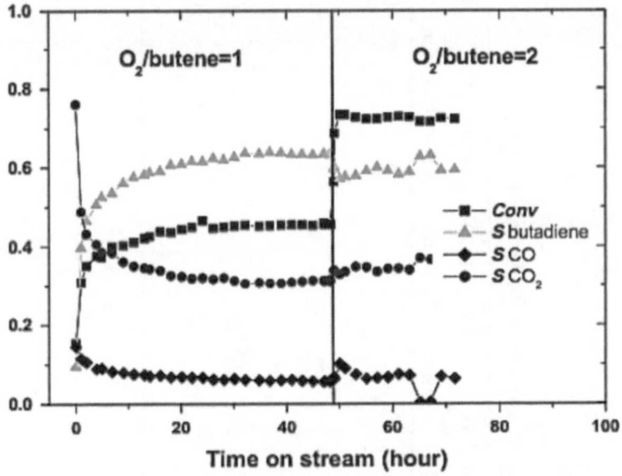


**FIGURE 1.4** Performance of CNTs in the ODH of 1-butene with different oxygen/butane molar ratios as a function of reaction time (■ butene conversion, ▲ selectivity to butadiene, ♦ selectivity to CO, ● selectivity to $CO_2$) [20].

*Source*: Reprinted with permission from Ref. [20]. © 2008 Elsevier.

The catalytic performance of the CNTs, as shown in Table 1.2, was found to be superior to that of AC and iron oxide. The butadiene formation rate obtained with CNTs is 10 times more than that with the AC, and the catalytic performance of CNT is also superior to that of iron oxide. The significant decrease in the BET area of the AC after the reaction was attributed to the carbon deposition and blockage of micropores, and the slight increase in the BET area of CNT was attributed to the combustion of amorphous carbon. TPO (temperature-programmed oxidation), XPS, and IR studies of the catalysts before and after the reaction indicated that the CNTs were highly functionalized during the reaction. The increase in the amount of oxygen-containing functional groups during the reaction suggested that quinone/hydroxyl functional groups could be the active sites for the ODH reaction.

**TABLE 1.2** The Catalytic Properties of Three Kinds of Carbon Materials and Iron Oxide Catalyst [20]

| Catalysts | $O_2$/Butene | Concentration (%) | $S_{butadiene}$ (%) | Yield (%) | BET Surface Area ($m^2/g$) | | Butadiene Formation Rate ($\times 10^{-5}$ mol h/$m^2$) |
|---|---|---|---|---|---|---|---|
| | | | | | Before Reaction | After Reaction | |
| CNT | 1 | 45 | 63 | 28 | 42 | 58 | 0.75 |
| | 2 | 72 | 60 | 43 | – | – | 1.11 |
| AC-1 | 1 | 31 | 42 | 13 | 837 | 311 | 0.062 |
| | 2 | 45 | 36 | 16 | – | – | 0.077 |
| AC-2 | 1 | 24 | 46 | 11 | 1,081 | 334 | 0.049 |
| | 2 | 36 | 38 | 14 | – | – | 0.062 |
| α-$Fe_2O_3$ [4] | 1 | 56 | 36 | 20 | – | – | – |

*Source*: Reprinted with permission from Ref. [20]. © 2008 Elsevier.

For comparison, two types of acid-washed ACs were used. One was supplied by the Alfa Aesar Co. (Germany), and the other was supplied by the COMBICAT research center (University of Malaya, Malaysia). CNT, AC-1, and AC-2 are the designations for CNTs and ACs (Alfa Aesar Co. and COMBICAT).

## 1.3.2 MODIFIED CNTS AS CATALYSTS

Modified CNTs have been successfully employed as catalysts in diverse chemical transformations, as discussed below.

### 1.3.2.1 OXIDATION

Wastewater originating from chemical, petrochemical, pharmaceutical, agricultural, and textile production plants frequently contains toxic, hazardous, and highly concentrated organic pollutants, which pose a severe threat to water bodies and human health. Therefore, the development of effective wastewater treatment technologies is crucial. WAO is one of the effective technologies which has been widely employed to oxidize organic compounds to $CO_2$, $H_2O$, and other biodegradable end products under high temperature (125–320°C) and pressure (0.5–20 MPa) using oxygen or air as an oxidant without any toxic emissions ($NO_x$, $SO_2$, dioxins, etc.). However, a very efficient process, severe operating conditions, and high operating costs limit the application of WAO in wastewater treatment. This has led to the development of the catalytic wet air oxidation (CWAO) process, which has been able to reduce the extreme process conditions to some extent (125–220C, 0.5–5 MPa), shorten the reaction time, decrease the operating cost, and enhance the oxidation efficiency. Heterogeneous catalysts appear to be a more promising and greener alternative for wastewater treatment without an additional separation step to remove the catalyst from the final treated effluent. In the last decades, various heterogeneous catalysts, including supported noble metals and transition metal oxides, have demonstrated good catalytic activity in the oxidation of several organic pollutants. Carbon materials that are fundamentally microporous are frequently less efficient for CWAO processes than mesoporous carbons with a large exterior surface area. Organic pollutants are rapidly transported to meso-and macropores, where they degrade at a faster pace than in micropores. As a result, MWCNTs are extremely appealing materials for CWAO, where the solution and, as a result, the dissolved contaminants are predicted to be more accessible to the carbon surface than with ACs, which are typically microporous. Significant mineralization of the pollutants and lower effluent toxicity was observed in several CWAO processes using CNTs as catalysts.

Acid-functionalized MWCNTs exhibited both high catalytic activity and good stability in the WAO of phenol [8]. MWCNTs, purified using 37% HCl solution to remove residual metal catalyst particles, were functionalized by treating with 67% $HNO_3$ + 98% $H_2SO_4$ (1:3 volume), sonicated for 20 min followed by refluxing the suspension at 50°C. The materials were cleaned with deionized water numerous times before being dried at 80°C overnight and pulverized. This acid treatment of MWCNTs increased the surface area and surface functionalization of MWCNTs by introducing carboxylic acid and quinone groups. Increasing the reflux time increased the concentration of the functional groups on the MWCNTs. After 120 minutes of reaction at 160°C and 2.0 MPa oxygen pressure with a starting phenol content of 1,000 mg/L, 100% phenol conversion, and 76% total organic carbon (TOC) abatement were obtained. Although phenol is totally removed after 120 min reaction in the CWAO over the functionalized MWCNTs, total mineralization of phenol was not achieved, indicating that some intermediates, i.e., the short chain carboxylic acids, mainly maleic/fumaric, malonic, oxalic, formic, and acetic acid are formed and not converted to $CO_2$ and $H_2O$ during the reaction. The increased activity of the functionalized MWCNTs was mostly due to surface functional groups (–COOH).

Ozone-treated MWCNTs have also been studied as catalysts in the WAO of phenol, nitrobenzene, and aniline at mild operating conditions (reaction temperatures of 155°C, total pressure of 2.5 MPa) in a batch reactor [21]. Commercial MWCNTs were purified with a 37% HCl solution by sonicating for 30 min at RT to remove any metal impurities that could possibly act as a catalyst in the reaction. The purified, dried, and powdered MWCNTs were further oxidized by ozone at RT. Ozonation improved the surface area of the catalyst and introduced oxygen-containing functionalities such as the carboxylic acid groups. The surface area of the functionalized MWCNTs gradually decreased with increasing ozonation time. In the CWAO of phenol, common contamination in industrial wastewater, over the functionalized MWCNTs, total phenol removal was obtained after 90 min run. The degradation efficiency of phenol increased with increasing the catalyst loading, reaction temperature, and initial phenol concentration.

In the CWAO of nitrobenzene, the presence of phenol improved the nitrobenzene removal, and 40% nitrobenzene was removed after 180 min. In the CWAO of aniline, 49% aniline conversion was obtained after 120 min.

Different chemical and thermal treatments were applied to MWCNTs in order to produce materials with different textural and chemical properties for application as catalysts in the CWAO of oxalic acid [10]. Original MWCNTs (CNT-O) were first treated with nitric acid at boiling temperature for 3 h in order to increase the nitrogen-containing groups on the surface, such as amide and amine groups (sample CNT-N). The nitric acid-treated samples were then treated with urea (sample CNT-NU) and was further given gas-phase thermal treatment (sample CNT-NUT). CNT-O, CNT-N, CNT-NU, and CNT-NUT samples were investigated as catalysts in the CWAO process, using oxalic acid as a model compound at 140°C and 40 bar of total pressure. The original MWCNTs (CNT-O) contained only a negligible amount of oxygen-containing surface groups (carboxylic acids, phenols, and some anhydrides, lactones, and carbonyl-quinones) found on the original material (CNT-O), while the highest amount of these groups was found on after the treatment with nitric acid. A slight increase of the H and N contents is also observed in CNT-N, possibly due to the incorporation of acid and nitrogen groups during the treatment. Nitrogen-containing groups are known to impart more basicity to carbon materials, which is beneficial for liquid-phase oxidation reactions. The urea-treated materials (CNT-NU) showed an additional increase in the nitrogen content and a decrease in the oxygen content, suggesting the loss of some carboxylic acid groups by the incorporation of nitrogen functional groups on the carbon material. The nitric acid treatment acidifies the original material, the urea treatment (CNT-NU) reduces the acidic character of the sample (by removing the carboxylic acid groups and increasing the N content), while CNT-NUT is the most basic sample and also the most active catalyst in CWAO of oxalic acid. In the absence of a catalyst, the oxalic acid was found to be poorly oxidized but was totally degraded in less than 30 min in the presence of MWCNTs. The rate of oxalic acid oxidation is influenced by the chemical characteristics of MWCNTs, with lower apparent initial first-order rate constants for MWCNTs with pronounced acidity. Catalyst recycling studies showed that the textural properties of MWCNTs are stable in cyclic CWAO experiments, but a decrease in their basic character leads to the reduction of their catalytic activity. Therefore, MWCNTs with more basic character are more active catalysts for CWAO than MWCNTs with a pronounced acid character. Their activity strongly depends on the stability of their surface chemistry.

Brominated SWCNTs [$(Br)_n$–SWCNTs] and MWCNTs [$(Br)_n$–MWCNTs] produced by the plasma-chemical technique using bromine vapor at 100 W power input and pressure of 9 Pa have demonstrated high catalytic activity in the liquid-phase aerobic oxidation of hydrocarbons [22]. $(Br)_n$–SWCNT was used as a catalyst in the oxidation of the model compound cumene using 2,2'-azobisisobutyronitrile (AIBN) as initiator at 60 (±0.02)°C and oxygen pressure of 20 kPa (air). $(Br)_n$–SWCNT and $(Br)_n$–MWCNTs were used as catalysts for the oxidation of petroleum naphthenic fraction (220–350°C) derived from commercial Azerbaijan (Baku) oils blend diesel cut at 135–140°C for several hours under continuous air current, passing through the liquid reaction mixture. The extent of oxidative conversion of oil hydrocarbons were measured in terms of the acid number (AN) of the reaction mixture (oxidate) and yields of synthetic (SPA) and oxy-synthetic (OSPA) petroleum acids. The acidic components (SPA + OSPA) were isolated from the oxidate by alkali (NaOH or KOH) treatment, followed by recovering SPA + OSPA with mineral acids. The SPA was separated from OSPA and unsaponifiable by light benzene or petroleum ether processing. The catalytic activity of $(Br)_n$–SWCNT was found to be markedly greater than the activity of the industrial catalysts, manganese salt of indigenous petroleum acids, used for the liquid phase petroleum hydrocarbons oxidation process.

Because its products, KA oil (a mixture of cyclohexanone (K) and cyclohexanol (A)), are starting materials for the synthesis of adipic acid and caprolactam, intermediates of nylon-6 and nylon-66 polymers, selective oxidation of cyclohexane (CyH) is an extremely important process in the modern chemical industry. Nitrogen-, phosphorous-, and boron-doped carbon nanotubes (N-CNTs, P-CNTs, and B-CNTs, prepared by a CVD method using xylene as carbon source and aniline-$NH_3$, triphenylphosphine and triethyl borate as nitrogen, phosphorous, and boron precursors, respectively, have been used as catalysts for the aerobic oxidation of CyH [23]. By tailoring the composition of reactants and reaction atmosphere, N-CNTs with nitrogen contents from 0% to 4.36% and P-CNTs with phosphorous contents from 0.55% to 5.14% were synthesized. N-and P-CNTs were discovered to be active in the liquid phase for the oxidation of CyH using molecular oxygen as an oxidant. Due to the improved electron transport of n-type dopants, it was established that N and P doping effectively boost the activity of CNTs in CyH oxidation.

N-CNTs synthesized from aniline in an $NH_3$ atmosphere had the highest mass-normalized activity of 761 mmol $g^{-1}$ $h^{-1}$, whereas P-CNTs had the highest surface-area-normalized activity of 28 mmol $m^{-2}$ $h^{-1}$. The effect of n- and p-type doping was also examined using B-doped CNTs (B-CNTs). Due to its electron shortage, B-doping has no effect on CNT activity. On the other hand, because structural flaws disrupt electron delocalization in the graphitic honeycomb network, fewer structural defects are required.

Nitrogen, as one of the most commonly used heteroatoms, can be incorporated into CNTs either by feeding nitrogen-containing precursors during the CNTs synthesis or by the post-treatment of CNTs with nitrogen-containing molecules. Nitrogen-doped CNTs, in the one-step aerobic oxidation of CyH to adipic acid, have shown selectivities as high as 60% at a conversion higher than 40% [24]. High activity was obtained over CNTs to yield a conversion of 34.3%, which is remarkably higher than those on AC, supported Au, and FeAlPO ($Fe^{III}$-substituted aluminum phosphate) catalysts. Since heteroatom doping is an effective way to enhance the performance of carbon materials, nitrogen-doped CNTs were synthesized by CVD with aniline as both a carbon and nitrogen source. The growth of N-doped CNTs was carried out in the $NH_3$ atmosphere to increase the N content and the proportion of pyridinic nitrogen, which may be responsible for the activity of N-doped carbon. The doped sample had a better selectivity for adipic acid and had a higher activity. The recyclability of the N-doped CNT catalyst was likewise satisfactory. During five cycling tests, there was almost no difference in both relative activities (defined as the conversion after 8 h reaction normalized by that of the first run) and adipic acid and KA oil selectivity. Compared with CNTs and N-doped, CNTs show similar adipic acid selectivity to FeAlPO, but poorer decarboxylated by-product selectivity. Because no metal is involved, carbon-catalyzed CyH oxidation is low-cost, the catalyst is easy to recover, and the reaction is resistant against pipeline fouling. Owing to their high activity, controllable selectivity, and outstanding recyclability, CNTs are promising catalysts for CyH conversion as well as single-step production of adipic acid.

MWCNTs functionalized with –C=O, via the oxidation with nitric acid following the procedure of Green et al., have been efficiently used as catalysts in the oxidation of aniline [9] and p-toluidine [25] with hydrogen peroxide ($H_2O_2$) using acetonitrile as a solvent to produce azobenzene and 4,4'-dimethylazobenzene as the product respectively. The initial product formed in the catalytic oxidation of aniline using these –C=O functionalized

MWCNTs is the nitrosobenzene which couples with aniline to produce the azobenzene. No such product is obtained in the absence of the functionalized CNTs. Also, unfunctionalized CNTs do not show the catalytic effect. The specificity observed here could be interpreted as due to the surface charge centers (such as C=O) present on the functionalized CNTs, which perhaps permits the entry of the reactants on/inside, the multi-walled surface.

Acetophenone (AcPO) is a commercially important product as it is used as an intermediate for the manufacture of some perfumes, pharmaceuticals, resins, alcohols, esters, and aldehydes. Commercially, AcPO is produced by the liquid-phase cobalt-catalyzed oxidation of ethyl benzene (EB) with molecular oxygen using acetic acid as the solvent and manganese or bromide species as promoters. However, the use of acidic solvents and homogeneous metal catalysts not only makes these processes environmentally unfriendly but also makes it difficult to separate the catalysts from the reaction mixture. CNTs as metal-free catalysts have exhibited excellent activity in the selective oxidation of EB to AcPO in the liquid phase with oxygen as the oxidant [26]. CNTs were subjected to oxidation and annealing treatments to change the surface structure of CNTs. With increasing oxidation, the carboxylic groups on the CNTs became more abundant. The catalytic activity of CNTs in EB oxidation reduced drastically as the surface carboxylic group content increased. However, the thermal treatment decreased the number of carboxylic groups on the catalyst surface, and as a result, the catalytic activity increased. Similar behavior was also observed on AC for the oxidation of EB. The low catalytic activity of AC was attributed to its long-range disordered structure and a high content of carboxylic groups on the surface. Compared with other metal-free catalysts such as N-hydroxyphthalimide (NHPI), CNTs were found to be more favorable for industrial applications because NHPI is a difficult homogeneous catalyst to separate. Moreover, compared with the commercial cobalt-based catalysts, CNTs performed much better than $Co(OAc)_2 \cdot 4H_2O$ (CoAc) under the same reaction conditions, which was attributed to the lower solubility of CoAc in acetonitrile. Although CoAc exhibited higher selectivity than CNTs with acetic acid as the solvent, the Co-based acetic acid homogeneous reaction systems required high operating expenses owing to serious corrosion and difficult separation. In addition, the CNTs revealed outstanding recyclability, and during six cycling tests, no evident difference in both the conversion of EB and selectivity of AcPO could be observed.

Hydrogen sulfide ($H_2S$), toxic and foul-smelling gas with detrimental effects on health and the environment, is mainly used as a precursor to elemental sulfur. The modified Claus process is the most significant desulfurizing process used in the industry, which selectively converts $H_2S$ into elemental sulfur. However, it is an energy-intensive process and requires post-treatment procedures to ensure high rates of desulfurization. A series of CNTs with different diameters made alkaline by impregnation with $Na_2CO_3$ has been used for the direct oxidation of $H_2S$ into sulfur at 30°C in the presence of $O_2$ and humidity [17]. The saturation sulfur capacity of alkaline CNTs was 1.86 g $H_2S$/g catalyst, which was about 3.9 times higher than that of a common commercial $H_2S$ oxidation catalyst based on AC. The large sulfur capacity of alkaline CNTs was attributed to the unique outside void of the CNT aggregates, which provided more space to store the solid sulfur produced. The introduction of $Na_2CO_3$, improved the hydrophilic and alkaline properties of CNTs, which are very important for low-temperature $H_2S$ catalytic oxidation. The optimum content of $Na_2CO_3$ loading was determined to be 20 wt.% in the CNTs. The catalytic performance of the alkaline CNTs was also dependent on the structure of the CNTs, and the SWCNTs with the smallest tube diameter exhibited the highest sulfur capacity (Figure 1.5).

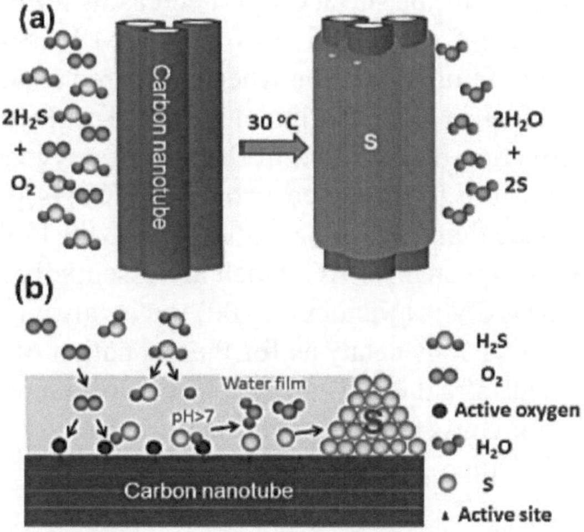


**FIGURE 1.5** The schematic representation of $H_2S$ oxidation over the alkaline CNTs: (a) the total reaction process; and (b) the detailed reaction step [17].

*Source*: Reprinted with permission from Ref. [17]. © 2011 Elsevier.

### 1.3.2.2 OZONATION

Commercial MWCNTs (Nanocyl 3100), oxidized with nitric acid followed by thermal treatment under nitrogen at different temperatures, have been studied as catalysts in the ozonation of sulfamethoxazole, a widely prescribed antibiotic which is frequently detected in both wastewater and drinking water, and is refractory to the conventional wastewater treatment processes [7]. The nitric acid oxidation and the thermal treatment produced MWCNTs with different levels of acidity/basicity. The nitric acid treatment increases the surface area and also the acidity by increasing the surface oxygen-containing groups, such as carboxylic acids, phenols, and carbonyls. The following thermal treatment selectively removes the groups introduced by oxidation with nitric acid, thereby decreasing the acidic nature with the increase in temperature, while slightly increasing the surface area. At 900°C, almost all oxygenated groups are removed, rendering a slightly basic catalyst with the largest surface area, which was found to be the most efficient in mineralizing sulfamethoxazole. These catalysts have also displayed high activity in the ozonation of oxalic and oxamic acids [6]. Generally, catalytic ozonation is favored by MWCNT with basic or neutral properties and high specific surface areas. The introduction of surface oxygenated electron-withdrawing groups reduces the electron density on the carbon surface, thus decreasing the catalytic activity of the material for the decomposition of ozone. MWCNTs significantly enhanced the extent of mineralization when compared to single ozonation. Recycling studies showed that the original MWCNT surface suffers a slight progressive deactivation due to the introduction of oxygenated groups on the surface by exposure to dissolved ozone. MWCNTs present a higher catalytic performance than AC for ozonation due to the high internal mass transfer resistances expected for AC which are essentially microporous.

Commercial MWCNT (Nanocyl 3100) have also been effectively employed as heterogeneous catalysts for the ozonation of organic pollutants such as herbicide atrazine [3] and oxalic acid in aqueous solutions and also in a semi-batch conventional stirred tank reactor for the mineralization of the herbicide, metolachlor [27].

### 1.3.2.3 REDUCTION

Although oxygen-containing functional groups can be introduced on CNTs through chemical functionalization methods, however, it is difficult to control the identity and quantity of oxygen functionalities by a traditional chemical modification method such as $HNO_3$ oxidation or ozone treatment. Moreover, these treatments are aggressive and have obvious influences on the surface structure of CNTs. A controllable modification method to obtain CNTs with identified oxygen functionalities and the integrated surface has been developed via the noncovalent functionalization of CNTs. Small organic molecules containing ketonic carbonyl groups, such as phenanthraquinone (PQ), anthraquinone (AQ), and anthrone (AR), were facilely supported on highly graphitized type of MWCNTs through noncovalent van der Waals and π-π interactions, and the resulting MWCNTs with carbonyl groups on their surface, especially quinone groups, showed high catalytic activity in the reduction of nitrobenzene to aniline [28]. The reduction of nitrobenzene is an important reaction because the main product, aniline, serves as a valuable intermediate and key precursor in the synthesis of pharmaceuticals, dyes, agrochemicals, pigments, and herbicides. The catalytic activity was found to be linearly related to the surface concentration of carbonyl groups, validating the active sites' identification. The turnover frequency (TOF) of highly graphitized MWCNT (HHT) functionalized with diketone groups (PQ and AQ) was higher than that with only one carbonyl group (AR); furthermore, the TOF of PQ functionalized HHT was also higher than that of AQ functionalized HHT because of the fully conjugated structure and high polarity of PQ molecules. PQ is a more powerful electron acceptor than AQ because PQ itself can be fully conjugated with a backbone π-electron system; however, the two aromatic rings in AQ are not fully conjugated with each other. Furthermore, PQ has significantly stronger polarity than AQ, with two carbonyl groups on the same side of the molecule, which may aid hydrazine adsorption and accelerate breakdown, thereby speeding up the reaction rate.

Although nitric acid and/or sulfuric acid oxidation of CNTs is an efficient method to prepare oxygen-functionalized CNTs, the harsh conditions, the corrosive acids as well as the nitrogen oxides generated during oxidation make the method environmentally unfriendly. In addition, thermal treatment of the oxidized CNTs is necessary to improve the activity as far as nitrobenzene reduction is concerned, leading to the consumption of extra energy. $H_2O_2$, as a green oxidant, has been used to functionalize CNTs under mild conditions. CNTs oxidized by $H_2O_2$ under different conditions were employed as metal-free catalysts in the reduction of nitrobenzene [29]. It was shown that the $H_2O_2$ functionalized samples exhibited better catalytic performance than that modified by nitric acid. The high activity of the $H_2O_2$ oxidized sample was ascribed to the relatively low content of the negative functionalities, such as the carboxylic group and anhydride. In addition, the $H_2O_2$ oxidized sample could also effectively catalyze the reduction of some substituted nitroarenes.

#### *1.3.2.4 HYDROGENATION/DEHYDROGENATION*

The presence of electron-rich nitrogen atoms in CNTs induces metal-like properties, and in turn, catalytic activity. Therefore, nitrogen-functionalized CNTs (NCNTs) can be used as metal-free catalysts, which are of great interest in the production of pharmaceuticals, health care, and fine chemicals. Avoiding precious metals not only lowers production costs but also makes it easier to acquire high-purity goods by eliminating time-consuming separation processes. Oxygen- and nitrogen-functionalized carbon nanotubes (OCNTs and NCNTs) were applied as metal-free catalysts in selective olefin hydrogenation of 1,5-cyclooctadiene [30]. CNTs were purified by washing with dilute nitric acid in order to remove residual growth catalysts used during CNT synthesis. Oxygen functionalization was achieved by treating the purified CNTs in $HNO_3$ vapor at 200°C to create oxygen-containing groups on the CNT surface. The obtained OCNTs were subsequently converted to NCNTs in flowing $NH_3$ (10% $NH_3$ in He, 50 $mL{\cdot}min^{-1}$) atmosphere at elevated temperatures. By adjusting the oxygen functionalization duration between 24 h and 120 h and the $NH_3$ post-treatment temperature between 200°C and 600°C, a variety of NCNTs with variable nitrogen content were created. At elevated temperatures, gas-phase treatments for oxygen and nitrogen functionalization increased surface flaws but did not produce structural damage in bulk. NCNTs had significantly higher activity in the hydrogenation of 1,5-cyclooctadiene than pure CNTs and OCNTs, as well as a higher selectivity for cyclooctene. The nitrogen-containing surface functional groups, as well as surface imperfections related to nitrogen species, are credited with favorable catalytic characteristics. Surface groups containing oxygen and surface imperfections generated by oxygen species, on the other hand, did not appear to play a significant role in hydrogenation catalysis. The optimal functionalization conditions in this work were found to be the treatment of CNTs in $HNO_3$ vapor at 200°C for 48 h and subsequently in $NH_3$ at 400°C for 6 h (Figure 1.6).

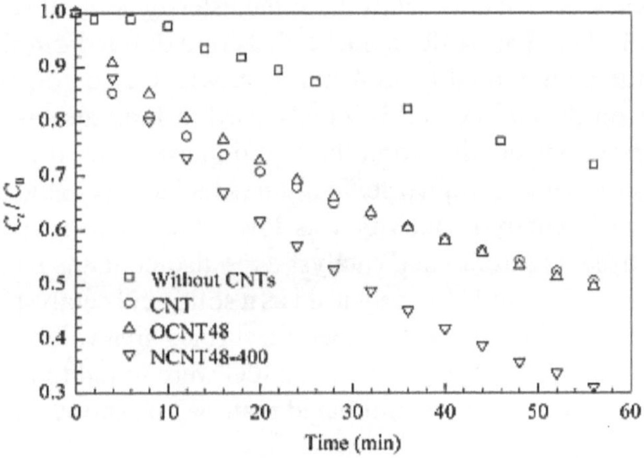


**FIGURE 1.6** 1,5-COD evolution as a function of time on stream without CNTs (with quartz sand) and using purified pristine CNTs, OCNT48, and NCNT48-400 as metal-free catalysts. $C_0$: initial 1,5-COD concentration; $C_t$: 1,5-COD concentration at reaction time *t* [30].

*Source*: Reprinted with permission from Ref. [30]. © 2013 Elsevier.

Styrene is a very important industrial product that is manufactured via the direct dehydrogenation of EB using a potassium-promoted iron catalyst in a very energy-consuming process (870–930°K.) Commercial CNTs have shown higher catalytic activity in the ODH of styrene than traditional catalysts at a far lower reaction temperature than the industrial process (>500°C). The catalytic activity of Fe-doped CNTs was found to be lower in the ODH of EB to styrene.

### 1.3.2.5 *ESTERIFICATION/TRANSESTERIFICATION*

Ethyl levulinate is a very promising compound for use as an oxygenated additive to fuels and is produced via the esterification of ethanol with levulinic acid, obtained via the acid hydrolysis of lignocellulosic biomass. Sulfuric acid is the most widely used inorganic acid catalyst for esterification reactions. Mineral acids, on the other hand, are unsuitable for use in the industry because they erode equipment and necessitate separation from the end product, which necessitates neutralization and waste management. As a result, heterogeneous catalysts that can be easily removed from the product and reused are preferred. Commercial MWCNTs (Nanocyl), sulfonated with concentrated sulfuric acid at six different sulfonation temperatures, were used as catalysts in the esterification of levulinic acid with ethanol [11]. The sulfonated MWCNTs exhibited similar catalytic activity to the commercial resin Amberlyst, which also contains sulfonic acid groups on its surface and is widely used in literature as a solid acid catalyst in reactions conducted in the liquid phase under mild conditions. The CNTs sulfonated at higher temperatures had a lower acid site density, so their specific catalytic activity was lower than that of Amberlyst-15, which was used as a reference catalyst as it also contains sulphonic acid groups on its surface and is widely used as a solid acid catalyst in the liquid phase under mild conditions. The specific surface area values obtained for CNTs sulfonated at 150, 180, 210, and 230C were found to be similar and were close to that of the non-sulfonated material, as shown in Table 1.3.

**TABLE 1.3** Chemical and Textural Properties of the Catalysts [11]

| Catalyst | Sg ($m^2$ $g^{-1}$) | Total Acidity ($0000000000 \mu mol\ NH_3\ g^{-1}_{solid}$) | Site Density ($\mu$mol $m^{-2}$) |
|---|---|---|---|
| CNT | 270 | 91 | <1 |
| CNT-150 | 255 | 504 | 2 |
| CNT-180 | 252 | 458 | 2 |
| CNT-210 | 256 | 491 | 2 |
| CNT-230 | 267 | 459 | 2 |
| CNT-250 | 305 | 320 | 1 |
| CNT-280 | 378 | 236 | <1 |
| Amberlyst-15 | 35 | 2,360 | 64 |

*Source*: Reprinted with permission from Ref. [11]. © 2014 Elsevier.

The modest reduction in the specific surface area seen could be owing to enhanced interactions between CNTs caused by sulfonic acid groups, which can create hydrogen bonds and so lower the pore size generated by nanotube entanglement. After the sulfonation step, the powder components became hard, dense solids. Nanotubes sulfonated at 250, and 280C had an increase in the specific surface area values, which may be a result of the opening of the CNT end caps due to the more severe treatment. Table 1.3 also shows that CNT-150, CNT-180, CNT-210, and CNT-230 had similar acidity values, while the nanotubes sulfonated at 250 and 280°C had lower values. The acidity of original CNTs may be due to the functional groups present on their surface (e.g., carboxyl and hydroxyl) that probably have been formed during its purification step after synthesis. Even at higher temperatures, treatment with sulfuric acid did not result in the degradation of the carbon nanotube structure. The activity results (Table 1.4) have shown that the nanotubes sulfonated below 250°C presented a specific activity higher than those sulfonated at 250 and 280°C. The amount of acidic sites determines the activity of the esterification reaction. In the esterification reaction, non-sulfonated CNTs have no activity. Recycling of the catalyst was unsuccessful as the levulinic acid strongly chemisorbs in the sulfonic groups incorporated to the CNTs surface, thus making the catalyst inactive.

**TABLE 1.4** Catalytic Performance of the CNT, CNT-TTT, and Amberlyst-15 [11]

| Catalyst | Conversion (%) | [a]Specific Reaction Rate ($\times 10^6$) | [b]TOF ($\times 10^2$) |
|---|---|---|---|
| Blank | 2 | – | – |
| CNT | 2 | – | – |
| CNT-150 | 55 | 1.5 | 1.3 |
| CNT-180 | 50 | 1.6 | 1.5 |
| CNT-210 | 54 | 1.7 | 1.5 |
| CNT-230 | 52 | 1.3 | 1.3 |
| CNT-250 | 40 | 0.8 | 1.3 |
| CNT-280 | 20 | 0.3 | 0.8 |
| Amberlyst-15 | 54 | 11.3 | 0.2 |

[a] *Specific reaction rate* = $mol_{EL}\ mol_{cat}^{-2}\ min^{-1}$;

[b] *TOF* = $mol_{EL}\ mol_{SO_3H}^{-1}\ S^{-1}$.

*Source*: Reprinted with permission from Ref. [11]. © 2014 Elsevier.

Sulfate-functionalized MWCNTs have demonstrated remarkable activity and stability in the esterification of glycerol with acetic acid at low temperatures under atmospheric pressure with a conversion of 75.9% after 2 h [13]. MWCNTs, purified by immersing in a 37% HCl solution, sonication for 20 min and kept overnight to remove metal catalyst particles, were used to prepare sulfate functionalized MWCNTs via an efficient and convenient method employing chlorosulphonic acid as a sulfonating agent instead of sulfuric acid. Acid-base titration with NaOH as a titrant was used to determine the quantity of surface sulfate groups, and the total number of acid groups was calculated to be 3.32 mmol/g. The yield of monoglyceride was 47.75%, while the diglyceride and triglyceride were 27.65 and 0.5%, respectively. The unmodified CNTs reached a conversion of 2.2% after the same time, with no effect on the conversion of acetic acid. The catalyst can be recycled and reused numerous times without losing its catalytic activity significantly. After 4 recycles, 72.7% conversion is still maintained.

A novel, simple, one-pot, deprotonation-carbometallation followed by an electrophilic attack of the bromotriethylamine was used to produce very homogeneous samples of $NEt_3$-MWCNT containing a high number of easily accessible amino groups. The catalytic activity of the $NEt_3$-MWCNT and the catalyst produced via the traditional oxidation-amidation route ($NH_2$-MWCNT) was evaluated for the transesterification of glyceryl tributyrate with methanol, which is considered as a test reaction for the production of biodiesel [16]. The major product was always found to be methyl butanoate, with the diglyceride and monoglyceride, which are reaction intermediates, accounting for less than a few percent of the total. The catalyst prepared by using the novel procedure gave a conversion of 77% after 8 hours, while the commercial hydrotalcite catalyst tested under the same conditions gave only a 51% conversion. The lowest conversion was obtained with the sample prepared by oxidation–amidation. Despite the fact that the grafted amino groups are basic enough to catalyze the transesterification reaction, this catalyst has low activity and a high deactivation rate. Acids and bases both catalyze transesterification processes. However, the proximity of both remaining acidic oxygen-containing groups and basic amino groups in this sample allows them to neutralize each other, as was observed by acid-base titration. As a result, fewer active sites are available to catalyze the transesterification step, resulting in reduced triglyceride conversion. This test reveals that the synthetic technique employed for covalent grafting of amino groups has a significant impact on the generated material's catalytic activity.

The catalytic activity of three catalysts prepared by the above novel grafting method with three different amines (triethylamine, ethylamine, and pyrrolidine, denoted as $Et_3N$-CNT, $EtNH_2$-CNT, and $C_4H_9N$-CNTs) have also been compared for correlating the different basicity of the amines to their catalytic activity in transesterification of glyceryl tributyrate at 60°C [31]. $Et_3N$-CNTs resulted in the most active catalyst with a conversion of 77% after 8 h. In comparison, the $EtNH_2$-CNTs catalyst reached a conversion of 55% after the same time, whereas $C_4H_9N$-CNTs showed a much lower conversion. The higher activity of $Et_3N$-CNTs compared to $EtNH_2$-CNTs is due to the higher basicity of triethylamine compared to ethylamine.

The reduced catalytic activity of C4H9N-CNTs is attributable to a lesser number of grafted amino groups (0.54 mmol $g^{-1}$) compared to the other two catalysts (1 mmol $g^{-1}$ for Et3N-CNTs and 0.99 mmol $g^{-1}$ EtNH2-CNTs). The reason for the lower amount of grafted pyrrolidine groups was ascribed to the different reactivity of 1-(2-chloroethyl)pyrrolidine hydrochloride in comparison to the Bromo derivatives used to graft the other two amines.

#### *1.3.2.6 MISCELLANEOUS*

Bisphenolic antioxidants play important roles in dealing with oxidative damage caused by reactive oxygen species. A facile and environmental-friendly method for the synthesis of bisphenolic antioxidants by the reaction of 2-*tert*-butyl-4-methylphenol and aldehydes using sulfonated MWCNTs (MWCNTs–$SO_3H$) as reusable and stable solid acid catalyst under solvent-free conditions giving high yields of the products in relatively short reaction times has been reported [14]. The total surface acidity of $SO_3H$ groups on the MWCNTs was 1.80 mmol $g^{-1}$.

Nitrogen-containing CNTs (NCNTs) have been used as reusable base catalysts in a Knoevenagel condensation reaction [32]. The CVD of C-and N-containing precursors (acetonitrile, pyridine) on a supported metal catalyst, namely Co or Ni supported on $SiO_2$, was used to make NCNTs with variable N/C compositions. The residual growth catalyst was chemically removed after NCNT production. The catalytic activity for the Knoevenagel condensation of ethyl—cyanoacetate with benzaldehyde was investigated in relation to the type of basic N-species in the NCNT. The catalytic activity of the samples for a Knoevenagel condensation is related to the concentration of pyridinic nitrogen present in the CNT, which could be tuned by the synthesis parameters of the catalyst. The creation of pyridinic nitrogen was favored at low temperatures, such as 823 K, whereas the formation of quaternary nitrogen was prominent at higher temperatures.

A series of imidazolium-based ionic liquids (ILs), bearing hydroxyl groups, with different alkyl chain lengths and counter anions ($Cl^-$, $Br^-$, and $I^-$) were synthesized and grafted on oxidized MWCNT surfaces via esterification reactions and utilized as highly efficient heterogeneous catalysts for the production of the cyclic carbonate, allyl glycidyl carbonate, via cycloaddition reaction of the epoxide, allyl glycidyl ether, and $CO_2$ [18]. In comparison with conventional heterogeneous catalysts that employ porous silica and polymer supports, the use of oxidized MWCNTs as supporting materials significantly enhanced the catalytic performance of immobilized ILs in cycloaddition reactions. The impacts of the type of the supported ILs were investigated, and it was discovered that supported ILs with a more nucleophilic anion and a thicker alkyl chain had higher activity and selectivity. In addition, these catalysts also demonstrated excellent reusability in cycloaddition reactions. Pristine CNTs and oxidized CNTs, however, did not show any catalytic activity toward the cycloaddition reaction. Therefore, the significant enhancement in the catalytic reactivity of CNT-ILs is mainly attributed to the combined effects of the intrinsic properties of oxidized CNT support and the immobilized ionic liquid.

## 1.4 CONCLUSION

Although the major use of CNTs in catalysis is as support, they have also been employed as catalysts in various chemical transformations, either as such (pristine) or after modifications by introducing surface functional groups via different oxidation/chemical treatments. The morphology of CNTs is found to be crucial in catalysis. CNTs can be modified with a variety of surface functionalities which play a significant role in catalysis. CNTs as catalysts have shown remarkable activity in oxidation, ozonation, dehydrogenation, hydroxylation, etc. In the last three decades, CNTs have been widely explored as catalysts and as supports in various fields such as wastewater treatment, pollution control, hydrogen storage, fuel cells, etc.; however, the full potential of CNTs in catalysis remains yet to be unraveled.

## KEYWORDS

- **carbon nanotubes**
- **chemical vapor deposition**
- **heterogeneous catalysis**
- **metal-free catalyst**
- **multi-wall carbon nanotubes**
- **oxidation**

## REFERENCES


1. Iijima, S., (1991). Helical microtubules of graphitic carbon. *Nature, 354*, 56–58.
2. Iijima, S., & Ichihashi, T., (1993). Single-shell carbon nanotubes of 1-nm diameter. *Nature, 363*, 603–605.
3. Fan, X., Restivo, J., Órfão, J. J. M., Pereira, M. F. R., & Lapkin, A. A., (2014). The role of multi-walled carbon nanotubes (MWCNTs) in the catalytic ozonation of atrazine. *Chem. Eng. J., 241*, 66–76.
4. Kang, Z., et al., (2006). Heterogeneous hydroxylation catalyzed by multi-walled carbon nanotubes at low temperatures. *Appl. Catal. A Gen., 299*, 212–217.
5. Soares, O. S. G. P., Gonçalves, A. G., Delgado, J. J., Órfão, J. J. M., & Pereira, M. F. R., (2015). Modification of carbon nanotubes by ball-milling to be used as ozonation catalysts. *Catal. Today, 249*, 199–203.
6. Goncalves, A. G., Figueiredo, J. L., Órfão, J. J. M., & Pereira, M. F. R., (2010). Influence of the surface chemistry of multi-walled carbon nanotubes on their activity as ozonation catalysts. *Carbon N. Y., 48*, 4369–4381.
7. Gonçalves, A. G., Órfão, J. J. M., & Pereira, M. F. R., (2013). Ozonation of sulfamethoxazole promoted by MWCNT. *Catal. Commun., 35*, 82–87.
8. Yang, S., Li, X., Zhu, W., Wang, J., & Descorme, C., (2008). Catalytic activity, stability, and structure of multi-walled carbon nanotubes in the wet air oxidation of phenol. *Carbon N. Y., 46*, 445–452.
9. Croston, M., et al., (2002). Conversion of aniline to azobenzene at functionalized carbon nanotubes: A possible case of a nanodimensional reaction. *Int. J. Nanosci., 1*, 285–293.
10. Rocha, R. P., Sousa, J. P. S., Silva, A. M. T., Pereira, M. F. R., & Figueiredo, J. L., (2011). Catalytic activity and stability of multi-walled carbon nanotubes in catalytic wet air oxidation of oxalic acid: The role of the basic nature induced by the surface chemistry. *Appl. Catal. B Environ., 104*, 330–336.
11. Oliveira, B. L., & Teixeira Da, S. V., (2014). Sulfonated carbon nanotubes as catalysts for the conversion of levulinic acid into ethyl levulinate. *Catal. Today, 234*, 257–263.
12. Sun, Z. P., Zhang, X. G., Liang, Y. Y., & Li, H. L., (2009). A facile approach towards sulfonate functionalization of multi-walled carbon nanotubes as Pd catalyst support for ethylene glycol electro-oxidation. *J. Power Sources, 191*, 366–370.
13. Ji, X., Chen, Y., Wang, X., & Liu, W., (2011). Sulfate-functionalized multi-walled carbon nanotubes as catalysts for the esterification of glycerol with acetic acid. *Kinet. Catal., 52*, 555–558.
14. Fareghi-Alamdari, R., Golestanzadeh, M., Agend, F., & Zekri, N., (2013). Synthesis, characterization, and catalytic activity of sulfonated multi-walled carbon nanotubes as heterogeneous, robust, and reusable catalysts for the synthesis of bisphenolic antioxidants under solvent-free conditions. *J. Chem. Sci., 125*, 1185–1195.
15. Yu, H., Jin, Y., Li, Z., Peng, F., & Wang, H., (2008). Synthesis and characterization of sulfonated single-walled carbon nanotubes and their performance as solid acid catalysts. *J. Solid State Chem., 181*, 432–438.
16. Tessonnier, J. P., Villa, A., Majoulet, O., Su, D. S., & Schlögl, R., (2009). Defect-mediated functionalization of carbon nanotubes as a route to design single-site basic heterogeneous catalysts for biomass conversion. *Angew. Chemie - Int. Ed., 48*, 6543–6546.

17. Chen, Q., et al., (2011). Alkaline carbon nanotubes as effective catalysts for H2S oxidation. *Carbon N. Y., 49*, 3773–3780.
18. Han, L., et al., (2012). Ionic liquids grafted on carbon nanotubes as highly efficient heterogeneous catalysts for the synthesis of cyclic carbonates. *Appl. Catal. A Gen., 429, 430*, 67–72.
19. Liu, Z. Q., Ma, J., Cui, Y. H., Zhao, L., & Zhang, B. P., (2011). Factors affecting the catalytic activity of multi-walled carbon nanotube for ozonation of oxalic acid. *Sep. Purif. Technol., 78*, 147–153.
20. Liu, X., Su, D. S., & Schlögl, R., (2008). Oxidative dehydrogenation of 1-butene to butadiene over carbon nanotube catalysts. *Carbon N. Y., 46*, 544–561.
21. Yang, S., Sun, Y., Yang, H., & Wan, J., (2014). Catalytic wet air oxidation of phenol, nitrobenzene, and aniline over the multi-walled carbon nanotubes (MWCNTs) as catalysts. *Front. Environ. Sci. Eng., 9*, 436–443.
22. Zeynalov, E., et al., (2013). Plasma-chemically brominated single-walled carbon nanotubes as novel catalysts for oil hydrocarbons aerobic oxidation. *Appl. Catal. A Gen., 454*, 115–118.
23. Cao, Y., et al., (2013). Nitrogen-, phosphorous- and boron-doped carbon nanotubes as catalysts for the aerobic oxidation of cyclohexane. *Carbon N. Y., 57*, 433–442.
24. Yu, H., et al., (2011). Selective catalysis of the aerobic oxidation of cyclohexane in the liquid phase by carbon nanotubes. *Angew. Chemie - Int. Ed., 50*, 3978–3982.
25. Croston, M., Langston, J., Sangoi, R., & Santhanam, K. S. V., (2002). Catalytic oxidation of p-toluidine at multi-walled functionalized carbon nanotubes. *Int. J. Nanosci., 1*, 277–283.
26. Luo, J., Peng, F., Yu, H., Wang, H., & Zheng, W., (2013). Aerobic liquid-phase oxidation of ethylbenzene to acetophenone catalyzed by carbon nanotubes. *ChemCatChem., 5*, 1578–1586.
27. Restivo, J., órfão, J. J. M., Armenise, S., Garcia-Bordejé, E., & Pereira, M. F. R., (2012). Catalytic ozonation of metolachlor under continuous operation using nanocarbon materials grown on a ceramic monolith. *J. Hazard. Mater., 239, 240*, 249–256.
28. Gu, X., et al., (2014). Noncovalent functionalization of multi-walled carbon nanotubes as metal-free catalysts for the reduction of nitrobenzene. *Catal. Sci. Technol., 4*, 1730–1733.
29. Wu, S., Wen, G., Schlögl, R., & Su, D. S., (2015). Carbon nanotubes oxidized by a green method as efficient metal-free catalysts for nitroarene reduction. *Phys. Chem. Chem. Phys., 17*, 1567–1571.
30. Chen, P., et al., (2013). Purified oxygen- and nitrogen-modified multi-walled carbon nanotubes as metal-free catalysts for selective olefin hydrogenation. *J. Energy Chem., 22*, 312–320.
31. Villa, A., Tessonnier, J. P., Majoulet, O., Su, D. S., & Schlögl, R., (2009). Amino-functionalized carbon nanotubes as solid basic catalysts for the transesterification of triglycerides. *Chem. Commun.*, 4405–4407. doi: 10.1039/b906123a.
32. Van, D. S., De Jong, K. P., & Bitter, J. H., (2006). Nitrogen-containing carbon nanotubes as solid base catalysts. *Chem. Commun.*, 4859–4861. doi: 10.1039/b610208e.

## CHAPTER 2

# Carbon Nanomaterials for Sensing Applications

D. RITHESH RAJ[1] and VINOD KUMAR NATHAN[2]

[1]*Department of Electronics and Communications, School of Electrical and Electronics Engineering, Sastra University, Thanjavur, Tami Nadu – 613401, India*

[2]*School of Chemical and Biotechnology, Sastra University, Thanjavur, Tami Nadu – 613401, India*

**Corresponding author. E-mail: ritheshraj@ece.sastra.edu*

## ABSTRACT

Carbon and its allotropes attracted immense research attention due to their unique electrical, optical, and structural properties. Recently, carbon-based nanomaterials like carbon dots (CDs), graphene, graphene nanodots, and carbon nanotubes (CNTs) have been widely explored for their sensing properties. Carbon-based nanomaterials turned out to be an appropriate candidate for fabricating sensing devices owing to their reliability and robustness. In this chapter, we present a rigorous review of recent advancements in the field of sensor technology using carbon-based materials. Specific attention is given to sensors developed using CDs, graphene quantum dots (GQDs), and graphene for optical, electrochemical, and environmental applications. This chapter aims to provide sufficient information for sensor development using carbon materials and their applications in different fields of study like biomedical sensors, Chemical sensors, and environmental sensors.

Carbon Composites: Composites with Nanotubes, Nanomaterials, and Graphene Oxide.
Eduardo A. Castro, Ann Rose Abraham, & A. K. Haghi (Eds.)

## 2.1 INTRODUCTION

Carbon and its allotropes are attracting immense research interest due to their unique optical, electrical, thermal, and structural properties [1]. Especially, graphene, graphene quantum dots (GQDs), carbon dots (CDs), and carbon nanotubes (CNTs) were rigorously investigated for their sensing properties [2–4]. In addition, the optical and electrical properties and favorable mechanical properties make the innovative material for robust and reliable sensor fabrication [5]. Carbon-based nanosensors have a crucial role in life science, chemical sensors, and the environment [6]. This chapter deals with recent developments in carbon and its allotropes-based nanosensors, properties fabrication, and their applications. The systematically presented properties, fabrication, and applications of carbon-based nanosensors will provide an overall idea, basic concepts, and new trends in the field new.

### 2.1.1 CARBON ALLOTROPES

Carbon and its allotropes were rigorously investigated and reported for electrical, optoelectronics, and sensing applications. Carbon allotropes with dissimilar crystal structures $sp^3$ to $sp^2$ hybridized, mixed states, and amorphous forms offer unique optical, electrical, and structural properties that make them an important class of materials for sensing applications [7]. $Sp^3$ hybridized carbon form like a diamond is of 3D cubic lattice while graphite is layered two-dimensional structures. Graphene is viewed as a two-dimensional building block of $sp^2$ hybridized forms of carbon [8]. Fullerene and low dimensional forms of carbon with irregular forms of layers. The elongated kind of fullerenes are nothing but CNTs [9]; it is few-layer graphene with a high aspect ratio. The structural properties of $sp^2$ and $sp^3$ structures, and the different optoelectronic properties of carbon allotropes make them indispensable candidates for different sensing applications.

### 2.1.2 UNIQUE PROPERTIES OF CARBON NANOMATERIALS

Carbon nanomaterials were broadly investigated, and reported their exceptional physicochemical properties. Electrical, electronic, and optical properties of carbon-based materials vary to a great extent while changing the structure, size, and composition [10, 11]. Also, the chemical complexes associated with carbon nanomaterials lead to the emergence of different properties. Exceptional electrical, electronic, and optical properties, together with low toxicity and biocompatibility, make them an inevitable candidate for sensing applications [12]. The carbon-based nanomaterials are classified based on the hybridization between carbon atoms. However, the physicochemical properties differ in size and composition. CDs and GQDs have excellent optical properties like luminescence, phosphorescence, and tunable fluorescence, together with biocompatibility, and high photostability [13]. CDs and GQDs are reported for their high quantum yield compared to conventional fluorescent tags [14]. Surface modifications result in tremendous changes in optical properties. CNTs are 1D variant of carbon-based nanomaterials that are soluble in water but functionalizing using different ligands makes them appropriate for biosensing applications as well [15]. Graphene has been one of the most investigated materials in the last decade because of its exceptional electronic and optical properties. Further, graphene is known for very good conductivity, thermal stability, and the existence of surface plasmon polaritons [16].

### *2.1.3 CARBON NANOMATERIALS: SYNTHESIS AND CHARACTERIZATIONS*

Early CNTs were derived from fullerene synthesis techniques like Kretschmer–Huffman method. Further, large-scale productions were reported using the ark discharge technique and laser ablation. Single-walled CNTs were synthesized with the help of metal nanoparticles as a catalyst using the chemical evaporation technique. Pyrolyzing metal carbines also can result in SWCNTs. Graphene was first synthesized by mechanically exfoliated from graphite by Geim and Novoselov using a simple scotch tape technique [17]. Scaling up of graphene production has been achieved by graphitization of silicon carbide wafers and monolayer graphene on metal carbides. Solution processing techniques also attracted attention due to their advantage in scalability.

## 2.2 TYPE OF CARBON-BASED NANOMATERIAL

### *2.2.1 CDS AND GQDS*

CDs made an impact on the biosensing and bioimaging research field due to their biocompatibility and distinctive fluorescence properties [18–20]. In addition, the electrochemical properties of CD make them a potential candidate for electrochemical biosensing applications [21, 22]. CD, composed of carbon and hydrogen with a diameter of less than 10 nm, mostly decorated the outer surface with hydroxyl carbonyl or carboxylic groups depending upon the preparation techniques [23]. The crystalline nature, size, and composition of the CD can be investigated using X-ray diffraction (XRD), high-resolution transmission electron microscope (HRTEM), and Fourier transforms infrared (FTIR), respectively. Raman Spectroscopy is the authoritative characterization technique to investigate the quality of the CDs. CDs-based systems are reported for their photoluminescence quenching properties posing new prospects in biosensor and bioimaging applications [18]. Electrochemiluminescence and proton adsorption behavior of CD were also exploited for sensing applications. CDs are investigated as fluorescent probes for bioimaging, owing to their blue and green fluorescence, biocompatibility, and photostability. GQDs are small fragments of graphene with different degrees of carbonization; like CDs, GQDs also possess exceptional fluorescent properties [24].

### *2.2.2 GRAPHENE*

Graphene, a single layer of tightly packed carbon atoms that forms a 2D honeycomb crystal lattice structure, has drawn a lot of interest due to its intrinsic features, especially in the energy storage industry [25]. Because of its fast-charging time, high strength-to-weight ratio, and wide surface area, graphene is commonly utilized as the anode in batteries and cells, as well as in supercapacitors [26]. The high electrical conductivity ($1.0\times10^8$ S/m), high melting point (4,510 K), high thermal conductivity (2,000–4,000 W $m^1$ $K^1$, 5,000 W/m K), and highest current density ($1.6\times109$ A/$cm^2$) of graphene, as well as its high electron mobility (2,00,000 $cm^2$ $V^1$ $s^1$ at electron density $2\times1{,}011$ $cm^{-2}$), make it ideal for electrochemical sensors, strain sensors, and electrical sensors [27, 28]. Graphene's physical as well as electrochemical properties have helped in the development of sensitive sensors and biosensors [85]. Electrochemical sensors made of graphene-based materials that are highly efficient, adjustable, biocompatible, and environmentally friendly are a difficult problem for computational chemistry [86]. The advantages of graphene-based nanoparticles over other nanoparticles include enhanced stability, thermal conductivity, corrosion and erosion resistance, and carrier mobility. Heat transfer, anti-infection therapy, defect sensors, medicinal, cosmetics, and energy harvesting systems are just a few of the uses for graphene-based nanofluids [87]. Among them, due to its unique properties, such as enhanced surface area, mechanical strength, and electro-conductivity, graphene nanoribbons are a highly developed kind of graphene with a wide range of uses [29].

Metals, metal oxides, quantum dots, organic polymers, semiconducting nanoparticles, as well as biomolecules can be easily mixed with graphene, GO, and rGO nanosheets to form a varied spectrum of graphene-based nanocomposites (NCs) with improved sensitivity to be employed for biosensor applications [88]. Due to a wrinkled surface roughness caused by a high density of surface defects, graphene has a stronger strength than CNTs, allowing for more contact with the polymer material than a rolled-up CNT [30]. Moreover, the biofunctionalization of graphene using carefully designed peptides was proven, as well as the production procedures for graphene-peptide NCs (monomer, nanotubes, as well as nanofibrils). The development of graphene-peptide NC-based biosensor architectures for fluorescent, electrochemical, electronic, and spectroscopic biosensing was discussed [31]. Functionalized graphene NCs, such as graphene along with its oxides or doped oxides, are extensively used as materials for high selectivity, sensitivity, as well as stability for developing sensors [32]. Novel sensing platforms based on pure and modified graphene fortified with other nanoparticles or polymers have been presented in recent years. To construct highly sensitive and selective biosensors, several of these biomolecules, namely DNA, enzymes, and antibodies, were employed. Based on graphene's chemical composition, strategies for attaching these biomolecules to its surface have been developed. Covalent bonding, such as connecting biomolecules via the 1-ethyl-3-(3-dimethyl aminopropyl) carbodiimide hydrochloride and N-hydroxysuccinimide (NHS) reactions, and physisorption are examples of these approaches [33].

### 2.2.3 CARBON NANOTUBES (CNTS)

Carbon nanotubes (CNTs) are a type of carbon allotrope with unique features that make them a useful product for a variety of nanotechnology applications, including electronics, optics, and medicines. They are an emerging class of nanostructured materials that could be used to construct chemical sensors that can monitor human as well as environmental health. Based on the number of sidewalls present and how they are distributed spatially, CNTs are classed as single-walled carbon nanotubes (SWCNTs) or multi-walled carbon nanotubes (MWCNTs) [34]. CNTs have piqued the interest of material scientists, physicists, chemists, and engineers for nearly three decades. CNTs have found use in a variety of industries due to their high aspect ratio, thermal capabilities along exceptional electrical and mechanical properties [35].

CNTs have a variety of multifunctional qualities, including high surface-to-volume ratios, excellent chemical capabilities, and programable physical properties. The discovery of the selectivity, sensitivity, and detection mechanisms of CNT-assisted chemical sensors, on the other hand, is a difficult undertaking [36]. CNTs are used to solve a variety of challenges, including environmental issues, pollution, biomedical applications, and more [37]. SWCNTs have been extensively explored for applications such as gas sensing due to their material and unique electrical properties [38]. SWNTS surface was modified in many studies with various sensing materials to improve sensing capabilities, including sensitivity and specificity (e.g., macromolecules, receptors, metallic nanoparticles, polymeric materials, etc.) [39]. A vast number of articles have focused on the use of CNTs as electrode components and the electrochemical behavior of these materials. The presence of contaminants, the use of SWCNTs or MWCNTs, and the chemical methods used all have an impact on the performance of modified sensors [89].

Although it has long been known that semiconducting single-wall carbon nanotubes (SWCNTs) have strong excitonic binding and produce light over a wide range of wavelengths, their use has been limited due to a low quantum yield and a high sensitivity to spectrum diffusion and blinking [40]. CNTs have accelerated the development of gas sensors based on CNT that are extremely sensitive to their surroundings. The presence of gases such as $O_2$, $NO_2$, $NH_3$, and a variety of other chemicals, as well as molecules, can either contribute or take electrons, changing the total conductivity. CNTs are suited for nanoscale gas-sensing materials because of their characteristics. Gas sensors have already been shown using conductive-based technologies. CNTs, on the other hand, have some drawbacks for gas sensor applications, including a slow recovery time, limited detection of gas, and sensitivity to humidity, as well as other gases. As a result, NCs, including polymer and CNTs, have attracted a lot of heed for gas sensing applications due to their better sensitivity over a spectrum of gas concentrations at ambient temperature than CNTs and polymeric materials used individually [41].

## 2.3 TYPE OF CARBON NANOMATERIAL-BASED SENSORS

### *2.3.1 CARBON-DERIVED MATERIALS FOR OPTICAL SENSING*

Unique optical properties of carbon-derived nanomaterials offer sensing solutions for detecting a wide variety of analytes. Optical sensors are known for their high sensitivity and rapid response, enabling detection up to the extent of single-molecule in biomedical applications. Also, they are used for the detection of analytes such as gas, chemicals, and environmental pollutants [42]. Optical sensing techniques include absorption, fluorescence, surface plasmon resonance, localized surface plasmon resonance, surface-enhanced Raman scattering, and FRET. The fluorescence properties of carbon-derived materials like SWCNT, CDs, GQDs, GO, and Reduced GO greatly vary surface properties, making them suitable candidates for optical sensing materials [43, 44]. Graphene is reported for its surface plasmon resonance, which is an excellent tool for sensing based on surrounding refractive index changes. Carbon-derived nanomaterials are a potential alternative for most conventional optical nanosensors. In this section, a brief review of the advantages and disadvantages of optical sensors based on carbon-based sensors is discussed.

Infrared fluorescence is very important in bioimaging as visible, and UV light always interferes with the autofluorescence of the biomolecules [45]. SWCNTs exhibit excellent fluorescence properties in the near-infrared region, which is well explored for bioimaging and biosensing applications [46]. Also, fluorescence properties are tunable with the chirality and size of the material [47]. In addition, SWCNT fluorescent tags are immune to photobleaching [48]. The tunable NIR fluorescence makes them a suitable candidate for in-vivo applications. Fluorescence intensity and wavelength drastically change with the change local environment, making them a suitable material for fabricating ultra-sensitive sensors. DNA hybridized SWCNTs are reported for target-specified detection of molecules as the presence of DNA leads to a change in the dielectric properties of the SWCNT [49]. This technique can be used for the detection of different complex fluids and molecules by monitoring fluorescence quenching [50]. SWCNTs are reported for the detection of proteins, RNA, DNA, bacteria, and small biomolecules [51, 90–94]. Also, SWCNTs are used for the fabrication of paper and optical fiber-based sensors for the sensitive detection of VOCs and other chemicals [95]. Aggregation and covalent bonding in the side walls leads to fluorescence quenching. This issue is tackled by noncovalent modification of SWCNT for preserving optical properties. Surfactants, polymers, and peptides are used for noncovalent modifications.

CDs have attracted immense research interest in the last decade due to their highly tunable photoluminescence, biocompatibility, and photostability. Also, CDs are very sensitive to the local environment leading to changes in their optical properties [52]. CDs were explored for their cation and anion-sensing properties, and the interaction between the surface group of CDs and ionic solvent led to a change in fluorescence [53, 96]. To overcome the specificity issues, CDs were functionalized with different ligands, such as transferrin folic acid (FA), also reported in the literature. The pH sensitivity of the CDs was exploited for developing pH sensors. Also, biosensors developed the detection of a wide variety of molecules, including amino acids, glucose, and ascorbic acid [54, 55]. Fluorescence ranging from UV to NIR makes them a suitable candidate for bioimaging. CDs were reported as a fluorescent tag for breast cancer cells owing to the simplistic cellular uptake of CDs [56]. *In vivo* imaging of tumor cells also reports using CDs as probs. CDs are reported for fluorescence-based pH sensors for monitoring cell pH conditions (Figures 2.1(a) and (b)) [57]. The fluorescent functionalized GQDs were reported for the detection of ferritin (Figures 2.1(c) and (d)) [58].

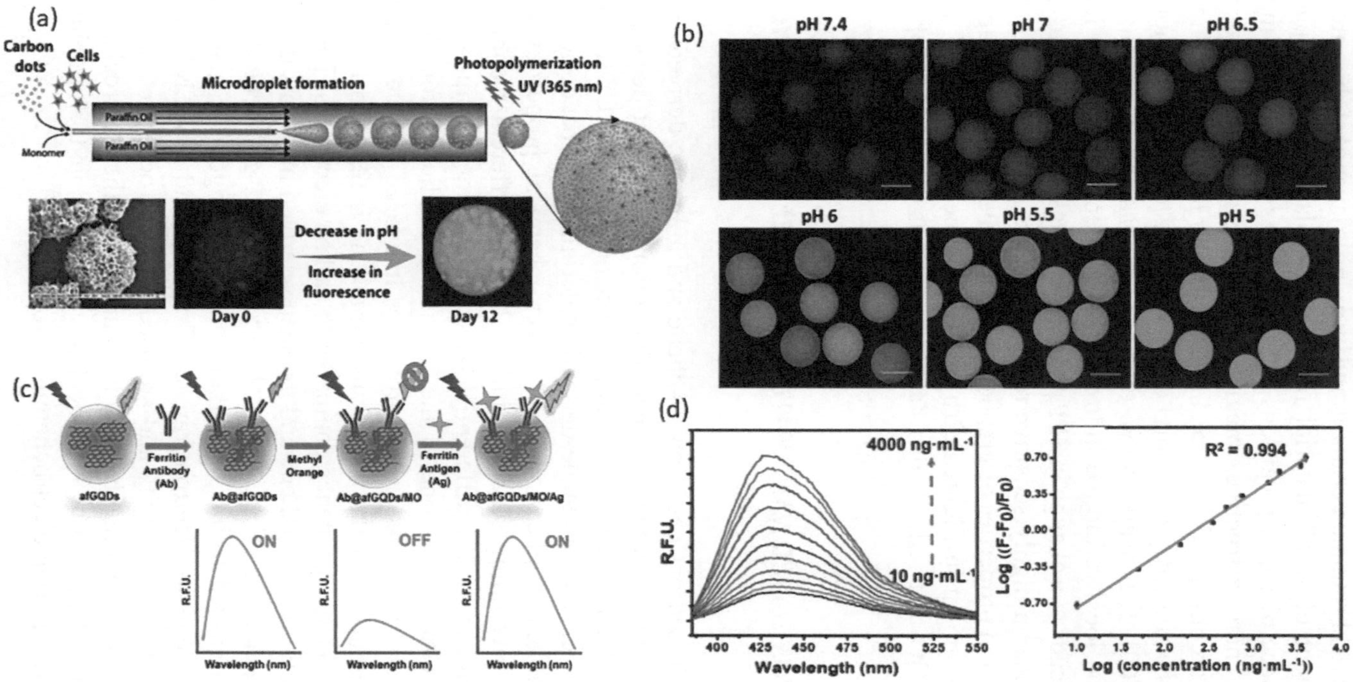


**FIGURE 2.1** (a) A schematic of carbon dots for fluorescent quenching-based cellular pH sensor. [Reprinted with permission from Ref. [57]; Copyright (2017), American Chemical Society]. (b) Fluorescence quenching of CDs with an increase in pH value. (c) GQDs for on and off biosensors. (d) Change in the fluorescence and linearity of GQD-based immunosensors. [Reprinted with permission from Ref. [58]; Copyright (2021), American Chemical Society].

Graphene is a 2D metamaterial formed by a single layer of carbon atoms, a very important class of material for terahertz plasmonics. Graphene possesses surface plasmon polaritons in the terahertz band, having many advantages in non-destructive imaging. Terahertz plasmonics has the potential to overcome the diffraction limit and use for imaging [59]. Figure 2.2(a) shows the chlorpyrifos-methyl detection using terahertz plasmonics of graphene [60]. It is well known that surface plasmon polaritons are sensitive to the local refractive index; any mass deposition in the local environment leads to a change in SPP and image things in the metasurface. Li et al. proposed a highly sensitive refractive index sensor based on the perfect absorption properties of graphene [61], and the schematic diagram of the work is shown in Figure 2.2(b). The graphene-derived nanomaterials such as GO and redGO also have tremendous potential in optical sensing [62].

### *2.3.2 CARBON-BASED NANOMATERIALS FOR ELECTROCHEMICAL SENSORS*

Electrochemical sensors based on graphene and graphene-derived nanomaterials have gained growing research attention in the last decade due to their extraordinary electrochemical performance [65, 66]. CDs, CNTs, GQDs, graphene, and graphene-derived materials have been explored for electrochemical applications [67]. Electrochemical properties were exploited for developing sensors with the help of different techniques, cyclic voltammetry (CV), differential pulse voltammetry, amperometry [68, 69], etc. Graphene is one of the promising candidates for electrochemical sensors due to its unique properties like large surface area and high electrical and thermal conductivity. Graphene is reported to have a lower charge transfer resistance compared to carbon electrodes. The electrochemical activity of graphene and graphene-derived materials depends on the edges per mass of the material. Graphene and graphene-derived materials are reported to have a higher number of edges per mass and, thereby, heterogeneous electron transfer. Graphene and MWCNTs show higher electrochemical activity and are used for electrochemical sensing of DNA, synthesis technique has an important role in the electrochemical properties of graphene, and CNTs mechanically exfoliated, and epitaxial grown reported different electrochemical properties [3].

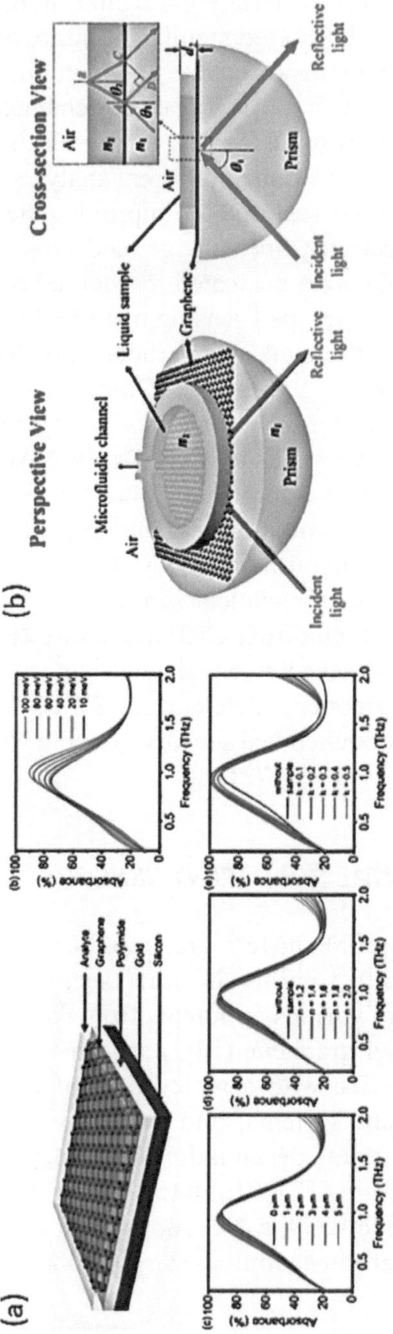


**FIGURE 2.2** (a) Graphene-based terahertz plasmonics for sensing applications. [Reprinted with permission from Ref. [63]; Copyright (2020), American Chemical Society]. (b) Perfect absorber-based graphene-based sensors for sensing applications. [Reprinted with permission from Ref. [64]; Copyright (2020), American Chemical Society].

GQDs are functionalized with different ligands to improve the electrochemical performance and reported significant improvement in the properties [2]. GQDs with metal nanoparticles and metal-organic frameworks, and other types of materials improve the electrochemical sensing properties [70]. GQDs have been used for developing electrochemical sensors for the detection and quantification of different analytes. Different material engineering techniques were reported for improving the sensor parameters such as sensitivity, selectivity, linear range, and limit of detection of the sensors. GQD-based NCs were evaluated for their electrochemical performance and successfully developed sensing material for different analytes. Composites of GQDs with metal nanoparticles, graphene, and polymers were reported for their sensing properties. Composites GQDs with Au-Pd alloys, SWCNT, NiO, and $MoS_2$ materials were reported for sensing biological analytes [97]. A composite of GQDs and MWCNT was reported for the electrochemical detection of dopamine [3]. Figure 2.3(a) and (b) show the scheme of the dopamine sensing mechanism and CV of different concentrations of dopamine on the GQD/MWCNT composite electrode. GQD-based label-free electrochemical sensors for the cardiac, immune sensor was reported for monitoring CRP biomarker (Figure 2.3(c)) [71]. Flexible graphene-based sensors for gas sensing were reported carbon nanomaterials such as graphene platelets and SWCNT were reported for the development of electrochemical sensors by using Nafion as an electrochemical sensor (Figure 2.3(d)) [72].

### *2.3.3 ENVIRONMENTAL POLLUTANT SENSORS*

Carbon-based nanomaterials have a great application in various sensor development, as described earlier in other sections. They are widely used for the detection of various pollutants from environmental samples. CNTs and laser-induced graphene (LIG) are carbon nanomaterials that are electrically conductive with a wide range of environmental applications, water treatment systems, and improved separation methods, including fouling-resistant desalination electrocatalytic as well as pollutant sensing platforms [73]. To detect harmful phenolic chemicals in the environment effectively, a NC doped with chromium (III) oxide nanoparticles coated carbon nanotubes ($Cr_2O_3$–CNT NC) was developed by Rahman et al. [74].

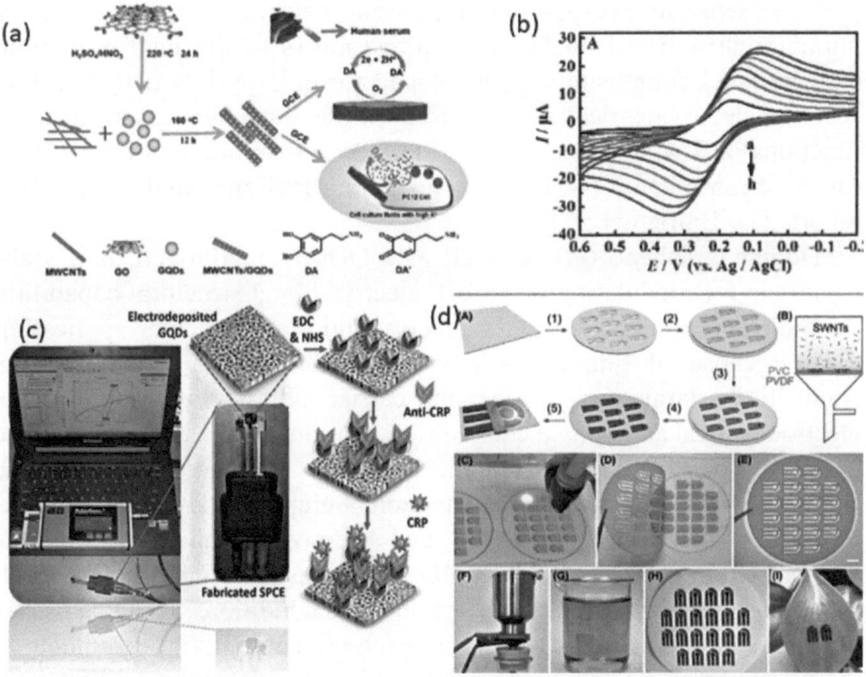


**FIGURE 2.3** Electrochemical sensors based on graphene. (a) Graphene nanowalls on PDMS substrate for sensors; (b) graphene quantum dots for biosensors [Reprinted with permission from Ref. [3]; Copyright (2020), American Chemical Society]. (c) SWCNT and MWCNT for sensor fabrication [Reprinted with permission from Ref. [71]; Copyright (2021), American Chemical Society]. (d) Graphene platelets and SWCNT as electrochemical sensors [Reprinted with permission from Ref. [72]; Copyright (2018), American Chemical Society].

Similarly, sensing materials were formed into printable inks using hybrid nanostructures of SWNTs functionalized with either a metalloporphyrin (MPor) or metal phthalocyanine (MPc) derivative. Sensing materials were ink-jet printed onto interdigitated platinum electrodes for each sensor in the array. Under various combinations of lighting conditions and specific sensing materials, the sensor array displayed adjustable sensitivity to $NO_2$, $SO_2$, and $O_3$ (in the range of sub-ppm to tens of ppm) [39]. To construct a phenolic chemical sensor that responds fast to selective 4-aminophenol (4-AP) in a buffer solution, polyaniline/graphene/carbon nanotubes (PANI/G/CNTs) NCs were deposited on a flat-silver electrode

[75]. In a separate investigation for pesticide detection, an electrochemical immunosensor based on $Fe_3O_4$-modified CNTs as the improved multi-enzyme label composite was developed for chlorpyrifos detection [76]. A novel electrochemical sensor based on the sonochemical synthesis of functionalized MWCNTs adorned 3D flower-like zinc oxide (ZnO) NC for the 4-nitrophenol (4-NP) detection from environmental samples was reported by Balram et al. [77].

Doping graphene (G), as well as GQDs with nitrogen, is a viable technique for modulating chemical, electrical, and structural capabilities for their exceptional features in energy and environmental applications [78]. Because of graphene's unique two-dimensional structure, high conductivity, large surface area, and carrier mobility, along with strong electrochemical and optical capabilities, substantial research on graphene-based sensors for $Hg^{2+}$ detection with high efficiency was reported. Similarly, for nitrate and nitrite detection, a graphene-based NC-modified electrode exhibits great sensitivity and selectivity, a quick reaction rate, and outstanding stability. The LOD is decreased to nano-molar levels, with good selectivity in a variety of aqueous conditions. RGO has strong conductivity and catalytic activity in general, making it more suitable for the development of composite materials and the detection of nitrate and nitrite using an electrochemical sensor [19] (Figure 2.4).

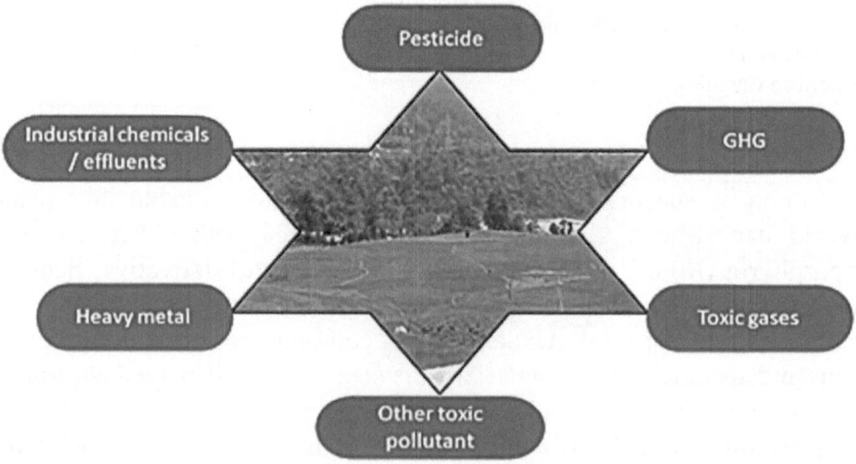


**FIGURE 2.4** Schematic of environmental sensing applications of carbon-based nanomaterials.

## 2.4 CONCLUSION

Carbon-based nanomaterials are proven candidates for sensing applications. Different allotropes of carbon, including graphene, quantum dots, CDs, SWCNT, MWCNT, and graphene oxides (GOs), were extensively investigated for their sensing properties. Carbon-derived materials offer new opportunities for developing optical sensors based on absorption fluorescence, surface plasmon resonance, and FERT. A wide variety of techniques were reported to exploit the optical features of carbon materials to develop sensors for biomolecules, ions, and gases. In addition, the electrochemical performance of carbon nanomaterials such as Cds, GQDs, graphene, and graphene-based materials makes them suitable candidates for electrochemical sensing. GQDs and their composites with different nanomaterials are used for the development of sensors based on amperometry and CV. Specifically, reports on optical, electrochemical, and environmental sensors based on carbon and its allotrope nanomaterials were reviewed, and analyzed their performance. Carbon-based optical nanosensors make using the fluorescence properties of CDs, GQDs, and GOs. Also, graphene exhibits surface plasmon polaritons leading to the fabrication of terahertz plasmonics-based sensors. Carbon nanomaterials attracted popularity in the last decade and still growing due to their potential electronic, electrical, and optical characteristics.

## KEYWORDS

- **carbon dots**
- **carbon nanotubes**
- **electrochemical sensors**
- **environmental sensors**
- **graphene**
- **graphene quantum dots**
- **optical sensors**

## REFERENCES


1. Tiwari, S. K., Kumar, V., Huczko, A., Oraon, R., Adhikari, A. D., & Nayak, G. C., (2016). Magical allotropes of carbon: Prospects and applications. *Critical Reviews in Solid State and Materials Sciences, 41*(4), 257–317.
2. Li, M., Chen, T., Gooding, J. J., & Liu, J., (2019). Review of carbon and graphene quantum dots for sensing. *ACS Sensors, 4*(7), 1732–1748.
3. Huang, Q., Lin, X., Tong, L., & Tong, Q. X., (2020). Graphene quantum dots/multi-walled carbon nanotubes composite-based electrochemical sensor for detecting dopamine release from living cells. *ACS Sustainable Chemistry & Engineering, 8*(3), 1644–1650.
4. Tajik, S., Dourandish, Z., Zhang, K., Beitollahi, H., Van Le, Q., Jang, H. W., & Shokouhimehr, M., (2020). Carbon and graphene quantum dots: A review on syntheses, characterization, biological and sensing applications for neurotransmitter determination. *RSC Advances, 10*(26), 15406–15429.
5. Ding, H., Hu, B., Zhang, B., Zhang, H., Yan, X., Nie, G., & Liang, M., (2021). Carbon-based nanozymes for biomedical applications. *Nano Research, 14*(3), 570–583.
6. Kurbanoglu, S., & Ozkan, S. A., (2018). Electrochemical carbon-based nanosensors: A promising tool in pharmaceutical and biomedical analysis. *Journal of Pharmaceutical and Biomedical Analysis, 147*, 439–457.
7. Vejpravová, J., (2021). Mixed sp2–sp3 nanocarbon materials: A status quo review. *Nanomaterials, 11*(10), 2469.
8. Ding, H., Wei, Y., Wu, Z., Tao, K., Ding, M., Xie, X., & Wu, J., (2020). Recent advances in gas and humidity sensors based on 3D structured and porous graphene and its derivatives. *ACS Materials Letters, 2*(11), 1381–1411.
9. Zhang, Q., Xie, X. M., Wei, S. Y., Zhu, Z. Z., Zheng, L. S., & Xie, S. Y., (2021). The synthesis of conical carbon. *Small Methods, 5*(3), 2001086.
10. Hu, S., (2016). Tuning optical properties and photocatalytic activities of carbon-based "quantum dots" through their surface groups. *The Chemical Record, 16*(1), 219–230.
11. Li, Z., Wang, L., Li, Y., Feng, Y., & Feng, W., (2019). Carbon-based functional nanomaterials: Preparation, properties, and applications. *Composites Science and Technology, 179*, 10–40.
12. Jović, D., Jaćević, V., Kuča, K., Borišev, I., Mrdjanovic, J., Petrovic, D., Seke, M., & Djordjevic, A., (2020). The puzzling potential of carbon nanomaterials: General properties, application, and toxicity. *Nanomaterials, 10*(8), 1508.
13. Liu, J., Li, R., & Yang, B., (2020). Carbon dots: A new type of carbon-based nanomaterial with wide applications. *ACS Central Science, 6*(12), 2179–2195.
14. Lei, Y., Yang, C., Hou, J., Wang, F., Min, S., Ma, X., Jin, Z., et al., (2017). Strongly coupled CdS/graphene quantum dots nanohybrids for highly efficient photocatalytic hydrogen evolution: Unraveling the essential roles of graphene quantum dots. *Applied Catalysis B: Environmental, 216*, 59–69.
15. Sajid, M. I., Jamshaid, U., Jamshaid, T., Zafar, N., Fessi, H., & Elaissari, A., (2016). Carbon nanotubes from synthesis to in vivo biomedical applications. *International Journal of Pharmaceutics, 501*(1, 2), 278–299.

16. Xiao, S., Zhu, X., Li, B. H., & Mortensen, N. A., (2016). Graphene-plasmon polaritons: From fundamental properties to potential applications. *Frontiers of Physics, 11*(2), 117801.
17. Novoselov, K. S., Fal, V. I., Colombo, L., Gellert, P. R., Schwab, M. G., & Kim, K., (2012). A roadmap for graphene. *Nature, 490*(7419), 192–200.
18. Shi, H., Wei, J., Qiang, L., Chen, X., & Meng, X., (2014). Fluorescent carbon dots for bioimaging and biosensing applications. *Journal of Biomedical Nanotechnology, 10*(10), 2677–2699.
19. Li, D., Wang, T., Li, Z., Xu, X., Wang, C., & Duan, Y., (2020). Application of graphene-based materials for detection of nitrate and nitrite in water—A review. *Sensors, 20*(1), 54.
20. Koutsogiannis, P., Thomou, E., Stamatis, H., Gournis, D., & Rudolf, P., (2020). Advances in fluorescent carbon dots for biomedical applications. *Advances in Physics: X, 5*(1), 1758592.
21. Liu, C., Chen, X., Zong, B., & Mao, S., (2019). Recent advances in sensitive and rapid mercury determination with graphene-based sensors. *Journal of Materials Chemistry A, 7*(12), 6616–6630.
22. Kour, R., Arya, S., Young, S. J., Gupta, V., Bandhoria, P., & Khosla, A., (2020). Recent advances in carbon nanomaterials as electrochemical biosensors. *Journal of The Electrochemical Society, 167*(3), 037555.
23. Liang, J., Liu, J., Yuan, X., Dong, H., Zeng, G., Wu, H., Wang, H., et al., (2015). Facile synthesis of alumina-decorated multi-walled carbon nanotubes for simultaneous adsorption of cadmium ion and trichloroethylene. *Chemical Engineering Journal, 273*, 101–110.
24. Hai, X., Feng, J., Chen, X., & Wang, J., (2018). Tuning the optical properties of graphene quantum dots for biosensing and bioimaging. *Journal of Materials Chemistry B, 6*(20), 3219–3234.
25. Gao, Y., (2017). Graphene and polymer composites for supercapacitor applications: A review. *Nanoscale Research Letters, 12*(1), 1–17.
26. Nag, A., Mitra, A., & Mukhopadhyay, S. C., (2018). Graphene and its sensor-based applications: A review. *Sensors and Actuators A: Physical, 270*, 177–194.
27. Pop, E., Varshney, V., & Roy, A. K., (2012). Thermal properties of graphene: Fundamentals and applications. *MRS Bulletin, 37*(12), 1273–1281. doi: 10.1557/mrs.2012.203.
28. Xu, Z., & Gao, C., (2015). Graphene fiber: A new trend in carbon fibers. *Materials Today, 18*(9), 480–492.
29. Shende, P., Augustine, S., & Prabhakar, B., (2020). A review on graphene nanoribbons for advanced biomedical applications. *Carbon Letters,* 1–11.
30. Kumar, A., Sharma, K., & Dixit, A. R., (2021). A review on the mechanical properties of polymer composites reinforced by carbon nanotubes and graphene. *Carbon Letters, 31*(2), 149–165.
31. Wang, L., Zhang, Y., Wu, A., & Wei, G., (2017). Designed graphene-peptide nanocomposites for biosensor applications: A review. *Analytica Chimica Acta, 985*, 24–40.

32. Xu, J., Wang, Y., & Hu, S., (2017). Nanocomposites of graphene and graphene oxides: Synthesis, molecular functionalization and application in electrochemical sensors and biosensors. A review. *Microchimica Acta, 184*(1), 1–44.
33. Peña-Bahamonde, J., Nguyen, H. N., Fanourakis, S. K., & Rodrigues, D. F., (2018). Recent advances in graphene-based biosensor technology with applications in life sciences. *Journal of Nanobiotechnology, 16*(1), 1–17.
34. Mohanta, D., Patnaik, S., Sood, S., & Das, N., (2019). Carbon nanotubes: Evaluation of toxicity at biointerfaces. *Journal of Pharmaceutical Analysis, 9*(5), 293–300.
35. Camilli, L., & Passacantando, M., (2018). Advances on sensors based on carbon nanotubes. *Chemosensors, 6*(4), 62.
36. Rasheed, T., Nabeel, F., Adeel, M., Rizwan, K., Bilal, M., & Iqbal, H. M., (2019). Carbon nanotubes-based cues: A pathway to future sensing and detection of hazardous pollutants. *Journal of Molecular Liquids, 292*, 111425.
37. Jain, N., Gupta, E., & Kanu, N. J., (2021). Plethora of carbon nanotubes applications in various fields–a state-of-the-art review. *Smart Science,* 1–24.
38. Vichchulada, P., Zhang, Q., & Lay, M. D., (2007). Recent progress in chemical detection with single-walled carbon nanotube networks. *Analyst, 132*(8), 719–723.
39. Su, H. C., Tran, T. T., Bosze, W., & Myung, N. V., (2020). Chemiresistive sensor arrays for detection of air pollutants based on carbon nanotubes functionalized with porphyrin and phthalocyanine derivatives. *Sensors and Actuators Reports, 2*(1), 100011.
40. He, X., Htoon, H., Doorn, S. K., Pernice, W. H. P., Pyatkov, F., Krupke, R., Jeantet, A., Chassagneux, Y., & Voisin, C., (2018). Carbon nanotubes as emerging quantum-light sources. *Nature Materials, 17*(8), 663–670.
41. Norizan, M. N., Moklis, M. H., Demon, S. Z. N., Halim, N. A., Samsuri, A., Mohamad, I. S., Knight, V. F., & Abdullah, N., (2020). Carbon nanotubes: Functionalization and their application in chemical sensors. *RSC Advances, 10*(71), 43704–43732.
42. Celasco, E., & Chaika, A. N., (2019). *Handbook of Graphene: Growth, Synthesis, and Functionalization* (Vol. 1). Hoboken NJ: John Wiley & Sons.
43. Abbas, A., Mariana, L. T., & Phan, A. N., (2018). Biomass-waste derived graphene quantum dots and their applications. *Carbon, 140*, 77–99.
44. Ravi, S., & Vadukumpully, S., (2016). Sustainable carbon nanomaterials: Recent advances and its applications in energy and environmental remediation. *Journal of Environmental Chemical Engineering, 4*(1), 835–856.
45. Yao, J., Yang, M., & Duan, Y., (2014). Chemistry, biology, and medicine of fluorescent nanomaterials and related systems: New insights into biosensing, bioimaging, genomics, diagnostics, and therapy. *Chemical Reviews, 114*(12), 6130–6178.
46. Li, H., Yan, X., Kong, D., Jin, R., Sun, C., Du, D., Lin, Y., & Lu, G., (2020). Recent advances in carbon dots for bioimaging applications. *Nanoscale Horizons, 5*(2), 218–234.
47. Nißler, R., Mann, F. A., Preiß, H., Selvaggio, G., Herrmann, N., & Kruss, S., (2019). Chirality enriched carbon nanotubes with tunable wrapping via corona phase exchange purification (CPEP). *Nanoscale, 11*(23), 11159–11166.
48. Heller, D. A., Baik, S., Eurell, T. E., & Strano, M. S., (2005). Single-walled carbon nanotube spectroscopy in live cells: Towards long-term labels and optical sensors. *Advanced Materials, 17*(23), 2793–2799.

49. Clendenin, J., Kim, J. W., & Tung, S., (2007). An aligned carbon nanotube biosensor for DNA detection. In: *2007 2nd IEEE International Conference on Nano/Micro Engineered and Molecular Systems* (pp. 1028–1033). IEEE.
50. Yang, C. J., Jockusch, S., Vicens, M., Turro, N. J., & Tan, W., (2005). Light-switching excimer probes for rapid protein monitoring in complex biological fluids. *Proceedings of the National Academy of Sciences, 102*(48), 17278–17283.
51. Antonucci, A., Kupis-Rozmysłowicz, J., & Boghossian, A. A., (2017). Noncovalent protein and peptide functionalization of single-walled carbon nanotubes for biodelivery and optical sensing applications. *ACS Applied Materials & Interfaces, 9*(13), 11321–11331.
52. Ren, J., Weber, F., Weigert, F., Wang, Y., Choudhury, S., Xiao, J., Lauermann, I., et al., (2019). Influence of surface chemistry on optical, chemical and electronic properties of blue luminescent carbon dots. *Nanoscale, 11*(4), 2056–2064.
53. Yan, F., Jiang, Y., Sun, X., Bai, Z., Zhang, Y., & Zhou, X., (2018). Surface modification and chemical functionalization of carbon dots: A review. *Microchimica Acta, 185*(9), 1–34.
54. Sahiner, N., Suner, S. S., Sahiner, M., & Silan, C., (2019). Nitrogen and sulfur-doped carbon dots from amino acids for potential biomedical applications. *Journal of Fluorescence, 29*(5), 1191–1200.
55. Wang, B., Liu, F., Wu, Y., Chen, Y., Weng, B., & Li, C. M., (2018). Synthesis of catalytically active multielement-doped carbon dots and application for colorimetric detection of glucose. *Sensors and Actuators B: Chemical, 255*, 2601–2607.
56. Wang, Z., Liao, H., Wu, H., Wang, B., Zhao, H., & Tan, M., (2015). Fluorescent carbon dots from beer for breast cancer cell imaging and drug delivery. *Analytical Methods, 7*(20), 8911–8917.
57. Chandra, A., & Singh, N., (2017). Cell microenvironment pH sensing in 3D microgels using fluorescent carbon dots. *ACS Biomaterials Science & Engineering, 3*(12), 3620–3627.
58. Garg, M., Vishwakarma, N., Sharma, A. L., & Singh, S., (2021). Amine-functionalized graphene quantum dots for fluorescence-based immunosensing of ferritin. *ACS Applied Nano Materials, 4*(7), 7416–7425.
59. Farmani, H., & Farmani, A., (2020). Graphene sensing nanostructure for exact graphene layers identification at terahertz frequency. *Physica E: Low-dimensional Systems and Nanostructures, 124*, 114375.
60. Xu, W., Huang, Y., Zhou, R., Wang, Q., Yin, J., Kono, J., Ping, J., et al., (2020). Metamaterial-free flexible graphene-enabled terahertz sensors for pesticide detection at bio-interface. *ACS Applied Materials & Interfaces, 12*(39), 44281–44287.
61. Li, C., Chen, G., Zhang, Y., Wu, F., & Wang, Q., (2020). Advanced fluorescence imaging technology in the near-infrared-II window for biomedical applications. *Journal of the American Chemical Society, 142*(35), 14789–14804.
62. Trusovas, R., Račiukaitis, G., Niaura, G., Barkauskas, J., Valušis, G., & Pauliukaite, R., (2016). Recent advances in laser utilization in the chemical modification of graphene oxide and its applications. *Advanced Optical Materials, 4*(1), 37–65.
63. Facure, M. H., Schneider, R., Lima, J., Mercante, L. A., & Correa, D. S., (2021). Graphene quantum dots-based nanocomposites applied in electrochemical sensors: A recent survey. *Electrochem, 2*(3), 490–519.

64. Li, C., Qiu, J., Ou, J. Y., Liu, Q. H., & Zhu, J., (2019). High-sensitivity refractive index sensors using coherent perfect absorption on graphene in the Vis-NIR region. *ACS Applied Nano Materials, 2*(5), 3231–3237.
65. Zuo, Y., Xu, J., Zhu, X., Duan, X., Lu, L., & Yu, Y., (2019). Graphene-derived nanomaterials as recognition elements for electrochemical determination of heavy metal ions: A review. *Microchimica Acta, 186*(3), 1–17.
66. Hou, X., Xu, H., Zhen, T., & Wu, W., (2020). Recent developments in three-dimensional graphene-based electrochemical sensors for food analysis. *Trends in Food Science & Technology.*
67. Liu, W., Li, M., Jiang, G., Li, G., Zhu, J., Xiao, M., Zhu, Y., et al., (2020). Graphene quantum dots-based advanced electrode materials: Design, synthesis and their applications in electrochemical energy storage and electrocatalysis. *Advanced Energy Materials, 10*(29), 2001275.
68. Guo, W., Pi, F., Zhang, H., Sun, J., Zhang, Y., & Sun, X., (2017). A novel molecularly imprinted electrochemical sensor modified with carbon dots, chitosan, gold nanoparticles for the determination of patulin. *Biosensors and Bioelectronics, 98*, 299–304.
69. Power, A. C., Gorey, B., Chandra, S., & Chapman, J., (2018). Carbon nanomaterials and their application to electrochemical sensors: A review. *Nanotechnology Reviews, 7*(1), 19–41.
70. Chuang, C. H., & Kung, C. W., (2020). Metal-organic frameworks toward electrochemical sensors: Challenges and opportunities. *Electroanalysis, 32*(9), 1885–1895.
71. Lakshmanakumar, M., Nesakumar, N., Sethuraman, S., Krishnan, U. M., & Rayappan, J. B. B., (2021). Fabrication of GQD-electrodeposited screen-printed carbon electrodes for the detection of the CRP biomarker. *ACS Omega.*
72. Song, M., Dang, L., Long, J., & Hu, C., (2018). Laser-cut polymer tape templates for scalable filtration fabrication of user-designed and carbon-nanomaterial-based electrochemical sensors. *ACS Sensors, 3*(12), 2518–2525.
73. Thamaraiselvan, C., Wang, J., James, D. K., Narkhede, P., Singh, S. P., Jassby, D., Tour, J. M., & Arnusch, C. J., (2020). Laser-induced graphene and carbon nanotubes as conductive carbon-based materials in environmental technology. *Materials Today, 34*, 115–131.
74. Rahman, M. M., Balkhoyor, H. B., & Asiri, A. M., (2017). Phenolic sensor development based on chromium oxide-decorated carbon nanotubes for environmental safety. *Journal of Environmental Management, 188*, 228–237.
75. Rahman, M. M., Hussein, M. A., Alamry, K. A., Al-Shehry, F. M., & Asiri, A. M., (2018). Polyaniline/graphene/carbon nanotubes nanocomposites for sensing environmentally hazardous 4-aminophenol. *Nano-Structures & Nano-Objects, 15*, 63–74.
76. Mehta, J., Bhardwaj, N., Bhardwaj, S. K., Tuteja, S. K., Vinayak, P., Paul, A. K., Kim, K. H., & Deep, A., (2017). Graphene quantum dot modified screen printed immunosensor for the determination of parathion. *Analytical Biochemistry, 523*, 1–9.
77. Balram, D., Lian, K. Y., & Sebastian, N., (2020). Ultrasound-assisted synthesis of 3D flower-like zinc oxide decorated fMWCNTs for sensitive detection of toxic environmental pollutant 4-nitrophenol. *Ultrasonics Sonochemistry, 60*, 104798.

78. Kaur, M., Kaur, M., & Sharma, V. K., (2018). Nitrogen-doped graphene and graphene quantum dots: A review on synthesis and applications in energy, sensors, and environment. *Advances in Colloid and Interface Science, 259*, 44–64.
79. Geim, A. K., & Novoselov, K. S., (2010). The rise of graphene. In: *Nanoscience and Technology: A Collection of Reviews from Nature Journals* (pp. 11–19).
80. Campuzano, S., Yáñez-Sedeño, P., & Pingarrón, J. M., (2019). Carbon dots and graphene quantum dots in electrochemical biosensing. *Nanomaterials, 9*(4), 634.
81. Liu, L., Anwar, S., Ding, H., Xu, M., Yin, Q., Xiao, Y., Yang, X., et al., (2019). Electrochemical sensor based on F, N-doped carbon dots decorated laccase for detection of catechol. *Journal of Electroanalytical Chemistry, 840*, 84–92.
82. Zhao, A., Zhao, C., Li, M., Ren, J., & Qu, X., (2014). Ionic liquids as precursors for highly luminescent, surface-different nitrogen-doped carbon dots used for label-free detection of $Cu^{2+}/Fe^{3+}$ and cell imaging. *Analytica Chimica Acta, 809*, 128–133.
83. Fahmi, M. Z., Haris, A., Permana, A. J., Wibowo, D. L. N., Purwanto, B., Nikmah, Y. L., & Idris, A., (2018). Bamboo leaf-based carbon dots for efficient tumor imaging and therapy. *RSC Advances, 8*(67), 38376–38383.
84. Hajati, Y., (2020). Tunable broadband multi-resonance graphene terahertz sensor. *Optical Materials, 101*, 109725.
85. Justino, C. I., Gomes, A. R., Freitas, A. C., Duarte, A. C., & Rocha-Santos, T. A., (2017). Graphene based sensors and biosensors. *TrAC Trends in Analytical Chemistry, 91*, 53–66.
86. Piras, A., Ehlert, C., & Gryn'ova, G., (2021). Sensing and sensitivity: Computational chemistry of graphene-based sensors. *Wiley Interdisciplinary Reviews: Computational Molecular Science, 11*(5), e1526.
87. Arshad, A., Jabbal, M., Yan, Y., & Reay, D., (2019). A review on graphene based nanofluids: Preparation, characterization, and applications. *Journal of Molecular Liquids, 279*, 444–484.
88. Krishnan, S. K., Singh, E., Singh, P., Meyyappan, M., & Nalwa, H. S., (2019). A review on graphene-based nanocomposites for electrochemical and fluorescent biosensors. *RSC Advances, 9*(16), 8778–8881.
89. Beitollahi, H., Movahedifar, F., Tajik, S., & Jahani, S., (2019). A review on the effects of introducing CNTs in the modification process of electrochemical sensors. *Electroanalysis, 31*(7), 1195–1203.
90. Landry, M. P., Vukovic, L., Kruss, S., Bisker, G., Landry, A. M., Islam, S., Jain, R., Schulten, K., & Strano, M. S., (2015). Comparative dynamics and sequence dependence of DNA and RNA binding to single walled carbon nanotubes. *The Journal of Physical Chemistry C, 119*(18), 10048–10058.
91. Gillen, A. J., Siefman, D. J., Wu, S. J., Bourmaud, C., Lambert, B. P., & Boghossian, A. A., (2019). The impact of exposed surface area on the response of SWCNT optical sensors. In: *ECS Meeting Abstracts* (No. 9, p. 720). IOP Publishing.
92. Sharifi, A. H., Zahmatkesh, I., Bamoharram, F. F., Shokouhi, T. A. H., Razavi, S. F., & Saneinezhad, S., (2020). Experimental measurement of thermophysical properties of alumina-MWCNTs/salt-water hybrid nanofluids. *Current Nanoscience, 16*(5), 734–747.

93. Shumeiko, V., Paltiel, Y., Bisker, G., Hayouka, Z., & Shoseyov, O., (2020). A paper-based near-infrared optical biosensor for quantitative detection of protease activity using peptide-encapsulated SWCNTs. *Sensors, 20*(18), 5247.
94. Kruss, S., Hilmer, A. J., Zhang, J., Reuel, N. F., Mu, B., & Strano, M. S., (2013). Carbon nanotubes as optical biomedical sensors. *Advanced Drug Delivery Reviews, 65*(15), 1933–1950.
95. Kaushik, B. K., & Majumder, M. K., (2015). *Carbon Nanotube-Based VLSI Interconnects: Analysis and Design* (pp. 1–14). New Delhi: Springer India.
96. Zhao, X., Lin, C. W., Wang, J., & Oh, D. H., (2014). Advances in rapid detection methods for foodborne pathogens. *Journal of Microbiology and Biotechnology, 24*(3), 297–312.
97. Facure, M. H., Schneider, R., Lima, J., Mercante, L. A., & Correa, D. S., (2021). Graphene quantum dots-based nanocomposites applied in electrochemical sensors: A recent survey. *Electrochem., 2*(3), 490–519.

## CHAPTER 3

# Carbon Nanotubes and Their Environmental Applications

I. S. VIDYALAKSHMI,[1] APARNA RAJ,[2] S. NEELIMA,[2] L. VIDYA,[2] and RIJU K. THOMAS[3*]

*[1]SB College, Changanacherry, Kottayam, Kerala, India*

*[2]School of Pure and Applied Physics, Mahatma Gandhi University, Kottayam, Kerala, India*

*[3]Bharata Mata College, Thrikkakara, Ernakulam, Kerala, India*

*[*]Corresponding author. E-mail: rijukthomas@gmail.com*

## ABSTRACT

Carbon nanotubes (CNTs) are the most incredible material in the fullerene family. The graphene sheets with a quasi-one-dimensional structure of nanometer scale diameter in the rolled-up form are called CNTs. Also, they are often called tubular fullerene. Due to their exceptional physical, chemical, mechanical, electronic, and thermal properties can do wonders in the various research areas of nanoscience. The significant aspects of CNTs, such as their lightweight, diminutive size with a high aspect ratio, fine tensile strength, and good conducting characteristics, make them a good candidate for various applications like biomedical, materials in electronics, sensors, field emission devices, energy storage, and conversion, etc. The most interesting fact is that they are environment-friendly materials, and this makes them the most beloved materials for environmental applications. In this book chapter, we present a general readership about the environmental applications of CNTs.

Carbon Composites: Composites with Nanotubes, Nanomaterials, and Graphene Oxide.
Eduardo A. Castro, Ann Rose Abraham, & A. K. Haghi (Eds.)

## 3.1 INTRODUCTION

The word "Nano" in nanotechnology is a Greek word, and it means dwarf. Scientific dealing at the nano level (atomic level) by using particular scientific instruments is acknowledged as nanotechnology, and the term nanotechnology was introduced by a Japanese scientist named Norio Taniguchi in 1974 [1]. The arrival of nanotechnology has brought a huge revolution in the area of research. Generally, materials having dimensions between 1 nm and 100 nm are called nonmaterial. More conveniently, we can say that a sheet of paper will be about $10^5$ nm thick while a single gold atom is about $\frac{1}{3}$ nm in diameter [2]. Hence, we can say that nanomaterials have a size larger than single atoms or molecules, but they are smaller than bulk materials. Their characteristic properties lie in between classical and quantum physics. The dimensionality of nanoparticles is classified into zero, one, and two. Since they process a low dimensionality, they have a large surface-to-volume ratio and better electrical and optical properties when compared with a bulk sample of the same material. All these properties take nanoparticles to a large area of research. Among all the elements, carbon has a much impression in the field of nanotechnology, and there are wide varieties of carbon-based nanomaterials.

Carbon, the first element of the $14^{th}$ group in the periodic table, is an exceptional and essential element in our world. It is known to be the sixth most common element in the universe, the $4^{th}$ most common element in our solar system, and the $17^{th}$ most common element in the earth's crust [3]. The relative abundance of carbon has been estimated about between 180 and 270 parts per million (ppm) [4]. Moreover, it is known to be the second most common element in the human body, which consists of about 18% of a human's body weight [5]. Carbon possesses a broad variety of metastable phases that can be produced near ambient circumstances and their roomy fields of kinetic stability, and this is one of the stupendous descriptions of carbon [6].

Even though the amount of elemental carbon present on the earth's crust is about 0.2% of the total mass of the earth, it exhibits implausible essential functions like bonding with other light elements [6]. This has paved the way for revolutionary research fields in subjects like chemistry and biology. The developments in carbon science led to the innovation of many such low dimensional allotropes of carbon nanostructures such as graphite, activated carbon (AC), carbon nanotubes (CNTs), and the $C_{60}$ family of buckyballs, poly-aromatic molecules, and graphene in various fields of nanoscience, material science and engineering technology [6]. The significant application of nanotechnology is that it can produce new materials with elite properties.

## 3.2 A BRIEF HISTORICAL CHEMICAL BACKGROUND OF SOME ALLOTROPES OF CARBON

For a very long period of time, there were only two natural crystalline allotropes of carbon were discovered, and they are graphite and diamond. However, both structures possess extremely different chemistry, especially in their structure and properties [7]. The covalent bonding between two carbon atoms may result in the formation of many new other allotropes of carbon [6]. The elements that are identical in their chemical properties and differ in their structural, as well as physical properties, are known as allotropes. There are several allotropes of carbon have been discovered, and they are graphene, Buckminsterfullerene, and CNTs and thus making carbon an element that contains a large number of allotropes [8–10] (Figure 3.1).

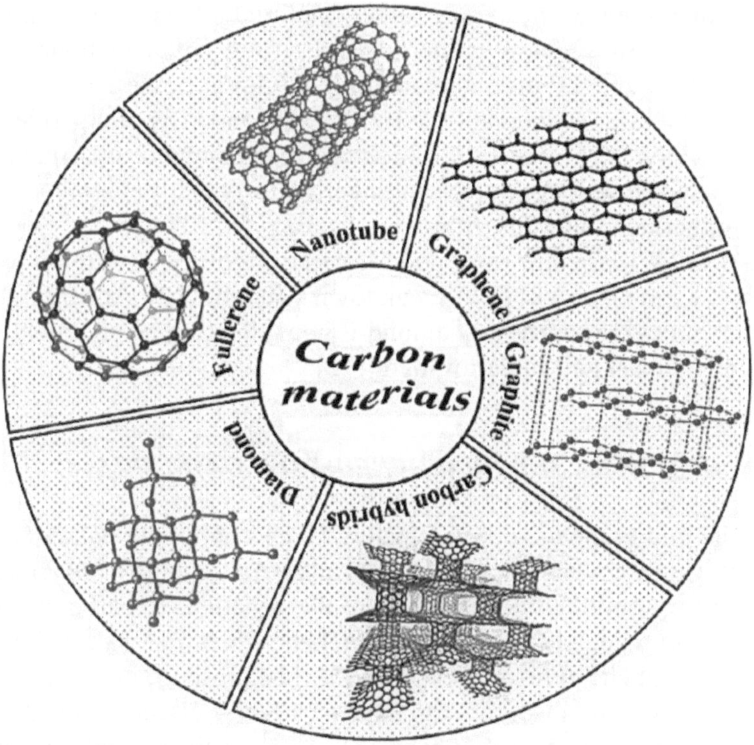


**FIGURE 3.1** Structure of carbon nanomaterials: Diamond, graphite, $C_{60}$, CNTs, graphene, and 3D graphene-CNT hybrid materials.

*Source:* Reprinted with Ref. [10]. https://creativecommons.org/licenses/by-nc/4.0/

### *3.2.1 GRAPHITE*

A layered structure consisting of six carbon atoms to form a ring arranged in widely spaced horizontal sheets is known as graphite. Since it has excellent electrical conductivity, it is mainly used for making electrodes in an electrical arc lamp. In graphite, the delocalization of $\pi$-electrons is responsible for the ability to conduct electricity. This phenomenon is absent in diamonds. In addition to this, graphite is known as the most stable form of carbon in standard conditions [6].

### *3.2.2 DIAMOND*

The most peculiar property of a diamond is its highest thermal conductivity when compared to any other material. This is determinedly endorsed by low phonon scattering and strong covalent bonding that held together the atoms. The natural diamond possesses a thermal conductivity of approximately 2,200 W/Mk, and it is about five times that of copper [11]. Due to the high thermal conductivity, diamond is broadly used in semiconductor engineering to prevent silicon and other semiconducting materials from overheating [6]. The other important properties of diamonds are good carrier mobility, saturated carrier velocities, and electrical field breakdown strength [12]. According to its physiochemical properties, Diamond is considered to be chemically and physically robust and radiation-hard [12].

### *3.2.3 GRAPHENE AND GRAPHENE-DERIVED MATERIALS*

Graphene and graphene-derived materials are commonly called Graphenoids. It was discovered by Andre Geim and Novoselov in 2004. Graphene is the primary building block of most of the structures of carbon materials [13]. The arrangement of piled-up graphene gives rise to graphite, and rolled-up graphene gives rise to CNTs [8]. The presence of defects, impurities, and structural disorders can have an adverse effect on the electrical and optical properties of graphene [8]. As a result, a single crystalline graphene thin film with high electrical and thermal conductivities is preferred for electronic applications [6].

### *3.2.4 ACTIVATED CARBON (AC)*

Activated carbon (AC) comes under the category of amorphous carbon due to its structure and properties. It has a microporous structure [6]. The materials that are composed of AC exhibit an excellent adsorption capacity for various chemical species. The significant property of AC is its high surface area to volume ratio, and therefore it is commonly used as an adsorbent in poisoning, reducing the cholesterol level plummeting internal gas flatulence, and many other useful applications [14, 15].

## 3.3 CARBON NANOTUBES (CNTS)

Fullerenes are often called molecular forms of carbon or carbon molecules. In 1985 three scientists, named H. W. Kroto, R. F. Curl, and R. E. Smalley, were discovered fullerenes and awarded with Nobel prize in chemistry in later. The family of fullerene consists of a number of atomic $C_n$ clusters (n>20), composed of carbon atoms on a spherical surface, and the carbon atoms are typically positioned on the vertices of the pentagons and hexagons [16]. Carbon atoms in fullerenes are in a $sp^2$-hybrid state and linked together by covalent bonds. The most common and the best-investigated fullerene is the $C_{60}$ fullerene, in which the spherical molecule is highly symmetric, and it consists of 60 carbon atoms situated at the vertices of 20 hexagons and 12 pentagons [16]. These closed $C_{60}$ fullerenes are informally called *Buckyballs*.

Another member of this fullerene which has a wide range of applications, is the CNTs. Being a member of the fullerene family, they are often called *buckytubes* [1]. As the word 'nanotube' demands in CNTs, the material consists of only two coaxial cylinders [6]. The discovery of CNTs thus made a revolution in the field of nanotechnology. CNTs were discovered independently by Iijima in 1991 and Benthune et al. in 1993. They are flawless cylinder-shaped macromolecules having a radius of a few nanometers and up to some micrometers in length. The walls of these tubes are made up of a hexagonal lattice of carbon atoms and capped by fullerene-like structures. According to their structure, they can be divided into two, multi-walled carbon nanotubes (MWCNTs) and single-walled carbon nanotubes (SWCNTs). A coaxial arrangement of two or more concentric cylindrical shells of graphene sheets around a central hollow area having a spacing between the layers will form a MWCNT, and they have a large diameter lies between 2 nm and 100 nm [2]. At the same time, a single cylinder graphite sheet held together by Van der Waals bonds will form a SWCNT and have a diameter of less than 2 nm [10]. Due to the narrow diameter, they exhibit excellent material properties like a high aspect ratio and large surface area. As a result, they can be approximated into one-dimensional nanostructures. The helicity and diameter of a CNT determine its electrical properties [2]. Their changing hybridization states and sensitivity of the structure to perturbations in synthesis conditions exploit their unique physical, chemical, and electronic properties [17]. This leads to innovations in new technologies and applications. Besides, these unique and fine properties offer impending advances in environmental systems from proactive (prevention of environmental degradation, optimizing energy efficiency) to retroactive (wastewater reuse, pollutant transformation) [17].

### 3.3.1 PROPERTIES OF CARBON NANOTUBES (CNTS)

It is reported that CNTs possess a large surface area, high aspect ratios, and remarkably high mechanical strengths [1]. They have a tensile strength that is 100 times greater than that of steel, and their electrical and thermal conductivity approaches that of copper [18].

#### 3.3.1.1 ELECTRONIC PROPERTIES OF CNTS

The electrical properties of CNTs are in a chiral form [1]. Each carbon atom in CNTs is covalently bonded to each other with $sp^2$ hybridization [19]. So, in each unit, the fourth valance electron remains free. Hence the delocalization of these free electrons over all atoms will occur, and this motion will help to attain the electrical nature of CNTs. Depending on the type of chirality, CNTs can be either conducting or semiconducting materials. Semiconducting SWNTs will be most likely to be of p-type semiconductors [20]. This electronic nature of CNTs can be applied to transistors and other switching applications, and the most recent application of CNT was an emitter [1].

#### *3.3.1.2 MECHANICAL PROPERTIES OF CNTS*

By considering their mechanical properties, CNTs were considered the strongest materials in nature. They have Young's modulus ranging from 270 to 950 GPa [1]. They also possess very high tensile strength ranges of 11–63 GPa. The covalent carbon bonds present in graphite are found to be the strongest ones in nature, and hence CNTs are considered to be the stiffest and toughest structure ever synthesized by scientists [1].

#### *3.3.1.3 THERMAL PROPERTIES OF CNTS*

Since the CNTs are very small in size, the quantum effects are very significant. Hence the low-temperature specific heat and thermal conductivity show direct evidence of the one-dimensional quantization of the phonon band structure in CNTs [21]. The factors affecting the thermal properties of CNTs are a number of phonon-active modes, the length of the free path for the phonons, and boundary surface scattering [22]. These properties may also depend on the atomic arrangement, the length and diameter of the tubes, the number of structural defects, and the morphology [23].

All these excellent properties of CNTs offer great promise in their environmental applications. They are used as sensors and sorbents for the detection and treatment of existing contaminants and thereby preventing pollution [24]. They are used for the storage of energy, such as in solar cells, used for the storage of hydrogen in hydrogen fuel cells, and for environment-friendly energy storage in the form of biofuels. Their water treatment applications include the treatment of wastewater and removing the heavy metals from the contaminated water, and they are also used for water transportation. Another important application is they are used as a sensor for the detection of toxic gases, and thereby they help to control air pollution. Another environmentally friendly application is that they are used for $CO_2$ capture. All these peculiarities and excellent properties make CNTs the most interesting nanomaterials to researchers.

## 3.4 ENVIRONMENTAL APPLICATIONS OF CARBON NANOTUBES (CNTS)

Nowadays, modern society is facing diverse public health implications as a result of environmental pollution. According to WHO, the global ambient air quality database (update 2018) estimates that 7 million people die every year due to exposure to fine particles in polluted air leading to diseases such as respiratory infections, heart disease, stroke, lung cancer, and chronic obstructive pulmonary diseases. Chemicals (natural/synthetic) that have adverse ecological effects on the environment and are not commonly monitored are called emerging pollutants (EP) [25]. EPs mainly include drugs, illicit drugs, pesticides, furans, dioxins, micro-plastics, perfluoroalkyl substances, Polycyclic aromatic hydrocarbons, and so on. They reach aquatic systems through discharge from wastewater treatment plants, sewage, runoff, etc. It can be bioaccumulated and biomagnified, causing adverse effects on the ecosystem. The occurrence of EP in the aquatic system has not been widely studied [26]. It may be due to the presence of organic compounds, lipids, and proteins that make them complex. Therefore, there is an urgent need for efficient materials for environmental pollution management [27]. Graphene/CNTs-based materials are promising candidates for environmental applications from sensing and monitoring to remediation [28]. New nanomaterials for the extraction of emerging organic contaminants (EOC) are CNTs, graphene-based materials, and metallic nanoparticles [25].

### 3.4.1 ENERGY APPLICATIONS

#### 3.4.1.1 HYDROGEN GAS STORAGE IN CNTS

The idea of hydrogen as a fuel has been a drastic increase in the production rate and utilization of technologies for gas. The reason is that hydrogen provides more energy than either gasoline or natural gas on the basis of its weight [29]. Among many ways of storing hydrogen, the most effective one is the use of CNTs. They can store hydrogen even at room temperatures (RTs). But the condition is that the capacities do not exceed ~0 wt.%. All in all, the reported capacities range from 0 to 60 wt.% [1]. These are categorized into three: (i) having an H/C ratio of 2 (~ 14 wt.% H) and are considered to be un-physical; (ii) consistent with expectations based on findings for activated and other, conventional high-surface-area carbons (~ 0 wt.%); and (iii) those are in the intermediate range (1–14 wt.%) [1]. According to the experimental results, the single-walled carbon nanotube structures can store about 8 wt.% of hydrogen.

Small pores in low pressure enhance the adsorption in pores and thus results in high storage [24], and the reverse effect will occur for high pressures. The excellent conductive properties, low mass densities, and strong physical properties of CNTs make them ideal and durable materials for fuel cell electrodes. In addition, the macrostructures formed by assembling the nanotubes possess a high surface area, and this makes them a suitable substitute for Pt catalysts and hydrogen adsorption [31]. A hydrogen fuel cell with CNT composite electrodes or membranes helps to reduce the use of rare and expensive catalysts [31]. However, further studies are being done in this field to get a perfect hydrogen-storing CNT structure (Figure 3.2).

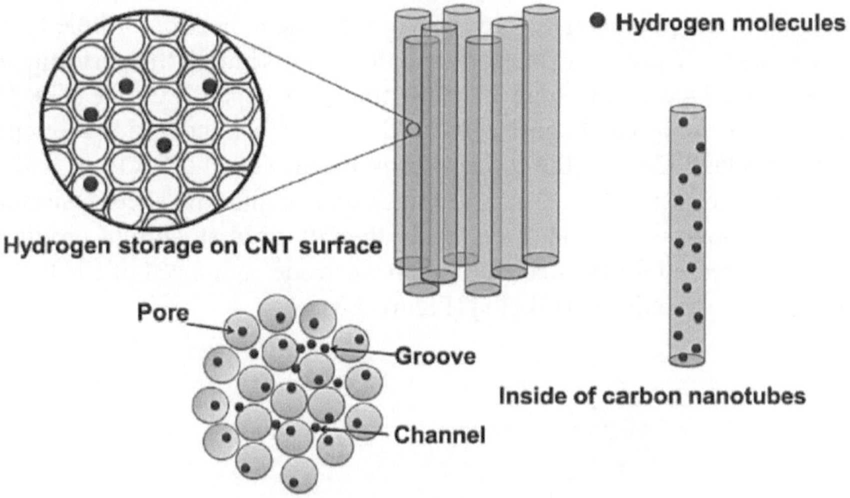


**FIGURE 3.2** Different storage sites of carbon nanotubes.

*Source:* Reprinted with permission from Ref. [31a]. © 2018 Elsevier.

### *3.4.1.2 CARBON NANOTUBES (CNTS) IN RENEWABLE ENERGY*

Nowadays, energy consumption is increasing across the world. So, long-term energy demand has become a challenge. Therefore, we should be aware of the requirement for advanced renewable energy technologies to protect the environmental balance. The application of CNTs in photovoltaic devices led to the use them in the solar cell was a major breakthrough. The device which absorbs photons from sunlight and converts them to electrical energy is called a photovoltaic device. The affordability and remarkable energy conversion of CNTs make them an important material in various solar cell structures, especially in silicon-based solar cells [32]. Their extremely small surface area enhances the massive absorption of photons for harvesting solar energy, while the mobility of charge transfer is facilitated by the delocalized π electron system [33]. Apart from silicon-based solar cells, there is another one known as an organic solar cell [31]. It consists of a conductive organic polymer like poly(3-octylthiophene) (P3OT), poly(3-hexylthiophene) (P3HT) or [6,6]-phenyl-C61-butyric acid methyl ester (PCBM) for light absorption and charge transfer [24].

The present research shows an increase in efficiency upon the incorporation of CNTs in the top electrode, the photoactive layer, and the back electrode of organic solar cells [35–37]. The role of CNT as a photoactive material in the photoactive layer provides efficient electron or hole transport at the CNT/polymer interface and thereby optimizes the performance of the cell. Due to the high electron transport capability of CNTs, the photoactive component constructed with P3OT/CNT shows a higher open circuit voltage [38]. In the top and back electrodes, the CNTs provide a large surface area for high optical transmittance and low sheet resistance to reduce power loss [31]. As a result, the rate of photocurrent produced will be increased for the design of a top electrode composed of CNT films and indium-tin-oxide (ITO) [17] (Figure 3.3).

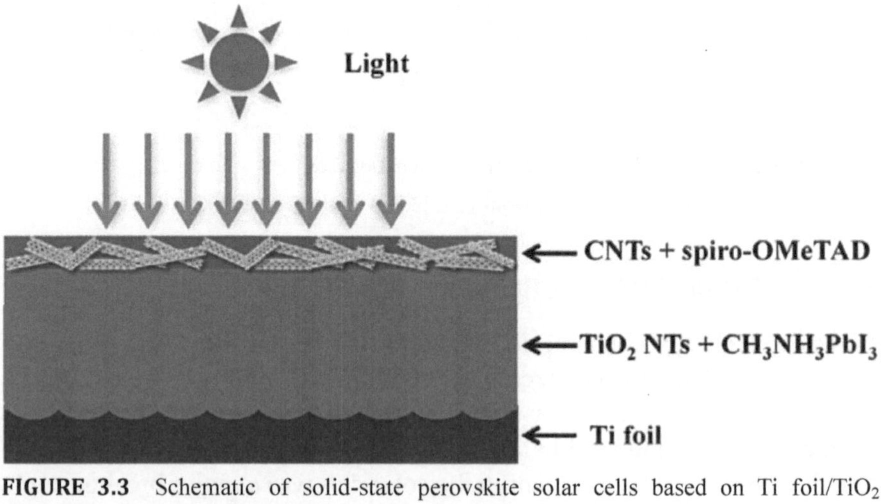


**FIGURE 3.3** Schematic of solid-state perovskite solar cells based on Ti foil/$TiO_2$ nanotubes and carbon nanotubes.

*Source:* Reprinted with permission from Ref. [17a]. © 2015 Elsevier.

### *3.4.1.3 CARBON NANOTUBES (CNTS) IN BIOFUEL CELLS*

The uses of cell phones, computers, and other electronic gadgets have become responsible for the large production of lithium batteries in the environment, which are absolutely non-environment-friendly chemicals. Hence the need for the clean production of electricity has stimulated the origin of new sources of sustainable and renewable energy without greenhouse gas emissions and environmental pollution. The increasing interest in environment-friendly technologies has opened doors to inventions in biotechnology that make specific useful products by using living organisms. A cell that produces electricity by making use of living organisms is known as a biofuel cell. Biofuel cells can be either microbial fuel cells (MFC) or enzymatic biofuel cells (EFC) [17]. MFC generates electric power by using its microbial catabolic activities. Taking complex organic waste and renewable biomass in wastewater as the source materials, MFCs can do better in wastewater treatments [40]. However, the research works on CNT electrodes are progressing. Due to their high conductivity and large surface area, the CNT electrodes are modified for increased power production [41]. The high specific surface area of CNT is very useful for the construction of highly porous 3-D CNT electrodes. All these peculiarities have made CNT become one of the important materials for the designing of electrochemical biosensors and biofuels [2]. In addition to it, attaching specific redox sites to these nanostructured electrodes through organic functionalization has tuned them very well, and this is very suitable for fixing proteins or catalyzing electrochemical reactions with enzymes or coenzymes [43]. This indicates that these CNTs can set up an electrical message with enzymes through electron transport to enzymes assured by electron hopping between immobilized redox centers. Since they are extremely small in size, CNTs can approach in close propinquity the prosthetic site of enzymes and hence attain direct electrical writing between the enzymes and the bulk electrode. As a result, modified CNT electrodes have prompted extensive attention to the invention of biofuel cells [43] (Figure 3.4).

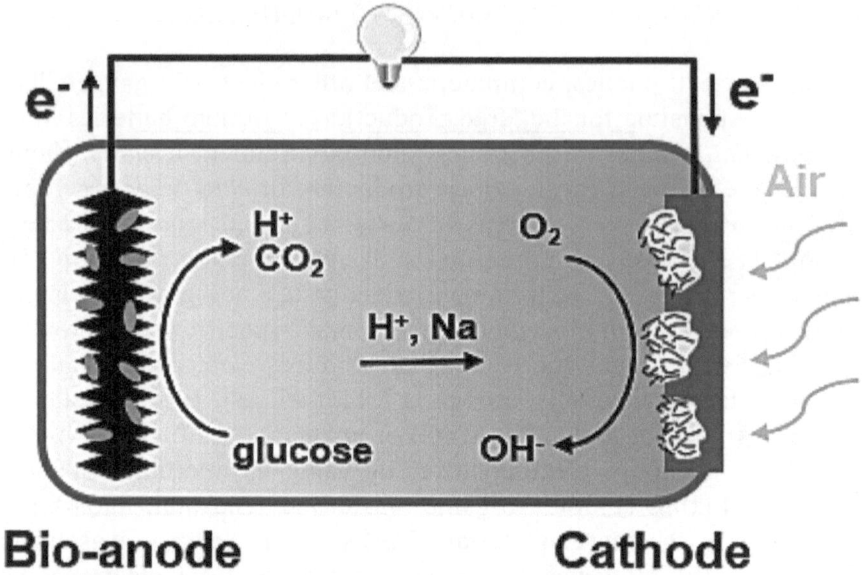


**FIGURE 3.4** Schematic setup of the microbial fuel cell.

*Source:* Reprinted with permission from Ref. [43a]. © 2019 Elsevier.

## 3.4.2 WASTE WATER TREATMENT APPLICATION OF CNTS

### 3.4.2.1 CNT SPONGE

Carbon nanotube sponge can be developed via chemical vapor decomposition methods using ferrocene as the precursor. A CNT sponge made from this technique possesses a randomly entangled 3-D structure and exhibits high porosity with minimum density. Oil-contaminated water can be cleaned by the CNT sponge. They float on the contaminated surface and will remove the oil by adsorbing it. Since the sponges are hydrophobic, they show a tendency to move to the oil film area and thereby to lead to exceptional floating-and cleaning capacity, which is very useful for spill clean-up. This oil-saturated CNT sponge can be regenerated easily through mechanical compression with concurrently the revival of important resources or straight burned in air devoid of destroying the sponge structure [24] (Figure 3.5).

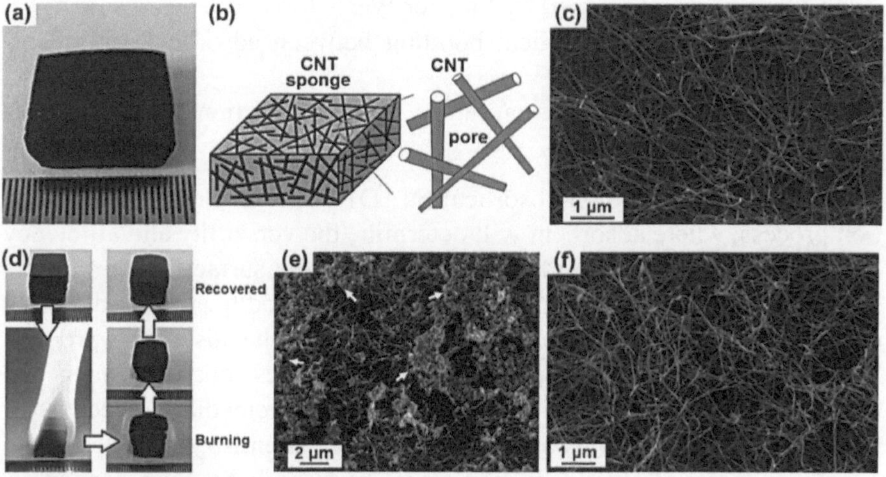


**FIGURE 3.5** CNT sponges: (a) photograph; (b) schematic of the pore structure; (c) SEM image of the CNT sponge; (d) burning and reuse of the CNT sponge. SEM images of (e) the surface; and (f) the core of the CNT sponge after burning.

*Source:* Reprinted with permission from Ref. [43b]. © 2011 Elselvier.

### *3.4.2.2 ADSORBENTS*

Adsorption is a surface phenomenon used for the elimination of pollutants. It is based on the interaction between absorbable solute and solid adsorbent. It is cost-effective and highly efficient in comparison with other methods. The general techniques for the purification of wastewater are ion exchange, chemical precipitation, reverse osmosis, electrolysis, coagulation/flocculation, ultra-filtration, and adsorption [44].

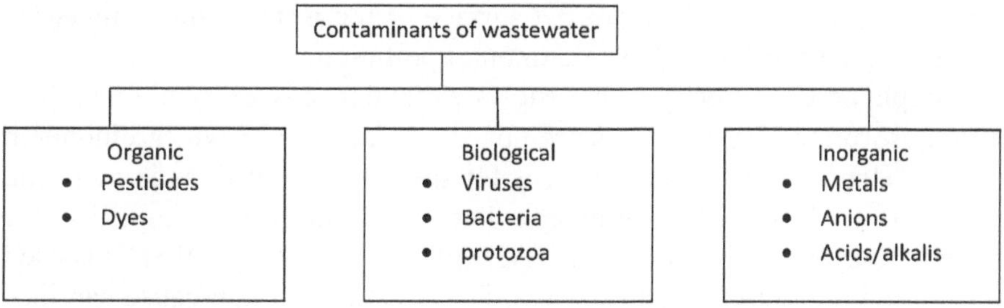


Adsorption can be:

- Physisorption, caused by Van der Waals forces;
- Chemisorption, chemical bonding between adsorbent and adsorbate; and
- Exchange adsorption, caused by charge attraction between adsorbate and the surface.

The importance of non-absorbent (NAD) comes to play in the adsorption process, where adsorbent will determine the versatility and efficiency of the adsorption process [45]. They have a large surface area, multiple sorption sites, tunable pore size, low-temperature modification, and surface chemistry that make them suitable candidates for the adsorption process. These nanomaterials can be nanowires, nanotubes, quantum dots, and nanoparticles. CNTs are important for wastewater remediation because of their high adsorption sites and tunable surface chemistry. An ideal nano adsorbent material should be non-toxic, have high sorption capacities, and have high accuracy for removing the contaminants even in their low concentrations. The surface should be easily reactivated. Recently eco-friendly synthesized $SnO_2$-CNT nanohybrid has been used as an efficient re-usable material for onsite water remediation. In this work, oxidative adsorption of As(III) on $SnO_2$-CNT nanocomposite (NC) was presented [46].

Non-biodegradable wastes have been increasingly released into water directly or indirectly, which causes serious toxic effects in the ecosystem [47]. Removal of such pollutants can be done using sorption, membrane filtration, or chemical precipitation. Effective adsorption of oil from water can also be achieved by new porous adsorbents that possess superhydrophobicity and super leophilicity. Super-hydrophobic and super leophilic hybrid foam of CNTs can be fabricated using a two-step chemical vapor deposition (CVD) technique. They can be selectively used to remove organic solvents and oils from the surface of the water, which otherwise cause long-term effects of environmental pollution.

Graphene CNT aerogel also shows excellent reusability and stability after adsorption. Organic dyes are non-degradable and have carcinogenic effects. Hence, a new development for the removal of dyes is of prime importance. Also, CNTs-graphene has good adsorption capacity for organic dyes. In work, MWCNT NAD called Co-MWCNT NAD's is used for removing methylene blue, which is a dyestuff that can cause carcinogenic and mutagenic effects on humans and aquatic life's-MWCNT NAD is fabricated using ultrasonic method [48].

The physical adsorption of hydrocarbon products (saturated and unsaturated) of Fischer-Tropsh (FT) synthesis on CNT-supported cobalt surface can be investigated using DFT (density functional theory). They found that in the FT process (4,4), CNT-Co nanocatalysts had greater efficiency in creating heavy hydrocarbons than light types [49].

In environmental engineering, CNTs are used for various adsorbent applications of the following species [50]:

- Heavy metals;
- Disinfection by-products;
- Endocrine disrupters;
- Biological contaminants;
- Natural organic matter and so on.

### *3.4.2.3 ROLE OF CNT IN WASTE WATER TREATMENT*

Wastewater coming from industrial, domestic areas, or agricultural sources consists of a huge amount of contaminants like heavy metal ions, some radioactive elements like Caesium and Strontium, etc. [51]. These pollutants are non-degradable, highly toxic, and carcinogenic, and hence they badly affect the quality of water as well as our health. So, sorbents with good sorption capability and high sorption efficiency can remove these contaminants. Having a high surface-active site-to-volume ratio and controlled pore size distribution, CNTs are very adaptable for wastewater treatments. Advanced studies discovered that the adsorption capacity of CNTs depends on both the surface functional group and the nature of the sorbate [17]. Generally, CNTs having a pH range of 7 to 10 are suitable for adsorption [52]. According to current technologies, CNTs are also used as nano filters to reduce the particle concentration in wastewater [53]. CNTs have shown extraordinary performance in transporting water regardless of their hydrophobic nature. Molecular dynamics simulations point out that the hydrophobic nature of CNTs pores builds weak interactions with water molecules and thereby facilitating a fast and nearly frictionless flow of water [54]. Recent studies show that CNT nano filters can remove the pathogenic microorganisms such as protozoa, bacteria, and viruses present in wastewater treatment, with microorganisms being reserved on the facade of CNT based on a depth filtration mechanism [55]. It can also remove the *E. coli* bacteria at low pressure using SWCNTs [17]. Moreover, the antimicrobial activity of CNTs helps to replace chemical disinfectants as a new efficient way of controlling microbial pathogens [56]. The application of CNTs in water disinfectant treatment avoids the creation of harmful disinfection by-products like trihalomethane, halo acetic acids, and aldehydes, as they are not strong oxidants and are comparatively static in water.

### 3.4.3 CNTS FOR CONTROLLING AIR POLLUTION

#### 3.4.3.1 CNTS AS GAS SENSOR

The metallic and semi-conductivity of CNT are greatly prejudiced by their one-dimensional cylindrical structure, such as size and chirality, although they possess unique and tunable electrical properties. The advantages, such as low power consumption, low operating temperature, and high sensitivity, will make CNTs dominant over other conventional metal-oxide gas sensors [57] (Figure 3.6).

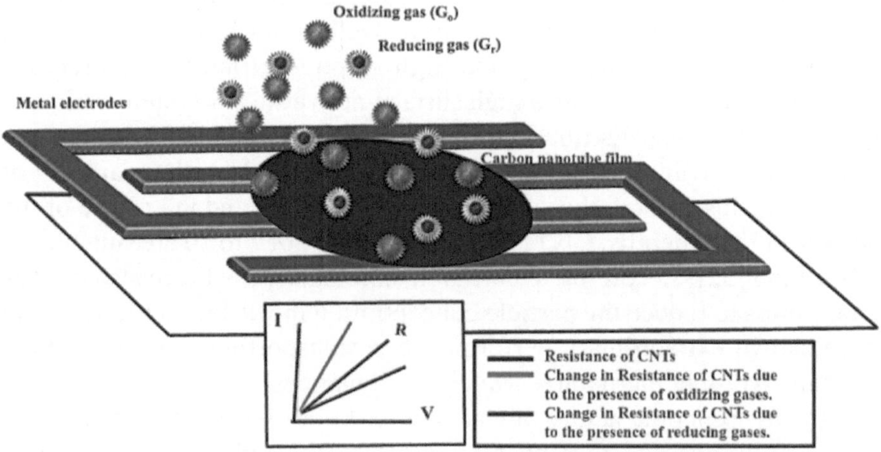


**FIGURE 3.6** Schematics of carbon nanotube (CNT) gas sensors for emissions from fossil fuel burning.

*Source:* Reprinted with permission from Ref. [57a]. © 2014 Elsvier.

A number of researchers have used the CNT-based gas sensor for the detection of various poisonous gases like Nitrogen Oxides ($NO_x$), Nitrogen Dioxide ($NO_2$), Ammonia ($NH_3$), and Sulphur Dioxide ($SO_2$) at RT [58–60].

#### *3.4.3.2 CARBON NANOTUBES (CNTS) AS THE ADSORBENT FOR THE $CO_2$ CAPTURE*

The combustion of fossil fuels in power plants is known as one of the most common technologies to generate electricity. Due to this combustion, a large amount of $CO_2$ will be produced. $CO_2$ is one of the greenhouse gases (GHGs) that cause global warming, and this leads to global climate change. A good remedy for this problem is the capture and storage of $CO_2$. The impregnation of amines onto solid materials is suitable for the adsorption of $CO_2$. Among them, CNTs are considered a suitable material for the impregnation of amines. Among the two classes of CNTs, SWCNTs show higher $CO_2$ adsorption capacity than MWCNTs [61].

### *3.4.4 ACTIVATED CARBON (AC) COATED WITH MULTI-WALLED CARBON NANOTUBES (MWCNTS) FOR REMOVING ANTIBIOTIC MATERIALS*

Antibiotics are partly metabolized in the human body, and they stay active after excretion. These materials can be seen in drinking water, surface water, sewage, and wastewater sources in various scale of amounts. The pharmaceutical wastes from the pharmaceutical industry and hospitals may come into contact with the water sources. This is a threat to both humans and all living organisms. This can result in high antibiotic resistance in the long term. One of the most common antibiotics is ciprofloxacin (CIP). It is present in a large amount of the environment, especially in wastewater. CIP in the aqueous solutions can be removed by AC supported with MWCNTs/AC. The factors affecting the removal of CIPs are the initial concentration of CIP, Adsorbent dosage, pH of the solution, temperature, and contact time [62]. They have a removal efficiency of 73%, and this made the MWCNT/AC as a potential tool for antibiotic removal.

### *3.4.5 CATALYTIC APPLICATIONS OF CNTS*

A catalyst is important for accelerating chemical reactions by reducing activation energy. Catalysts can be homogeneous or heterogeneous.

An optimal catalyst shows superior performance in mainly four areas: selectivity, activity, durability, and recoverability [63]. The activity of a catalyst expresses the number of molecules converted to product molecules by the catalyst per unit of time. Turnover frequency (TOF) is used to measure the activity of the catalyst. Homogeneous catalysts show a high TOF of 0.3 $s^{-1}$ or higher, while heterogeneous catalysts are in the range of 0.03 $s^{-1}$ or even lower. The lifetime of a catalyst is a measure of the number of cycles it can undergo without replacement. The turnover number (TON) is the number of products that can be formed by a given amount of catalyst. The recoverability of active catalytic substances is important. From the commercial point of view, it should be able to separate from the reaction mixture and reuse after the termination of the reaction. The selectivity, activity, durability, and recoverability are mainly influenced by catalyst size, shape, and surface composition. When the catalyst size comes in the nano range, the more active sites are accessible to the substrate. Graphene oxides (GOs) can be dispersed in CNT and Pt nanocatalysts. The resulting GOCNT-Pt nanocomposites show greatly enhanced Electrocatalysis and power density [42]. This so-prepared enzyme mimic exhibit catalytic performance comparable to protein peroxidase, with higher catalytic efficiency, environmental robustness, stability, and cost-effectiveness.

#### *3.4.5.1 WHY CNT IS SUITABLE FOR CATALYTIC ACTIVITY?*

Ideally, SWCNT is made of a perfect graphene sheet rolled up to a cylinder and closed at two sides (semi-fullerenes). The diameter varies between 0.4 nm and 2.5 nm, and the length can be several millimeters. The metallic particle residue from the production process can be seen inside the MWCNT cavity. Studies show that SWCNT behaves like quantum wires which have electron confinement along the tube axis. Mainly electronic properties are governed by tube diameter and helicity in which the graphene layer is rolled up. Electronic properties are drastically modified by defects in the system [30].

Alkalized $MoO_3$ nanocatalyst supported on CNT s is prepared by microemulsion technique. Its activity and selectivity in higher alcohol synthesis (HAS) have been assessed in fixed bed microreactor. In their previous work, they used the impregnation method to prepare CNT supported $MoO_3$-$k_2O$ HAS catalyst. CNT as catalyst support allow better metal dispersion and control the metal support interactions. The effect of molybdenum particle size on the catalyst properties, activity, and selectivity was also studied, supported by $MoO_3$-$k_2O$ nanocatalyst synthesized using microemulsion technique shows a decreased hydrocarbon selectivity from 32.4% to 24.5%, while the alcohol selectivity increased from 45.3% to 61.4%.

## 3.5 CONCLUSION

These environmental pollutants can be bioaccumulated and biomagnified, causing adverse effects on the ecosystem. Therefore, there is an urgent need for efficient materials for environmental pollution management. CNTs-based materials are promising candidates for environmental applications. With advancements in practical applications of CNT in environmental engineering, we can broaden the horizon of human life.

## ACKNOWLEDGMENT

Ms. S. Neelima acknowledges the Department of Science and Technology, the Government of India (DST-INSPIRE), for the award of JRF.

## KEYWORDS

- **adsorbent**
- **air pollution**
- **carbon nanotubes**
- **catalyst**
- **pollution**
- **wastewater treatment**

## REFERENCES

1. Ibrahim, K. S., (2013). Carbon nanotubes-properties and applications: A review. *Carbon Lett., 14*, 131–144.

2. Lahiff, E., Lynam, C., Gilmartin, N., O'Kennedy, R., & Diamond, D., (2010). The increasing importance of carbon nanotubes and nanostructured conducting polymers in biosensors. *Anal. Bioanal. Chem., 398*, 1575–1589.
3. Zhang, Y., & Yin, Q. Z., (2012). Carbon and other light element contents in the earth's core are based on first-principles molecular dynamics. *Proc. Natl. Acad. Sci. U. S. A., 109*, 19579–19583.
4. Allègre, C. J., Poirier, J. P., Humler, E., & Hofmann, A. W., (1995). The chemical composition of the earth. *Earth Planet. Sci. Lett., 134*, 515–526.
5. Pace, N. R., (2001). The universal nature of biochemistry. *Proc. Natl. Acad. Sci., 98*, 805–808.
6. Nasir, S., Hussein, M. Z., Zainal, Z., & Yusof, N. A., (2018). Carbon-based nanomaterials/allotropes: A glimpse of their synthesis, properties, and some applications. *Materials (Basel), 11*, 1–24.
7. Dai, L., Chang, D. W., Baek, J. B., & Lu, W., (2012). Carbon nanomaterials for advanced energy conversion and storage. *Small, 8*, 1130–1166.
8. Geim, A. K., & Novoselov, K. S., (2007). The rise of graphene. *Nat. Mater., 6*, 183–191.
9. Kroto, H. W., Heath, J. R., O'Brien, S. C., Curl, R. F., & Smalley, R. E., (1985). C60: Buckminsterfullerene. *Nature, 318*, 162, 163.
10. Hu, C., Qu, J., Xiao, Y., Zhao, S., Chen, H., & Dai, L., (2018). *Carbon Nanomaterials for Energy and Biorelated Catalysis: Recent Advances and Looking Forward.* https://doi.org/10.1021/acscentsci.8b00714.
11. Mildren, R., & Rabeau, J., (2013). *Optical Engineering of Diamond. Optical Engineering of Diamond.* doi: 10.1002/9783527648603.
12. Pierson, H. O., (1993). *Handb. Carbon, Graph. Diamonds Fullerenes*, 43–69. doi: 10.1016/B978-0-8155-1339-1.50008-6.
13. Choudhary, N., Hwang, S., & Choi, W., (2014). *Carbon Nanomaterials: A Review*, 709–769. doi: 10.1007/978-3-642-31107-9_37.
14. Mahmoudi, K., Hamdi, N., & Srasra, E., (2014). *Preparation and Characterization of Activated Carbon from Date Pits by Chemical Activation with Zinc Chloride for Methyl Orange Adsorption.*
15. Hernández-Montoya, V., & Bonilla-Petriciolet, A., (2012). *Lignocellulosic Precursors Used in the Synthesis of Activated Carbon: Characterization Techniques and Applications in the Wastewater Treatment.*
16. Zaytseva, O., & Neumann, G., (2016). Carbon nanomaterials: Production, impact on plant development, agricultural and environmental applications. *Chem. Biol. Technol. Agric., 3*, 1–26.
17. Ong, Y. T., Ahmad, A. L., Zein, S. H. S., & Tan, S. H., (2010). A review on carbon nanotubes in an environmental protection and green engineering perspective. *Brazilian J. Chem. Eng., 27*, 227–242. [a] Wang, X., Li, Z., Xu, W., Kulkarni, S. A., Batabyal, S. K., Zhang, S., Cao, A., & Helena, L., (2014). $TiO_2$ nanotube arrays based flexible perovskite solar cells with transparent carbon nanotube electrode. *Nano Energy,* 1–8. https://doi.org/10.1016/j.nanoen.2014.11.042.
18. Ebbesen, T. W., et al., (1996). Electrical conductivity of individual carbon nanotubes. *Nat., 382*, 54–56.

19. Eatemadi, A., et al., (2014). Carbon nanotubes: Properties, synthesis, purification, and medical applications. *Nanoscale Res. Lett., 9*, 1–13.
20. Amente, C., & Dharamvir, K., (2015). Thermally agitated self assembled carbon nanotubes and the scenario of extrinsic defects. *World J. Nano Sci. Eng., 5*, 17–25.
21. Hahm, M. G., Hashim, D. P., Vajtai, R., & Ajayan, P. M., (2011). A review: Controlled synthesis of vertically aligned carbon nanotubes. *Carbon Lett., 12*, 185–193.
22. Sinnott, S. B., Shenderova, O. A., White, C. T., & Brenner, D. W., (1998). Mechanical properties of nanotubule fibers and composites determined from theoretical calculations and simulations. *Carbon N. Y., 36*, 1–9.
23. Maeda, T., & Horie, C., (1999). Phonon modes in single-wall nanotubes with a small diameter. *Phys. B Condens. Matter, 263*, 479–481.
24. Bhanuse, U. M., Babar, A. B., & Ranveer, A. C., (2016). *Journal of Environmental Science, Computer Science and Engineering & Technology (JECET), 5*, 10–18.
25. Nouri, N., Khorram, P., Duman, O., Tunç, S., & Sereshti, H., (2020). Overview of nanosorbents used in solid phase extraction techniques for the monitoring of emerging organic contaminants in water and wastewater samples. *Trends in Environmental Analytical Chemistry, 25*. https://doi.org/10.1016/j.teac.2020.e00081.
26. Álvarez-ruiz, R., & Picó, Y., (2020). Analysis of emerging and related pollutants in aquatic biota. *Trends in Environmental Analytical Chemistry, 25*. https://doi.org/10.1016/j.teac.2020.e00082.
27. Mauter, M. S., & Elimelech, M., (2008). *Critical Review Environmental Applications of Carbon-Based Nanomaterials*, 5843–5859.
28. Fan, W. *Springer Briefs in Molecular Science Graphene-Carbon Nanotube Hybrids for Energy and Environmental Applications*.
29. Becher, M., *et al.,* (2003). Hydrogen storage in carbon nanotubes. *Comptes Rendus Phys., 4*, 1055–1062.
30. Tavasoli, A., Karimi, S., & Shoja, M., (2013). *Synthesis and Application of CNT-Supported $MoO_3$ -$K_2O$ Nanocatalyst Using Microemulsion Technique: Role of Nanoparticle Size on Catalyst Activity and Selectivity in Higher Alcohol Synthesis*, pp. 1–11.
31. Olusegun, A., et al., (2012). We are IntechOpen, the world's leading publisher of Open Access books built by scientists, for scientists TOP 1 %. Intech i, 38. [a] Singh, R., Srivastava, S., & Kumar, P., (2018). Nano-structures & nano-objects structural and surface modification of carbon nanotubes for enhanced hydrogen storage density. *Nano-Structures & Nano-Objects, 14*, 57–65. https://doi.org/10.1016/j.nanoso.2018.01.005.
32. Schmidt, J., & Peibst, R., (2018). *Surface Passivation of Crystalline Silicon Solar Cells: Present and Future*. doi: 10.1016/j.solmat.2018.06.047.
33. Kamat, P. V., (2007). *Meeting the Clean Energy Demand: Nanostructure Architectures for Solar Energy*, 2834–2860.
34. Janssen, R., (2005). *Polymer-Fullerene Bulk Heterojunction Solar Cells*. doi: 10.1557/mrs2005.6.
35. Pasquier, A. D., Unalan, H. E., Kanwal, A., Miller, S., & Chhowalla, M., (2005). Conducting and transparent single-wall carbon nanotube electrodes for polymer-fullerene solar cells. *Appl. Phys. Lett., 87*, 1–3.

36. Jeon, I., Matsuo, Y., & Maruyama, S., (2018). Single-walled carbon nanotubes in solar cells. *Top. Curr. Chem., 376*, 2–4.
37. Rowell, M. W., et al., (2006). Organic solar cells with carbon nanotube network electrodes. *Appl. Phys. Lett., 88*, 23–25.
38. Cataldo, S., Salice, P., Menna, E., & Pignataro, B., (2012). Carbon nanotubes and organic solar cells. *Energy Environ. Sci., 5*, 5919–5940.
39. Serp, P., Corrias, M., & Kalck, P., (2003). *Carbon Nanotubes and Nanofibers in Catalysis, 253*, 337–358. https://doi.org/10.1016/S0926-860X(03)00549-0.
40. Goh, K., & Chen, Y., (2017). Controlling water transport in carbon nanotubes. *Nano Today*. doi: 10.1016/j.nantod.2016.12.015.
41. Sharma, T., Mohana, R. A. L., Chandra, T. S., & Ramaprabhu, S., (2008). Development of carbon nanotubes and nanofluids based microbial fuel cell. *Int. J. Hydrogen Energy, 33*, 6749–6754.
42. Manuscript, A. *Nanoscale*. https://doi.org/10.1039/b000000x.
43. Cosnier, S., Holzinger, M., & Goff, A., (2014). Le. Recent advances in carbon nanotube-based enzymatic fuel cells. *Front. Bioeng. Biotechnol., 2*, 1–6. [a] Du, Y., Ma, F., Xu, C., Yu, J., Li, D., Feng, Y., & Zhen, L., (2019). Nano energy nitrogen-doped carbon nanotubes / reduced graphene oxide nanosheet hybrids towards enhanced cathodic oxygen reduction and power generation of microbial fuel cells. *61*, 533–539. https://doi.org/10.1016/j.nanoen.2019.05.001. [b] Gui, X., Li, H., Wang, K., Wei, J., Jia, Y., Li, Z., Fan, L., et al., (2011). Recyclable carbon nanotube sponges for oil absorption. *Acta Mater., 59*(12), 4798–4804. https://doi.org/10.1016/j.actamat.2011.04.022.
44. El-sayed, M. E. A., (2020). Science of the total environment nanoadsorbents for water and wastewater remediation. *Sci. Total Environ., 739*, 139903. https://doi.org/10.1016/j.scitotenv.2020.139903.
45. Olveira, S., Forster, S. P., & Seeger, S., (2014). *Nanocatalysis: Academic Discipline and Industrial Realities, 2014*.
46. Mohanta, D., & Nath, A., (2019). Environmentally benign fabrication of $SnO_2$ -CNT nanohybrids and their multifunctional efficiency as an adsorbent, catalyst and antimicrobial agent for water decontamination. *Sci. Rep.*, 1–19. https://doi.org/10.1038/s41598-019-49181-2.
47. Rocha, R. P., & Pereira, M. F. R., (2016). *Tuning CNT Properties for Metal-Free Environmental Catalytic Applications*. https://doi.org/10.3390/c2030017.
48. Çalımlı, M. H., (2020). Magnetic nanocomposite cobalt - multi-walled carbon nanotube and adsorption kinetics of methylene blue using an ultrasonic batch. *Int. J. Environ. Sci. Technol., 2020*(No. 0123456789). https://doi.org/10.1007/s13762-020-02855-1.
49. Shariatinia, Z., & Abdollahi-moghadam, M., (2018). Original article DFT computations on surface physical adsorption of hydrocarbons produced in the Fischer-tropsch synthesis on a CNT / Co nanocatalyst. *J. Saudi Chem. Soc., 22*(7), 786–808. https://doi.org/10.1016/j.jscs.2018.01.001.
50. Upadhyayula, V. K. K., & Gadhamshetty, V., (2010). *Appreciating the Role of Carbon Nanotube Composites in Preventing Biofouling and Promoting Biofilms on Material Surfaces in Environmental Engineering: A Review, 28*, 802–816. https://doi.org/10.1016/j.biotechadv.2010.06.006.

51. Wang, Y., Pan, C., Chu, W., Vipin, A. K., & Sun, L., (2019). Environmental remediation applications of carbon nanotubes and graphene oxide: Adsorption and catalysis. *Nanomaterials, 9.*
52. Rao, G., Lu, C., & Su, F., (2007). Sorption of divalent metal ions from aqueous solution by carbon nanotubes: A review. *Sep. Purif. Technol., 58*, 224–231.
53. Srivastava, A., Srivastava, O. N., Talapatra, S., Vajtai, R., & Ajayan, P. M., (2004). Carbon nanotube filters. *Nat. Mater., 3*, 610–614.
54. Noy, A., et al., (2007). Nanofluidics in carbon nanotubes. *Nano Today, 2*, 22–29.
55. Bohonak, D. M., & Zydney, A. L., (2005). Compaction and permeability effects with virus filtration membranes. *J. Memb. Sci., 254*, 71–79.
56. Savage, N., & Diallo, M. S., (2005). Nanomaterials and water purification: Opportunities and challenges. *J. Nanoparticle Res., 7*, 331–342.
57. Endo, M., Strano, M., & Ajayan, P., (2007). Potential applications of carbon nanotubes. in *Topics in Applied Physics, 111*, 13–61. [a] Mittal, M., & Kumar, A., (2014). Ac Ce p Te d Us T. *Sensors Actuators B. Chem., 2014.* https://doi.org/10.1016/j.snb.2014.05.080.
58. Ueda, T., et al., (2008). Development of carbon nanotube-based gas sensors for detection working at low temperature. *Phys. E Low-Dimensional Syst. Nanostructures, 40*, 2272–2277.
59. Wang, Y., & Yeow, J. T. W., (2009). A review of carbon nanotubes-based gas sensors. *J. Sensors, 2009*, 493904.
60. Suehiro, J., Zhou, G., & Hara, M., (2005). Detection of partial discharge in SF6 gas using a carbon nanotube-based gas sensor. *Sensors Actuators, B Chem., 105*, 164–169.
61. Osler, K., Dheda, D., Ngoy, J., Wagner, N., & Daramola, M. O., (2017). Synthesis and evaluation of carbon nanotubes composite adsorbent for $CO_2$ capture: A comparative study of CO2 adsorption capacity of single-walled and multi-walled carbon nanotubes. *Int. J. Coal Sci. Technol., 4*, 41–49.
62. Sharifpour, N., Moghaddam, F. M., Mardani, G., & Malakootian, M., (2020). Evaluation of the activated carbon coated with multi-walled carbon nanotubes in removal of ciprofloxacin from aqueous solutions. *Appl. Water Sci., 10*, 1–17.
63. Zubizarreta, L., (2020). *Sustainable Carbon as Efficient Support for Metal-Based Nanocatalyst: Applications in Energy, 2020.*

## CHAPTER 4

# Graphene Oxide-Based Nanocomposites for Wastewater Treatment

SANDIPAN BHATTACHARYA,[1] PRIYA BANERJEE,[2] and PAPITA DAS[1,3]

[1]*Department of Chemical Engineering, Jadavpur University, Kolkata – 700032, West Bengal, India*

[2]*Department of Environmental Studies, CDOE Rabindra Bharati University, Kolkata – 700091, West Bengal, India*

[3]*School of Advanced Studies in Industrial Pollution Control Engineering, Jadavpur University, Kolkata – 700032, West Bengal, India*

**Corresponding author. E-mail: prya_bnrje@yahoo.com*

## ABSTRACT

Pharmaceuticals and personal care products (PPCP) are widely used for medicinal and cosmetic purposes. Due to their extensive use in recent times, PPCPs are emerging as a prominent group of micropollutants with near universal presence. PPCPs are frequently encountered as a trace pollutant in almost every water body (lakes, rivers, and oceans). PPCPs are considered to be a reason for concern due to their persistent nature, ability to bioaccumulate in bodies of the organism, and their ability to induce a physiological effect even at a very low concentration. Recent studies have reported the adverse effects of PPCPs on different forms of life. Traditional water treatment plants are not capable of complete removal of these PPCPs from contaminated water. Therefore, advanced water

Carbon Composites: Composites with Nanotubes, Nanomaterials, and Graphene Oxide.
Eduardo A. Castro, Ann Rose Abraham, & A. K. Haghi (Eds.)

treatment methods are needed to effectively remove these PPCPs from water. In this review work, a detailed discussion is provided about novel graphene oxide (GO)-based nanocomposites (NCs) reported in recent studies and their application in water treatments like adsorption, advanced oxidative processes, and membrane filtration. This study also includes a comprehensive and precise analysis of the advantages and disadvantages of each method discussed herein. Factors affecting the PPCP removal efficiency of these NCs and methods of process optimization have also been discussed herein.

## 4.1 INTRODUCTION

Pharmaceuticals and personal care products (PPCP) have turned up as emerging contaminants during recent times by virtue of their widespread use and corresponding poor removal by conventional water treatment plants [1]. As a consequence of which, PPCPs are denoted as contaminants of global concern [2] and, more recently, as agents of global change [3]. In a broad sense, the term PPCP is used to denote any substance which has a medicinal and/or cosmetic value for humans and other life organisms [4]. PPCP are emanating as substances of concern due to their ability to invoke toxic response in certain aquatic organism or conversely to trigger resistance among certain pathogens [5]. Many PPCPs may themselves rapidly dispel in the environment [4], but by virtue of their constant and widespread use, they are continually added to the ecosystem and hence act as pseudo-persistent materials [2], which in turn can have a significant impact on organisms [4].

There are many routes by which the PPCPs can enter the ecosystem, namely sewage treatment plants (STP), wastewater treatment plants (WWTPs), runoff and landfill leaching, industrial services, hospitals, aquaculture facilities [4, 6]. Although generally found at very low concentrations (ng/L to μg/L level) but still the ubiquitous presence of PPCPs in waters all over the globe can be attributed to their ineffective removal by conventional wastewater or sewage treatment systems. The contamination of freshwater by PPCP can happen in many ways – one of them is the uptake of PPCP by human body through consumption which is then subsequently eliminated as faeces and urine into the septic tank or sewage. A high ratio of PPCP which is utilized by humans is excreted back in their original form while the rest are eliminated in a conjugated or metabolized form [5]. After the preliminary sewage treatment (which is often not sufficient to eliminate the PPCPs), this water can be potentially utilized (along with the bio-solids) as sludge for agricultural purposes through where the PPCPs can move back into the circulation [6]. Some of the other ways by which PPCPs are released back into the environment are untreated household effluent, wastewater from the manufacturing plant to the WWTP from where it is released into the environment in the form of sludge, hospital effluent, disposal of unused medicines and agricultural runoff [4, 6]. Like humans, the waste generated from domesticated animals also contains traces of veterinary drugs that are administered to them. These drugs find their way back into the environment when these animal wastes are used for various agricultural purposes or when they are released into water bodies without proper treatment [6]. The daily bathing and washing activities also result in the release of externally applied personal care products like shampoos, soap, toothpaste, lotion, sunscreens, etc., into the environment. Additionally, recreational activities like swimming and water sports also contribute to the release of PPCPs in the environment [4].

As mentioned before PPCPs are considered as contaminants of global concern due to their omnipresence and biologically active nature [5]. As mentioned before, the concentrations at which PPCPs are generally found in the environment are in the parts per billion (μg/L) to parts per trillion (ng/L) range [7], but still, even at that concentration the PPCPs are adept at causing unintended toxic effects on non-target organisms [6]. PPCPs have a tendency to persist and bioaccumulate in the environment and by virtue of these properties can impart various toxic effects to a wide variety of organisms, some of which are described below.

One of the most significant cases of PPCP-induced toxicity on the environment is the case of poisoning of three species of Indian vultures of the genus *Gyps* by the pharmaceutical diclofenac (DCF) [8]. DCF is a non-steroidal anti-inflammatory drug (NSAID) which was applied to the cattle to treat inflammation, fever, and body pain. After the cattle died, they were consumed by the vultures, signifying the route by which the vultures were exposed to the drug [9]. However, the drug DCF was found to have a nephrotoxic effect on the vultures resulting in their death by renal failure, and the autopsy of the dead vultures showed necrosis of kidney tissues and extensive visceral gout [8]. DCF has also exhibited a toxic effect on Brown trout (*Salmo trutta*) at a concentration of 5–50 μg/L, by affecting their gill, kidney, and immunity [6]. PPCPs have also been observed to alter the natural livelihood of various organisms, for example, Oxazepam, a common antidepressant was found to change the feeding behavior of European perch (*Perca fluviatilis*). Low concentrations of the serotonin uptake inhibitor (SRI) fluoxetine (3 μg/L) have been shown to render the Tadpoles (*Bufo arabicus*) more vulnerable to predation by dragonfly larvae (*Anax imperator*) [2]. PPCPs are also generally present in the environment as mixtures and not as individual components. And while the individual concentration of PPCPs may be significantly low but in combination these PPCPs can still exhibit a significant toxic effect, which is illustrated in the example that carbamazepine (CBZ), an antiepileptic drug and Clofibric acid, a lipid-lowering agent exhibited a much stronger effect on *Daphnia magna* in combination than it did individually at the same concentrations [6]. Emergence and metamorphosis of aquatic insect Diptera happened prematurely and at an accelerated rate when it was exposed to the 20 μg/L of Fluoxetine and Citalopram [2] and on exposure to amphetamine, the cumulative insect emergence increased by up to 89% [10]. When the arthropod crab (*Carcinus maenus*) was exposed to a mixture of CBZ, Ibuprofen, Caffeine, and Novobiocin at 0.1–50 μg/L for 28 days, it exhibited a significant decrease in Lysosomal membrane stability [11]. Studies reporting the effect of PPCP in the environment have also reported their effect on aquatic plants, such as the toxic effect of Oxolinic acid on *Daphnia magna*, toxic effect of ciprofloxacin (CIP) to green algae as well as toxic effect on Fluoroquinolone antibiotic on duckweed (*Lemna minor*), green algae (*Pseudokirchneriella subcapitata*), cyanobacterium (*Microcystis aruginosa*), fathead minnow (*Pimephales promelas*) and the crustacean (*Daphnia magna*) [6]. Pharmaceuticals like Fluoroquinole affect the replication of chloroplast so the plants are generally more sensitive towards them [9]. Another important ecological phenomenon connected to the presence of PPCPs in the environment is antibiotic resistance in some pathogenic bacteria resulting in serious complications in terms of treating infectious caused by them. Conversely, detrimental effect on some other bacterial species is also observed [6]. Therefore, it can be unequivocally agreed that trace PPCP contaminants, in aquatic environment or elsewhere have damaging effects on the environment. And as standard STPs and WWTPs are not efficient enough to effectively remove these trace pollutants from water, then alternative methods of water treatment should be adapted for this purpose.

Numerous methods have been evaluated for the removal of recalcitrant trace pollutants like PPCPs from water [1, 4, 12]. For this review, the emphasis would be on graphene oxide (GO) based nanocomposites (NCs) and their efficacy to effectively remove the PPCPs from water. GO is the oxidized form of graphene (a 2-D material with atomic thickness) with functional groups like hydroxyl, epoxy, and carboxyl groups on its surfaces and edges and with very large surface area [13]. GO also has highly negative charge density and exists in the solution as a single layer [13]. GO has been used extensively for the removal of PPCPs from water either as an adsorbent [13–16] or in the form of NCs [17–20]. A comprehensive database regarding the synthesis of different GO-based composites and their application in the removal of pharmaceuticals is given in this review chapter.

## 4.2 PPCP REMOVAL USING NOVEL GO NANOCOMPOSITES (NCS)

### *4.2.1 GO-POLYMER NANOCOMPOSITE (NC)*

Membrane technology has emerged as a very efficient method for the purpose of removing PPCPs from water due to its simple operation, simple operational setup, and ability to remove a wide range of materials [21]. But at the same time, contemporary membrane technologies are very much dependent on size exclusion principle for effective membrane filtration, thereby resulting in a huge energy demand [22]. Therefore, it is a challenge to design a membrane which is at the same time functionally efficient and cost effective [21]. As mentioned earlier, GO has generated considerable interest in recent times due to its 2-Dimensional unique structure, modifiable physicochemical properties, chargeability, high strength, large surface area and antimicrobial properties and high dispersibility in water and organic solution [21]. The combination of these properties renders GO to be a highly suitable material in the synthesis of polymer membranes for the treatment of PPCP-laden wastewater.

Oh et al. [23] synthesized a GO/PAH (poly(allylamine)hydrochloride) membrane by alternatively depositing positively charged PAH and negatively charged GO on the alternative sides of a hPAN (hydrolyzed polyacrylonitrile) support layer. The hPAN support layer was synthesized by dissolving PAN (polyacrylonitrile) and lithium chloride (LiCL) were mixed in N,N-dimethylformamide (DMF) in a ratio followed by heating and subsequent casting of the membrane. Then the membrane was hydrolyzed with NaOH solution. Then the h-PAN membrane was coated with PAH and GO on its alternate sides by a dip-coating method. The addition of GO allows the creation of a membrane with tailored charge distribution which enables efficient removal of both positive and negative charged particles. Also, the designed distribution of charges within the membrane layer without changing the overall surface charge of the membrane, enables the membrane to efficiently adsorb the charged particles at the desired spots without compromising its anti-fouling abilities. Also, the coating of the membrane with GO decreases its pore spots thereby enabling greater removal of contaminants by size exclusion. Lou et al. [21] synthesized a GO membrane on a polyethersulfone (PES) ultrafiltration membrane as a support, and small molecular diamines like butanediamine (BBD), ethylenediamine (EED), and p-phenylenediamine (PPD) were utilized as cross-linking agents to investigate the effect of diamine in GO membrane. The PES membrane was pre-treated first by washing, then by dipping it in a solution of dopamine hydrochloride and of Tris(hydroxymethyl) aminomethane acetic acid for 48 hours. A uniform dispersion solution with GO and binary amines BBD, EED, and PPD were prepared by sonication, which was then set aside for 5 hours for the reaction between the GO and the monomers to take place. After that, a certain amount of the dispersion was diluted to 100 mL, after which it was filtered into the PES membrane with a filtration of self-assembly method. After that, the membrane was dried and dipped in methanol for 12 hours to remove the excess untreated monomers. After which, the film was dried at room temperature (RT) for later use. The reaction of the amine molecules with the hydroxyl and carboxyl bonds of GO increased its hydrophobicity and decreased the permeability of the membrane. Unlike the traditional nano-filtration membrane where the filtration mainly takes place by electrostatic repulsion, in case of this GO-amine membrane the removal of PPCPs takes place by adsorption and molecular sieving. As greater hydrophobicity of the GO-amine membrane ensures greater removal of the PPCPs therefore incorporating the amine groups in the GO membrane ensures higher removal of PPCP molecules. Chen et al. [24] synthesized GO membrane modified with β-cyclodextrin for high percentage removal of Bisphenol-A (BPA). The GO dispersed in a suspension of DI water and to 20 ml of this solution, 2 gm of mono-(6-amino-6-deoxy)-β-cyclodextrin

(AmβCD) was added. The pH of the solution was kept at 9 with the help of borax buffer in order to enhance the speed of the reaction. The reaction took place at 40°C for 18 hours under continuous stirring and the β-cyclodextrin GO (CDGO) nanosheets were synthesized by the reaction between the amino groups of AmβCD and epoxy group of GO nanosheets. This resultant solution was then purified by centrifugation. After this, the finished product was obtained by drying it by lyophilization. The CDGO nanosheets were converted to CDGO membrane by suction filtration on Cellulose ester (CN-CA) membrane substrate by the process described by Xi et al. [25, 26]. After the completion of the membrane synthesis, the CN-CA membrane was peeled off. The synthesized CDGO membranes consisted of 2-D nanochannels on which the β-cyclodextrin molecules acted as adsorption sites. The imparting of β-cyclodextrin inside the GO nanosheets resulted in the formation of wave-like structures in the membrane that increased its surface area and interlayer spacing. As a consequence of which the dimension of the 2-D nanochannels becomes remarkably larger than that of conventional membranes. Due to this increased surface area and larger 2-D nanochannels, the CDGO membranes exhibit significantly greater water flux and increased adsorption capacity in comparison to other conventional membranes. Moreover, these CDGO membranes also exhibited high reusability (with almost 100% efficiency) even after repeated use after washing with ethanol.

### *4.2.2 GO-CERAMIC NANOCOMPOSITES (NCS)*

Ceramic membranes are widely used for the water treatment process due to their chemical and thermal stability, high mechanical strength, and high-water flux. However, two main drawbacks which are hindering the wide-scale utilization of ceramic membranes are low separation efficiency and membrane fouling [27]. In order to overcome this obstacle, the Ceramic membranes are modified with different types of nanoparticles in order to change their surface wetting behavior and improve its separation and anti-fouling properties. Among the different types of nanoparticles tested, GO has emerged as a premier choice by virtue of the functional groups present on its surface, which provides it with chemical stability, hydrophilicity, and the ability to uniformly disperse in water [27].

Chu et al. [28] synthesized a GO-coated Ceramic ultrafiltration membrane for the removal of pharmaceuticals Ibuprofen and Sulfamethoxazole along with natural organic matter and inorganic salts from water. For the synthesis of the GO-coated ceramic membrane, a vacuum filtration via phase inversion method was utilized. In brief, GO was dispersed as a solvent in Deionized (DI) water for 1 hour with sonication at 40–50 Hz. Followed by which the solution was completely dissolved in water by centrifugation and was used as a stock solution. As per the preliminary study of Chu et al. [29, 30], the 10 nm thickness of GO membrane exhibited the best anti-fouling and water permeability properties. Therefore, the thickness of the GO membrane was decided to be kept at 10 nm. In order to control the thickness of the GO membrane, the required concentration of the GO suspension was obtained by dilution of the stock solution. The Ceramic membrane used for this study was brought commercially from Starlitech Co (Kent, WA, USA). The GO suspension was deposited on the Ceramic membrane by filtering it by using the Millipore system. After the GO deposition, the membrane was washed with ultrapure water to wash off any non-bonded or weakly bonded GO particles from the membrane surface. Following which the GO-coated ceramic membrane was heated at 100°C for 2 hours to increase the bond strength of the GO particles with the membrane surface. The uniform GO coating on the ceramic membrane was due to the strong hydrogen bond between the carboxylic group of the GO membrane and the hydroxyl group of the ceramic membrane. The GO-coating of the ceramic membrane imparted it with decreased pore size, higher hydrophilicity, and build-up of negative surface charge. These properties in turn resulted in greater size exclusion, reduced hydrophobicity and strong electrostatic repulsion resulting in greater removal of the target organisms. This made the GO-coated ceramic membrane superior to pristine ceramic membranes for the purpose of removing components like pharmaceuticals, natural organic matter, and inorganic salts from water. Thus, it can be stated that coating of the ceramic membranes by GO resulted in greater efficiency, thereby proving this to be a viable method of membrane modification.

Kang et al. [27] modified a ceramic β-SiALON membrane with GO by coating the β-SiALON membrane with GO by a dip coating method. The GO was dispersed in DI water as a suspension, and the β-SiALON membrane was immersed in the GO suspension at 75°C for 1 hour under ultrasonication to coat the GO particles on its surface. After that, the GO-coating the β-SiALON membrane was dried and then sintered in the furnace for 4 hours at 600°C. The modification of the β-SiALON membrane by GO resulted in smaller pore size, greater anti-fouling, and higher separation performance. Thus, modification of the ceramic β-SiALON membrane by GO increased the efficiency of the membrane.

### 4.2.3 GO-METAL/METAL OXIDE NANOCOMPOSITE (NC)

GO acts as a suitable additive in the synthesis of metal/metal NC by virtue of its characteristics and properties like very high surface area (2,630 $m^2$/gm), presence of a large number of active sites on its surface, and bases, delocalized π-electrons systems on its surface, high chemical stability, high hydrophilicity, and high dispersibility in water [21, 31]. The addition of GO to metal/metal oxide NC can increase the adsorptive ability of the NC [32, 33] or increase the photocatalytic ability of the NC [20, 34] or both [35].

In a recent study, synthesized a reduced GO and Iron (II, III) oxide ($Fe_3O_4$) NC [32]. The GO-$Fe_3O_4$ NC was synthesized by a simple stirring method. $FeCl_3$.6 $H_2O$ and $FeCl_2$.4 $H_2O$ was dissolved in water under acidic condition. To this solution, a solution of GO was added and the pH of the solution was raised to a basic range through the addition of sodium hydroxide (NaOH). The solution was stirred for about 30 mins and the GO-$Fe_3O_4$ NC thus synthesized had a magnetic nature and was separated from the solution by using a magnet. The composite was then washed and dried for future use. The synthesized GO-$Fe_3O_4$ was able to effectively remove the drug phenazopyridine from water and also showed high reusability. Nodeh et al. [33] synthesized a NC with GO, $Fe_3O_4$ and Silica ($SiO_2$). GO was used for its adsorption ability, whereas $Fe_3O_4$ in this case was used for higher water permeability, to prevent sheet aggregation and to enable separation of the composite with a magnet. Silica ($SiO_2$) was used in order to sustain the adsorbent in an acidic medium. The NC was synthesized by adding the $Fe_3O_4$ and $SiO_2$ NC into a solution of GO which was then sonicated for 30 mins. Then NaOH was added into this mixture to form a black solution. The solution was stirred for 5 hours, following which the supernatant was thrown away and the precipitate was washed and dried at 80°C for a whole day. The GO-$Fe_3O_4$-$SiO_2$ NC was used to remove the pharmaceutical Naproxen which it did with greater efficiency and in lesser time than other magnetic-carbonaceous adsorbents and natural and functionalized nano-clay. Although adsorbents like porous sugarcane bagasse had greater adsorption efficiency but it took greater time. Also, the magnetic nature of the GO-$Fe_3O_4$-$SiO_2$ NC helped in their more efficient separation.

Nawaz et al. [20] synthesized a porous 3-dimensional GO and titanium dioxide ($TiO_2$) aerogel for the degradation of CBZ. The addition of GO to the $TiO_2$ is supposed to provide it with greater surface area, greater adsorption ability, chemical stability, delayed recombination of electron-hole pairs, and greater photocatalytic activity. The NC was synthesized by adding the $TiO_2$ (at different ratios) to the uniform GO dispersion under sonication. To this solution, a certain amount of L-ascorbic acid was added and then the solution was kept in a glass vial at 95°C for 1 hour. After the synthesis, the NC was extracted followed by washing with DI water several times. Then the NC was freeze-dried and stored for future use. The bonding of GO with $TiO_2$ was strengthened by the presence of presence of functional groups on the surface of GO, whereas the addition of L-ascorbic acid gave the solution uniformity. The addition of GO with $TiO_2$ had a synergistic effect on the removal of CBZ, which was about 20% more than that observed in the case of only $TiO_2$. Overall a $TiO_2$-GO NC having a ratio of 2:1 of $TiO_2$ and GO was able to degrade 99% of CBZ under a UV light with wattage 13.5 W/m$^2$ and with frequency in the range 300–400 nm. Sani et al. [34] synthesized a GO and zinc oxide (ZnO) NC with Gallium (Ga) and Sulphur (S) as dopants by an innovative wet chemical method. GO was mixed with ZnO in order to reduce the electron recombination capability of ZnO thereby increasing its photocatalytic ability. Doping with Ga resulted in extending of adsorption edge towards the visible region, improving photogenerated charge separation and improving oxygen vacancy whereas doping with S narrowed down the bandgap resulting in greater photocatalytic activity. The NC was synthesized by dispersing ZnO in ethanol with continuous stirring under basic conditions at constant temperature (70°C) allowing the ZnO to completely hydrolyze. Then a dispersion of GO in double distilled water was added to this solution. Then this solution was stirred and sonicated twice. Then the solution was transferred into a Teflon lined autoclave where it was hydrothermally treated at 140°C for 24 hours. Then the NC was extracted by decanting the supernatant followed by washing and drying for future use. The Ga and S were added to the NC by adding required amounts of $Ga(NO_3)_3.xH_2O$ and thiourea ($CH_4N_2S$) to the ZnO solution. The synthesized NC was called $Ga_xS_y$@ZG with the 'x' and 'y' denoting the weight percentage of Ga and S in the NC, respectively. As per the results of the study, the combination of GO along with the metal dopants had the desired effect and the $Ga_1S_5$@ZG NC was 8 times more efficient than the photocatalytic activity of pristine ZnO under similar experimental conditions. The NC was able to achieve complete degradation and 51% mineralization of the pharmaceutical Acetaminophen at a concentration of 50 mg/L.

In another study, authors reported the synthesis of a NC consisting of GO and nano-scale zero-valent iron (nZVI) by a wet chemistry method [35]. The nZVI particles have the ability to degrade the PPCPs by acting as Fenton catalysts through the generation of $Fe^{2+}$ ions by slow dissolution. These $Fe^{2+}$ ions combine with $H_2O_2$ to form reactive oxygen species, which degrade the PPCPs. However, the downside of using only nZVI is that they aggregate among themselves due to their magnetic forces and van der Walls attraction forces thereby decreasing their surface reactivity. Therefore, in order mitigate this problem a NC was synthesized with GO and nZVI where the GO will support the nZVI particles. The GO also improved the catalytic activity of the nZVI by providing fast electron transfer. The NC was prepared by dispersing the GO in water by sonication followed by addition of $FeCl_3.6\ H_2O$ into the same mixture. A $NaHB_4$ solution was added to the mixture in order to simultaneously reduce the GO and $FeCl_3$. After the addition of $NaBH_4$, the solution was kept stirring for 30 minutes at 300 rpm. The whole experiment was conducted under an inert environment at RT under continuous nitrogen gas flow. After the completion of the synthesis, the NC was washed with DI and ethanol followed by drying in a vacuum oven in order to store it for further experiments. As per the results of the study, GO-nZVI NC showed very high removal of the PPCPs by adsorption and advanced oxidation, higher than what is exhibited individually by its parent compounds GO and nZVI. Also, the combined adsorption and advanced oxidation process exhibited by GO-nZVI showed higher removal when compared to only adsorption.

## 4.3 APPLICATION OF GO NANOCOMPOSITES (NCS) FOR PCPP REMOVAL

GO NC-based processes reported in recent studies for PCPP removal from aqueous phases have been enlisted in Table 4.1. Different processes for the same have been discussed in detail as follows.

**TABLE 4.1** Removal of PPCPs by Different Systems by GO-based Nanocomposite

| SL. No. | Components of Nanocomposite | Compound Targeted | Method of Treatment | Major Findings | References |
|---|---|---|---|---|---|
| 1. | Zinc (Zn) and Iron (Fe) double-layered hydroxide modified with GO and coated on Glass Surface (ZnFe-LDH/GO/GS). | Phenazopyridine hydrochloride (PhP) | Photocatalysis | 60.01% removal of PhP was observed under an initial PhP concentration of 15 mg/L, pH of 8 and 3 sheets of Zn-Fe LDH/GO/GS.<br><br>Degradation of PhP was due to oxidative species in the order of $O_2^- >$ •OH> $h^+$. Adding 1 mmol/L of Potassium Persulfate increased the degradation efficiency to 93.95%. 60% of COD was removed after 300 minutes of irradiation denoting the mineralization of PhP. | [36] |
| 2. | Palladium nanoparticle coated GO (Pd NPs-GO) | Ibuprofen | Advanced oxidation by ultrasonic irradiation | The highest degradation of Ibuprofen was achieved under the condition of pH 3, PdNPs-GO dosage of 0.8 gm/L, Ibuprofen concentration of 30 mg/L and ultrasonic irradiation time of 50 minutes at 35 kHz. The main species responsible for the oxidation reaction were $•O_2^-$ and •OH.<br><br>Complete degradation of Ibuprofen found in Hospital effluents at concentration of 0.003 mg/L, 0.011 mg/L, 0.006 mg/L and 0.005 mg/L were observed under this optimized pH (pH 3) and PdNPs-GO dosage (0.8 gm/L) after 5 minutes of ultrasonication irradiation. | [37] |
| 3. | UiO-66_GO with 5% GO loading. Applied on a nano-filtration (NF) membrane with pressure assisted self-assembly (PASA) method. | Carbamazepine (CBZ) and diclofenac (DCF) | Nanofiltration | 3% of UiO-66_GO nanocomposite was dispersed in water (5%, 10%, and 15%) which was then loaded onto a NF membrane by PASA method. The UiO-66_GO NF membrane with 10% nanocomposite loading exhibited good water flux, good adhesion, and stability due to the interaction between GO and pollutant and uniform dispersal of UiO-66_GO on the nanofiltration membrane. Hence the UiO-66_GO NF membrane with 10% nanocomposite was used for future studies.<br><br>Using the UiO-66_GO (10%) NF membrane, the steady state rejection of CBZ was found to be 70% and that of DCF was found to be 93%. The removal of the pharmaceuticals was determined by size exclusion and charge repulsion. | [38] |

**TABLE 4.1** *(Continued)*

| SL. No. | Components of Nanocomposite | Compound Targeted | Method of Treatment | Major Findings | References |
|---|---|---|---|---|---|
| 4. | GO-$TiO_2$-based ceramic membrane | Naproxen (NAP), carbamazepine (CBZ) and diclofenac (DCF). | Membrane filtration and photocatalysis | Coating of the ceramic membrane by the GO-$TiO_2$ nanocomposite increased both the rejection rate and photocatalytic activity of the ceramic membrane.<br>Without UV radiation, the GO-$TiO_2$-coated ceramic membrane of the DCF and NAP showed 80% and 70% removal, respectively, while CBZ exhibited 13% removal.<br>Under UV irradiation, DCF and NAP was degraded up to 100%, whereas CBZ was degraded up to 90% within 20 minutes of radiation. The coating of the ceramic membrane by GO-$TiO_2$<br>Enhanced the degradation efficiency by four times and reduced the electrical energy per order ($E_0$) by an order of 10. | [39] |
| 5. | Ag (Silver)/AgBr (silver bromide)/GO nanocomposite | Diclofenac sodium | Photo-catalysis | The Ag/AgBr/GO nanocomposite was synthesized by an ultrasonic method.<br>The optimized dosage of Ag/AgBr/GO nanocomposite was found to be 0.03 gm/L and the concentration of DCF was taken as 25 mg/L. The optimum pH for the photocatalytic experiment was the natural pH of DCF (pH 6.2). Under these conditions, Ag/AgBr/GO nanocomposite was able to degrade 95% of DCF and 93% COD under 6 min of sunlight. | [40] |

### 4.3.1 ADSORPTION

Among the various process utilized for the purpose of removing pharmaceuticals from water, adsorption by GO remains one of the most popular choices [9]. GO has been utilized both as an individual and as an adsorbent, or it has been employed as an additive in a NC to enhance its adsorption ability. The reason for utilizing GO as an adsorbent is its significantly higher surface area compared to some of the other adsorbents like activated carbon (AC), carbon nanotubes (CNTs), clays, and montmorillonite [9]. The presence of functional groups like epoxy, carboxyl, and hydroxyl groups on the surface of GO also enhances its adsorption ability [41]. Furthermore, the functional groups present on the surface of GO can be replaced by other functional groups or molecules, which further increases their adsorption ability [9, 42]. As mentioned before, GO in itself has been used for the purpose of removing pharmaceuticals from water in many cases. For instance, Balasubramani et al. [43] used GO for the purpose of removing the recalcitrant pharmaceutical Metformin from water, where the highest adsorption efficiency was found to be 122.61 mg/gm. Banerjee et al. [41] used GO for the purpose of removing Ibuprofen from water. The highest removal of Ibuprofen was observed to be 98.17% at experimental conditions of GO dosage of 1 gm/L, pH 6, agitation speed of 180 rotations per minute, Ibuprofen concentration of 6 mg/L, a temperature of 308 K, and reaction time of 60 minutes. However, GO has also been widely reported to be utilized in combination with other materials for the purpose of removing pharmaceuticals from water. For instance, as per the work of Deng et al. [44], GO was used in combination with Attapulgite clay and $Fe_3O_4$ particles for the purpose of adsorbing Propranolol from water. GO was used as the dispersion medium onto which ATP was distributed and $Fe_3O_4$ was added to the mixture in order to impart a magnetic property to the composite so that it can be easily separated out after the treatment process. Furthermore, the functional groups present on the surface of GO also help in the uptake of Propranolol by virtue of $\pi$-$\pi$ bonding and hydrogen bonding [44, 45]. The removal of Propranolol was found to be about 99%, and the adsorption was reportedly faster and higher than other adsorbents at the neutral pH range. Naskar et al. [46] synthesized a NC consisting of reduced GO-coupled Gadolinium (Gd) doped $ZnFe_2O_4$ by virtue of a low-temperature, one-pot synthesis method. From the characterization study, it was found that the $ZnFe_2O_4$ was evenly dispensed over the rGO layers. The Gd doped (5%) rGO-$ZnFe_2O_4$ NC was able to remove 86% of the original concentration of the antibiotic drug Levofloxacin from water. Tabrizian et al. [47] synthesized a GO-supported Iron (Fe)/Copper (Cu) bimetallic nanoparticle for the purpose of removing the tetracycline (TC) from water. GO was utilized as the supporting material for the dispersion of Fe/Cu nanomaterial composite. The application of GO increased the dispersivity, reduced the agglomeration and thereby increased the sustainability of the bi-metallic nanoparticle. The GO-Fe/Cu nanoparticle exhibited 2.7 times increase in adsorption ability as compared to non-GO supported Fe/Cu NCs as reported in previous literature [47]. Also, the GO-Fe/Cu NC was able to retain its magnetic property and was easily separated from the reaction medium after the reaction. Under optimum conditions, the NC was able to remove nearly 100% TC under about 15 minutes.

### 4.3.2 ACTIVE OXIDATIVE PROCESSES

Among the different methods that have been applied for the purpose of removal of pharmaceuticals from water, the active oxidation process (AOP) remains a very popular choice. In broad terms, the procedure of AOP involves the generation of highly reactive chemical species which can then degrade the target compounds to harmless and benign products [48]. The reactive oxygen-containing radicals with at least one unsaturated electron pair bond like $HO^{\bullet}$, $O_2^{\bullet-}$, $HO_2^{\bullet}$, and $RO^{\bullet}$ are mostly used for this purpose [49]. GO is utilized as a part of these AOP mainly in the form of composites for its chemical and physical attributes such as chemical stability, capacity for high charge transport, and very high surface area [50].

#### 4.3.2.1 FENTON OXIDATION

Fenton reaction can be described as a process by virtue of which $OH^{\bullet}$ radicals can be generated by using a mixture of any Iron (Fe) containing salt and hydrogen peroxide ($H_2O_2$) where the Iron salt is mainly used as the catalyst. The mechanism by which the $OH^{\bullet}$ radicals can be generated is elaborated in Eqns. (1) and (2) [49].

$$Fe^{2+} + H_2O_2 = Fe^{3+} + ?OH^- + OH^o \quad (1)$$

$$Fe^{3+} + H_2O_2 = Fe^{2+} + HO_2^o + H^+ \quad (2)$$

Kong et al. [51] synthesized five double metal cross-linked alginate/graphene hydrogels for the purpose of degrading TC from water. Among the different types of metals used for the study, the alginate-GO-iron-cerium (AG-Fe-Ce) NC exhibited the best result. The role of GO in the mechanism of the degradation of the NC was to act as an adsorbent by virtue of the π-π interaction enabled by the oxygen-containing functional groups present on its surface. Also, during the degradation process, the highly conductive nature of GO enabled the transfer of free electrons between the surface of the Ag-Fe-Ce NC and the active iron sites, which possibly could have provided more reactive sites for $H_2O_2$ activation.

As per the works of Yu et al. [52], a GO-Iron Oxide ($Fe_3O_4$-GO) NC was synthesized by a co-precipitation followed by a hydrothermal method where a solution of $FeCl_2$.4 $H_20$ and solution of $FeCl_3$.6 $H_2O$ was added to different amounts of GO (5, 10, and 15% by weight) were hydrothermally treated under basic pH conditions. The $Fe_3O_4$-GO NC thus formed were used to degrade phenol under a UV light of 500 W with an irradiance of 100 $mW/cm^2$. Under optimal conditions, it was found that the $Fe_3O_4$-GO NC were able to degrade 98.8% phenol and 81.3% total organic carbon (TOC) from water after 120 minutes of reaction time. The utility of using GO in this NC can be given as follows, i.> it increased the surface area of the NC thereby increasing the rate of the mass transfer of pollutants toward the reactive sites on the surface of the photocatalyst ii.> Enhanced adsorption by virtue of π-π interaction iii.> Greater electron transfer between GO and $Fe_3O_4$. Moreover, the $Fe_3O_4$-GO was found to exhibit a high degree of reusability.

Another interesting use of the Fenton oxidation reaction for removing Pharmaceuticals is presented in the work of Shan et al. [22]. In this work, a hydrophilic and strengthened 3D reduced GO /nano-$Fe_3O_4$ hybrid hydrogel was used for the removal of two very common and widely used pharmaceuticals, namely CIP and TC. After the adsorption process, the NC was regenerated by a Fenton-like reaction where $H_2O_2$ was added to the NC. The $H_2O_2$ acted with the $Fe_3O_4$ salt to furnish $OH^\bullet$ radicals which oxidized almost the whole amount of the pharmaceuticals that was adsorbed on the surface of the NC. As a result of which, the NC was regenerated and reused for about 10 times with minimal loss in its effectivity. During the adsorption-desorption study for regeneration of the NC, the weight loss percentage of Fe was found to be about 1%. This low loss of Fe could be attributed to the presence of GO which has anti-corrosion effect due to its chemical stability and isolation effect [53]. Thus, due to the presence of GO, the reusability potential of the NC was highly increased.

#### *4.3.2.2 OZONOLYSIS*

Like Fenton oxidation, ozonation is also an advanced oxidation procedure (AOP), which is commonly used for the purpose of degrading pharmaceuticals from water [49].

Li et al. [54] synthesized a NC consisting of sea urchin-like three-dimensional $\alpha$-$MnO_2$ and reduced GO for the purpose of eliminating bisphenol A (BPA) from water in the presence of Ozone. The NC was prepared by a facile hydrothermal procedure. From the characterization study, it was found that the NC consisted of Mn, O, and C in the ratios at which they were taken. Moreover, from the BET adsorption and desorption study with nitrogen, it was observed that the $\alpha$-$MnO_3$/GO had a higher surface area (109 $m^2$/gm) and lower average pore size diameter as compared to pristine $\alpha$-$MnO_3$. This suggests that the addition of GO to the $\alpha$-$MnO_3$ increased its surface area, which is beneficial for the purpose of catalysis. It was found that under optimal conditions, the $\alpha$-$MnO_3$/GO was highly effective for the purpose of degrading the BPA from water and the addition of GO increased the surface area and enhanced the electron transfer for more effective catalytic degradation of BPA.

Checa et al. [55] synthesized a GO-$TiO_2$ NC by a sol-gel method and used it to mineralize the popular drug Primidone under an aqueous condition under a 425 nm LED visible light and Ozone. Due to the presence of GO in these NCs, the band gap of $TiO_2$ was reduced. As a result of which, under optimized conditions, i.e., in the presence of both LED irradiation and Ozone, a 0.75% GO-$TiO_2$ NC showed significant higher mineralization (83%) as compared to only $TiO_2$ (70%).

Jothinathan and Hu [56] synthesized different GO-based NCs like GO/$TiO_2$, GO/$Fe_3O_4$ and GO/Ti/$Fe_3O_4$ respectively. The GO/$TiO_2$ NC was synthesized by a microwave-based hydrothermal procedure whereas GO/$Fe_3O_4$ and GO/Ti/$Fe_3O_4$ was synthesized by mixing $Fe_3O_4$ with GO and GO/Ti under ultra-sonication (to exfoliate the GO particles and to convert the carboxylic groups on the surface of GO to carboxylate ions) followed by stirring and then drying. The NCs were used for degrading two PPCPs, the pharmaceutical Ibuprofen and the chemical p-chlorobenzoic acid. For the p-chlorobenzoic acid degradation study, it was found that the Ozone ($O_3$) decomposition followed by •OH generation was highest for the GO/Ti/$Fe_3O_4$ NCs followed by GO/$TiO_2$ and then by GO/$Fe_3O_4$. As a result of which highest removal of p-chlorobenzoic acid was exhibited by the GO/Ti/$Fe_3O_4$ NC. From this experiment it was inferred that GO decomposes $O_3$ at a greater rate than the NCs however, it does not produce high rates of •OH radicals due to GO acting as a •OH scavenger. This problem was mitigated by the synthesis of NCs. For the degradation of Ibuprofen, three systems were used, namely only $O_3$, $O_3/H_2O_2$ and GO/$Fe_3O_4$. From there, it was observed that the highest removal of Ibuprofen was exhibited by GO/$Fe_3O_4$ (85%) followed by $O_3$/GO and $O_3$ alone. Therefore, from this experiment it was found that surface-modified GO NCs generated sufficient amount of •OH radicals in presence of Ozone and hence can be used as a medium for enhanced AOP reactions in the future pending a few more developments.

Sheydaei et al. [57] synthesized an N-$TiO_2$/GO/Titan grid sheet composite by coating nitrogen-doped $TiO_2$ (N-$TiO_2$) and GO on a titan grid sheet by following an electrophoretic deposition (EPD) procedure. The synthesized composite showed photocatalytic activity under visible light. The NC was used for the degradation of the pharmaceutical Cetirizine by light-assisted photocatalytic ozonation. In the experimental setup, photocatalysis was performed by lighting LED lamps (with irradiation power of 7.45 $W/m^2$) on the surface of the N-$TiO_2$/GO/Titan grid sheet composite, whereas Ozone was generated by an Ozone generator. The simultaneous photocatalysis, as well as the ozonation process, resulted in the increased generation of hydroxyl radicals resulting in higher degradation of Cetirizine. Thus, from the experimental results, it was inferred that the combined synergistic effect of photocatalysis and ozonation was a far more effective process for degrading Cetirizine than individual removal mechanisms like adsorption, ozonation, and photocatalysis.

#### *4.3.2.3 PHOTOCATALYSIS*

Among the different types of AOPs that have been employed for the purpose of removing pharmaceuticals from water, photocatalysis remains a popular choice [58]. The process of photocatalysis generally involves the participation of a semiconductor material that can generate radical species upon being activated by light [59]. But traditional photocatalysts generally have very low quantum yields by virtue of their very wide band gaps, making it impractical to use them individually for real-time applications [59]. As a consequence of which, these photocatalysts are generally modified by combining them with other materials in order to render them more applicable for utilization.

Eleftheriadou et al. [60] developed a novel polymer-supported GO-$TiO_2$ NC for the purpose of degrading pharmaceuticals in water. $TiO_2$ is well known as an efficient photocatalyst with high photocatalytic activity in the UV region; however, it suffers from several drawbacks like high energy band gap, low quantum yield, and fast recombination of electron pairs that affect its performance under the visible spectrum of light. In order to overcome this disadvantage, $TiO_2$ is combined with GO. The GO helps to amplify the photocatalytic ability of $TiO_2$ by increasing the surface area for adsorption and by delaying electron-hole recombination. Also, $TiO_2$ has a tendency to agglomerate, thereby reducing its efficiency and reusability. In order to mitigate this problem, the GO-$TiO_2$ was immobilized on a polymer named poly(L-lactic acid) (PLLA) which was an extensively used biopolymer. The PLLA films were synthesized by a phase inversion method and the GO and $TiO_2$ were added to PLLA in the form dispersed solution followed by which the PLLA-GO-$TiO_2$ NC films were formed. The PLLA-GO-$TiO_2$ films consisted of 1% GO and 10, 25, and 50% of $TiO_2$, respectively. The PLLA-GO-$TiO_2$ was used for the photocatalytic degradation of nine widely used antibiotics. The photocatalytic experiments were conducted under solar simulator lamps. From the results of the experiment, it was found that the PLLA-$TiO_2$-GO NC with 50% $TiO_2$ exhibited the highest effectivity. The increase in the dosage of $TiO_2$ accelerated the degradation of the antibiotics as more reactive species were produced and a higher surface area was available for photodegradation. Also, the adverse effects that could have been caused by the increased dosage of $TiO_2$, like light scattering and increased turbidity, were nullified by the immobilization of $TiO_2$ on the thin films. The PLLA-$TiO_2$-GO NC also exhibited high efficiency for degrading the antibiotics in real-time wastewater effluent, although the rate of degradation was lower as compared to the antibiotic solutions in the ultrapure water due to the presence of various interferences like organic waste products and various inorganic ions. The PLLA-$TiO_2$-GO NC also exhibited high reusability potential, being able to be reused four times without any noticeable decrease in efficiency. A toxicity study was performed with Daphnia Magna, and from there, it was evaluated that the acute toxicity imparted by the antibiotics decreased after their photocatalytic degradation, thus making photocatalysis a viable method for the treatment of the antibiotics.

Moztahida et al. [61] synthesized an rGO-loaded $Fe_3O_4$ NC by a co-precipitation method with the intention of degrading the anticonvulsant drug CBZ under visible light. Magnetite nanoparticles consisting of Fe(II) and Fe(III) can operate as effective photocatalysts even under visible light due to their very low band gap. The addition of GO helps in increasing the surface area for adsorption and for decreasing the recombination rate of the photogenerated electrons. Also, the addition of reduced GO has been reported to shorten the time for photocatalyst activation. For this particular experiment, rGO-$Fe_3O_4$ NC with three different ratios of GO loading (1% by weight, 10% by weight, and 20% by weight) were synthesized. The experiments were conducted under a Xenon Arc Solar lamp (150 W). From the optimization study it was observed that highest effectivity was exhibited by rGO-$Fe_3O_4$ NC with 10% GO loading, catalyst dose of 0.5 gm/L and pH of 6.5. CBZ was mainly degraded by the $OH^-$ radicals produced by the photocatalytic activity of the magnetite nanoparticles. Under optimal conditions, the rGO-$Fe_3O_4$ was able to degrade 98.7% of CBZ. In a toxicity study following the degradation, the treated water showed no significant toxic effect whereas the untreated water exhibited a 25% increase in the toxic effect.

Fu et al. [58] synthesized an rGO-$ZnIn_2S_4$ NC by a simple hydrothermal method and evaluated its efficiency for the purpose of degrading Naproxen, a widely used analgesic and antipyretic medicine. $ZnIn_2S_4$ is a ternary metal chalcogenide semiconductor. Its use has evolved in recent times and among other things, $ZnIn_2S_4$ has also been used for the degradation of pollutants. $ZnIn_2S_4$ exhibits photocatalytic activity under visible range. rGO, on the other hand, can provide support to the $ZnIn_2S_4$ and can act as an acceptor and mediator of electrons generated by photocatalysis. The rGO-$ZnIn_2S_4$ NCs were prepared with four different mass ratios of GO, i.e., 0.5%, 1%, 2.5%, 5%, and 8%. The photocatalytic experiment was conducted in the presence of a 300 W xenon lamp with a cut-off filter at 420 nm. The intensity of the bulb was 115 $mW/cm^2$. The rGO-$ZnIn_2S_4$ with 1% GO exhibited the highest efficiency and $h^+$ and $\bullet O_2^-$ were shown to be the main reactive species behind the degradation of Naproxen molecules. Under optimum conditions, the rGO-$ZnIn_2S_4$ NC with 1% GO were able to degrade 99.4% Naproxen in 60 minutes, making the NC a feasible option for the purpose of degrading similar organic contaminants in the future.

Lin et al. [19] synthesized a series of $TiO_2$-rGO coated side glowing optical fibers (SOFs) with various ratios of GO loading by a polymer-assisted hydrothermal method. GO was added with the intention of decreasing the band gap of $TiO_2$ and also to act as an electron transporter/acceptor, thereby increasing the lifetime of the electron-hole pairs generated on the surface of $TiO_2$. The $TiO_2$-rGO was coated onto SOFs in order to facilitate their recovery after treatment and to reduce the wastage of incident radiation by liquid adsorption or incident ray scattering. The effectivity of $TiO_2$-rGO coated SOFs were evaluated by their ability to degrade three well-known pharmaceuticals, namely Ibuprofen, CBZ, and sulfamethoxazole. The $TiO_2$-rGO-coated SOFs with 2.7% GO loading exhibited the highest degradation efficiency. $TiO_2$-rGO (2.7%) degraded 81% Ibuprofen, 54% CBZ, and 92% Sulfamethoxazole after 180 min of high-pressure UV exposure. This was higher than the amount degraded by Degussa-P25 coated SOF and synthesized $TiO_2$ coated SOFs, thereby signifying that the addition of GO helped to enhance the photocatalytic of $TiO_2$. The reason for the enhanced photodegradation due to the addition of GO was due to increased surface area for adsorption, lowering of band gap for activation of $TiO_2$, thereby enabling its photoactivation by visible light, and due to the formation of a heterojunction interface in the $TiO_2$-rGO matrix which reduced the rate of recombination of electrons thereby increasing the photocatalytic efficiency. However, there lies an optimum dosage of GO for achieving the highest efficiency of $TiO_2$. If more GO is added beyond that point, then the surface of $TiO_2$ may get unintentionally covered by GO lowering the availability of active sites for catalysis. The $TiO_2$-rGO (2.7%) coated SOFs were used for degrading Ibuprofen under visible light. And although the degradation of Ibuprofen was found to be significantly less under visible light (18%) as compared to UV light (81% for high-pressure UV lamp, 41% for low-pressure UV lamp) it was still higher than that of only $TiO_2$-coated SOF. Since $TiO_2$ is mostly photoactivated by UV light therefore from this experiment, it is again proved that the addition of GO decreased the band gap of $TiO_2$. From this experiment, it was also inferred that the rate of degradation of pharmaceuticals was positively correlated quantum flux in the case of both UV and white light.

Therefore, the rGO-$TiO_2$ coated SOF could exhibit much higher efficiency when exposed to higher quantum flux of solar light. Thus, from the result of this study, it can be inferred that with necessary effort a photocatalyst can indeed be synthesized, which could be optimally functional under solar/visible light range.

### 4.3.3 MEMBRANE FILTRATION

Among the various water treatment methods that has emerged in recent times for the purpose of removing pollutants from water, membrane technology has emerged as one of the most effective processes due to its inert nature and effective regeneration ability. By applying the membrane filtration technique, water purification can be performed up to the ionic level. Another added advantage of membrane filtration for water purification is that the membrane filter only allows aqueous particles to pass through them, thereby separating out any other particles like undissolved solids, gas, or microorganisms from water [62]. As per some previous literature, the addition of GO enhances the hydrophilicity of modified membranes, increases the water flux, and amplifies the rejection rate of pharmaceuticals [21].

As mentioned earlier, Lou et al. [21] synthesized a GO-based membrane on a polyethersulfone (PES) support. Three amine groups, namely, EED, BBD, and p-phenylenediamine (PPD), were used as cross-linking agents, to evaluate the effect of these diamines on the GO-based membrane. The membrane was synthesized by a pressure-assisted self-assembly method, with these diamines acting as the cross-linking agents. Ibuprofen, Gemfibrozil, and Triclosan were chosen to be the target chemicals for this study. The GO imparted the membrane with high surface area, chargeability, mechanical strength, and anti-bacterial properties. Moreover, the functional groups present on the surface of GO helps in increasing the flexibility and adaptability of GO. In terms of interlayer spacing of membranes, it was found that as the length of EDA molecule was short therefore the interlayer-spacing of GO-EDA membrane was less than GO-BDA and GO-PPD membrane. Although the size of BDA and PPD were same, GO-PPD was bigger than GO-BDA due to steric hindrance of Benzene rings. The membranes were used to remove the three hydrophobic pharmaceuticals from water by adsorption and membrane filtration. In the case of adsorption, greater the interlayer spacing, greater the stearic hindrance, higher the hydrophobicity, and higher the removal. Thus, the GO-PPD membrane exhibited the highest removal of PPCPs by adsorption. Alternatively, in the case of membrane filtration, the shorter the membrane layer spacing, the greater the steric hindrance and the higher the removal rate. As a result of which, GO-EDA exhibited the highest removal by membrane filtration. GO-EDA was able to remove 99.7% triclosan, 73.59% gemfibrozil, and 47.2% gemfibrozil under a flux of 5.27 $L/m^2/h/bar$. Also, adsorption only has a positive effect on the removal rate in the initial stages of filtration, after which the removal of the PPCPs is mainly dependent on molecular spacing and stearic hindrance.

Li et al. [63] synthesized a GO ultrathin membrane supported on top of a cross-linked Matrimid substrate by a pressure-assisted filtration method. GO was utilized in the synthesis of the membrane due to its well-known separation ability, whereas Matrimid was selected because it cross-links easily and is resistant to harsh solvents. The width of the GO layer on top of the ultrathin membrane was about 70 nm. Due to the addition of GO, the pore size of the substrate membrane decreased considerably. As a result of which, the pure water permeability (PWP) and molecular weight cut-offs (MWCO) of the Matrimid-GO membrane reduced considerably when compared to both the pristine Matrimid membrane and cross-linked Matrimid membrane. The solvent for dissolving the drugs was Isopropanol, and the pressure was 15 bar with the original concentration of the drugs being 50 mg/L. From the result of the membrane filtration study, the Matrimid-GO membrane were able to remove 65.8% of TC, 82.67% of Rifampicin, 84.27% of Spiramycin, and 92.21% of Roxithromycin. Around 95.34% of Vitamin B-12 and 98.44% of Lecithin were also removed under the same conditions. The membrane also performed satisfactorily in three 7-day stability tests with 50 mg/L solution of Vitamin 12 in Isopropanol and ethanol and a 50 mg/L solution of Lecithin in Hexane. During the course of the test, the membrane exhibited a good rejection rate with a reasonable flux in all three solvents. This result confirmed the stability of GO in a Matrimid-based laminar composite membrane against various types of polar and non-polar solvents and under a good amount of pressure (15 bar). Thus, it can be inferred that the GO-Matrimid membrane thus synthesized can be utilized for industrial purposes.

Wang et al. [64] synthesized a rGO-Carbon Nanotube (CNT) intercalated membranes with a polyvinyl fluoride (PVDF) membrane with a pore size of 0.22 μm as support. Three distinct methods were taken to synthesize these membranes. For membrane 1, rGO and CNT were sonicated together in the form of a suspension in ultrapure water with a probe sonicator for their proper dispersion, followed by which it was injected into the PVDF membrane. For membrane 2, rGO and CNT were first sonicated separately, followed by which the suspensions were mixed together and sonicated again, followed by which it was added to the PVDF membrane. For membrane 3, rGO and CNT were sonicated separately in their respective suspension, followed by which they were sequentially added to the PVDF membrane. For benchmarking purpose, membrane 4 was synthesized by sonicating a suspension of Multi-Walled CNT (MWCNT) in ultrapure water, followed by their addition onto the PVDF support. From the characterization study, it was inferred that the r-GO and CNT particles were almost uniformly distributed in the case of membrane 2 and consequently, membrane 2 exhibited the highest efficiency in terms of membrane permeability (4,454 L/m$^2$/h/bar), anti-fouling property and PPCP (Acetaminophen, Triclosan, Caffeine, and Carbendazim) removal (76–100%). Thus, from this experiment, it was inferred that the nanostructures in rGO-CNT heavily influenced the performance of the membrane and hence manipulation of rGO-CNT nanostructures can be done for future studies in order to extract the optimal performance of the membranes.

With the aim of removing BPA from water by a cost effective and sustainable method, Nasseri et al. [65] synthesized a polysulfone (PSF)/GO membrane by a phase inversion method. Other than PSF, the casting solution also consisted of polyvinylpyrrolidone (PVP) and DMF. Two ratios of GO are taken for the synthesis of membrane (0.4% and 1%) and another was synthesized without GO. The addition of GO greatly increased the water permeability of the membrane, with PSF, PSF/GO (0.4%) and PSF/GO (1%) membrane exhibiting pure water flux of 226 L/m$^2$ h, 449 L/m$^2$ h and 512 L/m$^2$ h under operating pressure of 2 bar. The highest removal of BPA was exhibited by the PSF/GO (0.4%) membrane, which can be attributed to the highest zeta potential exhibited by this membrane (in comparison to PSF and PSF/GO (1%) membrane). The increased zeta potential can enhance the repulsion between the surface of the membrane and negative particles like BPA and thus can result in increased rejection of BPA by the membrane. This PSF/GO (0.4%) membrane was used for further optimization study by central composite design (CCD) of response surface methodology (RSM) system. From the result of that experiment, it was found that the ideal conditions under which the highest removal of BPA (93%) was exhibited by the PSF/GO (0.4%) membrane were input pressure of 1.02 bar, operating time of 10.6 minutes, initial concentration of BPA of 7.5 mg/L and pH of experimental solution of 5.5. The addition of GO can also decrease the fouling of the membrane by microorganisms and humic acid by increasing the hydrophilicity of the membrane. Thus, from this study, it was observed that this PSF/GO (0.4%) membrane was able to remove BPA from water under cost-effective and sustainable conditions, and hence further studies can be done on the feasibility of upscaling the production of this type of membranes for the purpose of removing similar kinds of pollutants from water.

## 4.4 PROCESS PARAMETERS AFFECTING PPCP REMOVAL FROM WASTE WATER

From the literature surveyed, it was found that various parameters were responsible for the removal of PPCPs from water by GO-based NCs. Parameters also tended to change with the change in the system that was being employed at that time. However, among them, some parameters like pH of the experimental solution, dosage of adsorbent, and percentage of GO in NC were checked in almost all the systems due to their consequential effect on the removal of PPCPs. Therefore, the effect of these parameters on the removal efficiency of GO-based NCs have been elaborated in this section.

### *4.4.1 PH OF THE SOLUTION*

The pH of the experimental solution had a significant effect on the removal efficacy of the GO-based NCs [9]. The change in pH plays an important role in determining the surface properties of both the adsorbate and adsorbent molecules [9]. The $pH_{pzc}$ (point of zero charge) and Zeta potential of the adsorbent and pKa (dissociation constant) of the adsorbate plays a crucial role in the adsorption of PPCP by adsorbates [52]. For example, in the case of the work of Deng et al. [44], the pH of the experimental solution had a significant effect on the removal of Propranolol from water by GO-Attapulgite-$Fe_3O_4$ NC. At lower pH, the abundance of

$H^+$ ions competed with Propranolol for adsorption on the surface of the GO-Attapulgite-$Fe_3O_4$ NC. When the pH increased the concentration of $H^+$ ions decreased thereby facilitating adsorption. Also at the pH range of 5.5–10, some secondary amine group present with Propranolol (with pKa > pH 9) attained a protonated form and formed a bond with the $COO^-$ groups present on the surface of GO. As a consequence of which highest removal was observed across the pH range of 5.5–10. When the pH of the experimental solution increased beyond 10, the amine groups became deprotonated resulting in a decrease in the rate of adsorption. The pH of the experimental solution also has an effect on the stability of the NC, as exhibited by the work of Tabrizian et al. [47]. The GO-Fe/Cu nanoparticle exhibited the highest removal (almost 100%) at the pH range of 5–6, which is due to the stability of the nanoparticles, the charge of the nanomaterials and contaminants, and as well as the bond between the nanoparticles, and the contaminants. At a lower pH, the GO-Fe/Cu nanoparticle might themselves get partially oxidized, thereby reducing their adsorption capacity, whereas at pH close to 7, there might be some repulsion between GO-Fe/Cu nanoparticle and their target pollutant (TC) due to acquiring of similar charges. As the stability of the GO-Fe/Cu nanoparticle increased gradually up to pH 6, the highest adsorption of TC by the GO-Fe/Cu nanoparticle was also observed at around pH 6. The pH also has a significant effect on the advanced oxidation processes with GO-based NC. For instance, GO-NC-based Fenton reaction is interrupted at higher pH because at pH greater than 4, the $H_2O_2$ gets decomposed into water ($H_2O$) and Oxygen ($O_2$), whereas at lower pH, the $H^+$ ions compete with the target pollutants for adsorption on the surface of the NC [51]. During the removal of PPCPs by Ozonation, higher removal is observed at a higher pH range because, reportedly, the decomposition rate of Ozone is facilitated at higher pH [54, 57].

### *4.4.2 DOSE OF THE ADSORBENT*

The dosage of the adsorbent/photocatalyst utilized had a significant effect on the percentage removal of PPCPs. In the case of the adsorption reactions, the removal of the respective target chemicals increased with the increase in adsorbent dosage, mainly due to the availability of more active sites for adsorption [44, 46, 47]. However, each adsorption experiment exhibited a saturation point beyond which no further removal of pharmaceuticals was observed, even with the increase in adsorbent dosage. This is known as reaching of the adsorption equilibrium point and is generally attributed to the saturation of active sites on the surface of adsorbents [44, 46] or to the agglomeration of adsorbent particles which in turn diminishes their uptake adsorbates [55]. In the case of photocatalysis reactions, the addition of photocatalysts over a certain dose may increase the rate of adsorption of pollutants, but in turn, can impart a shielding effect on the incoming radiation, thereby reducing the rate of photocatalysis [20, 61]. Thus this factor is also kept in consideration while determining the optimum dosage of the catalyst for the photolysis experiment [52, 55].

### 4.4.3 DOSAGE OF GO IN NANOCOMPOSITE (NC)

The utility of GO in the GO-based NC varies as per the functionality of the NC. During the preparation of an adsorbent or catalyst, GO is mainly utilized as a supporting material [44, 46, 47]. During the synthesis of catalyst, GO is mainly utilized for its superior adsorption ability, chemical stability, and ability to act as an electron acceptor, thereby delaying the electron-hole recombination [20, 22, 52]. Moreover, the combination of GO with the photocatalyst $TiO_2$ extends the photoactivation region of $TiO_2$ into the visible region [57, 60], thereby enhancing its photocatalytic ability even in the visible region of the spectrum. GO is utilized in membrane synthesis for its chemical stability, charge separation, and size exclusion properties and for its ability to reduce membrane fouling by increasing membrane permeability [21, 63–64].

From the literature reviewed, it can be termed that the success of a GO-based NC is highly dependent on determining the correct GO content. The appropriate content of GO in a particular NC is related to the function of the NC and it is mostly determined experimentally by varying the loading of GO over a particular range. For example, as per the work of Yu et al. [52], the correct loading of GO in the synthesis of a GO-$Fe_3O_4$ NC intended for the purpose of enhanced Fenton oxidation of phenol was found to be 5%. The maximum degradation of phenol was recorded at that GO loading and when the loading of GO was increased any further the effectivity of the NC decreased due to $\pi$-$\pi$ stacking of GO. Nawaz et al. [20] synthesized GO-$TiO_2$ for the purpose of degrading CBZ. Four ratios of GO-$TiO_2$ were taken namely 1:1, 1:2, 1:3, and 1:4 and the highest effectivity was exhibited by the ratio of 1:2. At higher $TiO_2$ dosage the degradation of CBZ was not adequate whereas the higher GO dosage was hindering the UV light from entering the solution. Nasseri et al. [65] synthesized a PSF/GO nanocomposite membrane with two amounts of GO (0.4% by weight and 1% by weight). The PSF/GO membrane with 1% GO exhibited greater permeability of water whereas the PSF/GO membrane with 0.4% GO exhibited greater exclusion of BPA. Thus, it can be said the correct quantity of GO in the GO-based NC is an important parameter that has to be determined on the basis of knowledge gained during the experiments.

## 4.5 PROCESS OPTIMIZATION CONCERNING THE REMOVAL OF PPCP BY GO-BASED NANOCOMPOSITES (NCS)

Normally under batch study, one experimental factor is varied across a particular range to observe its effect on a particular process operation whereas the other factors are kept constant. As a result of which this process is considered to be time-consuming and cumbersome [66]. Moreover, the most desirable outcome of a particular experiment is obtained when it is carried out under a set of optimized conditions and not a single optimized parameter [66]. Therefore, with that in mind, various analogical programings are being utilized in recent times for process optimization with a greater understanding of the inter-parameter relationships resulting in a reduced number of experiments and hence reduced costing and time consumption [66]. RSM and artificial neural network (ANN) are two very popular optimization tools which has been utilized widely for assessing and subsequently enhancing the collective effect of different experimental parameters on a particular system [67–69]. The optimization section with reference to the removal of PPCPs by GO-based NCs shall be expounded with reference to these two processes.

### *4.5.1 RESPONSE SURFACE METHODOLOGY (RSM)*

Response surface methodology (RSM) can be defined as a set of statistical techniques and mathematical models which is used for the purpose of optimizing a set of experimental parameters termed independent variables, by virtue of performing a set of experiments (termed runs) with different variations of these parameters in order to acquire the desired outcome of a process [41]. Analysis of variance (ANOVA) is used for ascertaining the most relevant model for the optimization of the concerned process [41], whereas the optimum experimental conditions (within their given range) are decided by the Derringers desirability function [41].

### 4.5.2 ARTIFICIAL NEURAL NETWORK (ANN)

An artificial neural network (ANN) can be defined as a statistical tool that has been developed on the basis of the knowledge that has been derived from the biological nervous system [13]. ANN generally consists of three interconnected parallel layers, an input layer that consists of independent variables, a middle layer that consists of the hidden neural network, and an output layer that consists of the dependent variable [13, 18]. The three layers are joined by means of various transfer functions [66]. The ANN model employs various algorithms for the optimization of the experimental data after normalization. The best-fitting algorithm model for the experimental data is determined by the coefficient of correlation ($R^2$). The training of the ANN model is dependent on the optimal number of neurons in the hidden layer, which has to be determined by the trial-and-error method. In order to confirm the optimum output from the ANN model, the performance target and ramp constant of the ANN model has to be kept constant [66].

In multiple studies concerning the removal of PPCP by GO-based NC, RSM, and ANN has been utilized as an optimization parameter either singly or in combination. Balasubramani et al. [70] synthesized a GO-Cellulose nanogel composite which was utilized for removing the anti-depressant drug flupentixol (FPL) from water by adsorption. The adsorption process was optimized by a particle swarm optimization (PSO) algorithm-based ANN model with a Box-Behnken design. The experimental parameters that were varied were pH of the experimental solution, adsorbent dosage, initial concentration, and temperature. The maximum removal of FPL (99%) by adsorption onto the surface of GO-Cellulose NC was observed at experimental parameters of pH-5.35, GO-Cellulose NC dosage of 116.4 mg/gm, Initial concentration of FPL at 47.8 mg/L and temperature of 21.5°C. Thus, the collective effect of different experimental parameters and their optimized values for the highest removal of FPL was determined by this optimization process.

Magnetic GO NC was synthesized by a co-precipitation route by mixing $FeCl_3.6H_2O$ and $FeCl_2.4H_2O$ with GO under a nitrogenous atmosphere with constant stirring [71]. The NC was used for removing methadone (an opioid agonist) from water. The process was optimized by a Box-Behnken design. The validity of the model was checked by ANOVA, which gave a $R^2$ value of 0.9696, probability (p) value of less than 0.001, and standard deviation of +/–0.8, thereby denoting that the experimentally obtained result was concurrent with the predicted result. The optimized condition under which the highest removal was observed were pH=6.2, adsorbent dose of 0.0098 gm, contact time of 30 minutes, and temperature of 295.7 K. The highest removal of methadone was observed to be 87.2 mg/gm.

In another study, a GO-$TiO_2$ NC was synthesized, which was used for the removal of Phenol under UV radiation [118]. First, a batch study was performed to optimize various experimental parameters under which the highest removal of phenol was exhibited by the GO-$TiO_2$ NC. For the batch study, the highest removal of phenol was observed to be 97% at a photocatalyst dosage of 200 mg/100 ml and a reaction time of 120 minutes. The process was further optimized by both RSM and ANN methods. For the RSM study, the experimental parameters that were considered were pH of the experimental solution, dosage of GO-$TiO_2$, and contact time of the reaction. From the RSM study, the highest removal of phenol (96.2%) was observed at a photocatalyst dosage of 200 mg/100 ml, pH 5 and reaction time of 115 min. This result was in accordance with the result of the batch study. The $R^2$ value for the RSM model was found 0.9571. In the case of the ANN model, the best fit was exhibited by the Levenberg Marquardt Backpropagation algorithm with a $R^2$ value of 0.987. Therefore, in this study by Ganguly et al. [18], the removal of Phenol by GO-$TiO_2$ was observed under a batch study condition followed by its optimization by the RSM and ANN methods. The results of optimization study matched with the optimum result from the batch study to check the correlation and validity of the two systems.

## 4.6 CONCLUSION

In this present review, the use of GO-based NC for the purpose of removing PPCPs from water has been discussed with reference to recently published literature. The aim was to do an exhaustive review regarding the recent developments about the synthesis and application of GO-based NCs for the purpose of removing PPCPs from water. All the different systems under which the PPCPs can be removed by GO-based NCs, like adsorption, advanced oxidation, and membrane filtration has been discussed in this chapter. As per the findings of this review, it can be stated that by virtue of its unique set of physical and chemical characteristics, GO was able to enhance the efficiency of almost all the NCs to which it was added. The effect of various experimental parameters on the removal capacity of GO-based NC was also elucidated in this review. The use of optimization tools like RSM and ANN for optimizing the experimental parameters for enhancing the removal capacity GO-based NC was also discussed in this chapter. Thus, the credibility of GO for the purpose of removing recalcitrant and emerging pollutants (EPs) from the aqueous medium was re-established through this chapter. Future studies could be aimed at the synthetization of more efficient NC with a further optimized dose of GO for the purpose of removing different types of emerging and recalcitrant pollutants from water.

## KEYWORDS

- **adsorption**
- **advanced oxidation**
- **emerging pollutants**
- **membrane filtration**
- **process optimization**
- **sewage treatment plant**
- **wastewater treatment plants**

## REFERENCES

1. Wang, J., & Wang, S., (2016). Removal of pharmaceuticals and personal care products (PPCPs) from wastewater: A review. *J. Environ. Manage., 182*, 620–640. https://doi.org/10.1016/j.jenvman.2016.07.049.

2. Richmond, E. K., Grace, M. R., Kelly, J. J., Reisinger, A. J., Rosi, E. J., & Walters, D. M., (2017). *Pharmaceuticals and Personal Care Products (PPCPs) are Ecological Disrupting Compounds (EcoDC).*
3. Bernhardt, E. S., Rosi, E. J., & Gessner, M. O., (2017). *Synthetic Chemicals as Agents of Global Change*. https://doi.org/10.1002/fee.1450.
4. Yang, Y., Sik, Y., Kim, K., Kwon, E. E., & Fai, Y., (2017). Science of the total environment occurrences and removal of pharmaceuticals and personal care products (PPCPs) in drinking water and water / sewage treatment plants: A review. *Sci. Total Environ., 596, 597*, 303–320. https://doi.org/10.1016/j.scitotenv.2017.04.102.
5. Balakrishna, K., Rath, A., Praveenkumarreddy, Y., & Siri, K., (2017). A review of the occurrence of pharmaceuticals and personal care products in Indian water bodies. *Ecotoxicol. Environ. Saf., 137*, 113–120. https://doi.org/10.1016/j.ecoenv.2016.11.014.
6. Ebele, J. A., Abdallah, M. A., & Harrad, S., (2017). Pharmaceuticals and personal care products (PPCPs) in the freshwater aquatic environment. *Emerg. Contam., 3*, 1–16. https://doi.org/10.1016/j.emcon.2016.12.004.
7. Dai, G., Wang, B., Huang, J., Dong, R., Deng, S., & Yu, G., (2015). Chemosphere occurrence and source apportionment of pharmaceuticals and personal care products in the Baiyun River of Beijing, China. *Chemosphere, 119*, 1033–1039. https://doi.org/10.1016/j.chemosphere.2014.08.056.
8. Cuthbert, R., Ichard, J., Taggart, M. A., Sarkar, M., Sharma, A., Das, A., Kulkarni, M. D., et al., (2016). Continuing mortality of vultures in India associated with illegal veterinary use of diclofenac and a potential threat from nimesulide. *Oryx, 50*, 104–112. https://doi.org/10.1017/S003060531500037X.
9. Khan, A., Wang, J., Li, J., Wang, X., & Chen, Z., (2017). The role of graphene oxide and graphene oxide-based nanomaterials in the removal of pharmaceuticals from aqueous media: A review. *Environ. Sci. Pollut. Res*. https://doi.org/10.1007/s11356-017-8388-8.
10. Lee, S. S., Paspalof, A. M., Snow, D., Richmond, K., Rosi-marshall, E. J., & Kelly, J. J., (2016). *Occurrence and Potential Biological Effects of Amphetamine on Stream Communities*. https://doi.org/10.1021/acs.est.6b03717.
11. Prichard, E., & Granek, E. F., (2016). Effects of pharmaceuticals and personal care products on marine organisms: From single-species studies to an ecosystem-based approach. *Environ. Sci. Pollut. Res*., 22365–22384. https://doi.org/10.1007/s11356-016-7282-0.
12. Tomul, F., Arslan, Y., Kabak, B., Kendüzler, E., Lima, E. C., & Nguyen, H., (2020). Peanut shells-derived biochars prepared from different carbonization processes: Comparison of characterization and mechanism of naproxen adsorption in water. *Science of the Total Environment, 726*. https://doi.org/10.1016/j.scitotenv.2020.137828.
13. Banerjee, P., Sau, S., Das, P., & Mukhopadhyay, A., (2015). Ecotoxicology and environmental safety optimization and modelling of synthetic azo dye wastewater treatment using graphene oxide nanoplatelets: Characterization toxicity evaluation and optimization using artificial neural network. *Ecotoxicol. Environ. Saf., 119*, 47–57. https://doi.org/10.1016/j.ecoenv.2015.04.022.

14. Banerjee, P., Das, P., Zaman, A., & Das, P., (2016b). Application of graphene oxide nanoplatelets for adsorption of Ibuprofen from aqueous solutions: Evaluation of process kinetics and. *Process Saf. Environ. Prot., 101*, 45–53. https://doi.org/10.1016/j.psep.2016.01.021.
15. Bhattacharya, S., Banerjee, P., Das, P., Bhowal, A., Majumdar, S. K., & Ghosh, P., (2020). *Removal of Aqueous Carbamazepine Using Graphene Oxide Nanoplatelets: Process Modeling and Optimization*, 8.
16. Bhattacharyya, S., Glass, C., Banerjee, P., Bhattacharya, S., Karan, R., & Rathour, S., (2018). *Comparative Assessment on the Removal of Ranitidine and Prednisolone Present in Solution Using Graphene Oxide (GO) Nanoplatelets*. https://doi.org/10.5004/dwt.2018.23170.
17. Alim, N. S., Lintag, H. O., & Yuliati, L., (2016). Photocatalytic removal of phenol over titanium dioxide- reduced graphene oxide photocatalyst photocatalytic removal of phenol over titanium dioxide- reduced graphene oxide photocatalyst. *Mater. Sci. Eng., 107*. https://doi.org/10.1088/1757-899X/107/1/012001.
18. Ganguly, P., Sarkhel, R., Bhattacharya, S., Das, P., & Saha, A., (2020). Integral approach of treatment of phenolic wastewater using nano-metal coated graphene oxide in combination with advanced oxidation. *Surfaces and Interfaces, 21*, 100660. https://doi.org/10.1016/j.surfin.2020.100660.
19. Lin, L., Lin, L., Wang, H., & Xu, P., (2016). *Immobilized $TiO_2$-reduced Graphene Oxide Nanocomposites on Optical Fibers as High Performance Photocatalysts for Degradation of Pharmaceuticals*. https://doi.org/10.1016/j.cej.2016.04.024.
20. Nawaz, M., Miran, W., Jang, J., & Lee, D. S., (2017). Environmental one-step hydrothermal synthesis of porous 3D reduced graphene oxide / $TiO_2$ aerogel for carbamazepine photodegradation in aqueous solution. *Applied Catal. B, Environ., 203*, 85–95. https://doi.org/10.1016/j.apcatb.2016.10.007.
21. Lou, Y., Tan, F. J., Zeng, R., Wang, M., Li, P., & Xia, S., (2020). Preparation of cross-linked graphene oxide on polyethersulfone membrane for pharmaceuticals and personal care products removal. *Polymers (Basel), 12,* 1921.
22. Shan, D., Deng, S., Wang, Y., & Huang, J., (2018). *Environmental Science Nano Typical Pharmaceuticals,* 1650–1660. https://doi.org/10.1039/c8en00422f.
23. Oh, Y., Armstrong, D. L., Finnerty, C., Zheng, S., Hu, M., & Torrents, A., (2017). Understanding the pH-responsive behavior of graphene oxide membrane in removing ions and organic micropollutants. *J. Memb. Sci., 541*, 235–243. https://doi.org/10.1016/j.memsci.2017.07.005.
24. Chen, Z., Liu, Z., Hu, J., Cai, Q., Li, X., & Wang, W., (2020). β -cyclodextrin-modified graphene oxide membranes with large adsorption capacity and high flux for efficient removal of bisphenol A from water β -cyclodextrin-modified graphene oxide membranes with large adsorption capacity and high flux for efficient rem. *J. Memb. Sci., 595*, 117510. https://doi.org/10.1016/j.memsci.2019.117510.
25. Xi, Y., Hu, J., Liu, Z., Xie, R., Ju, X., Wang, W., & Chu, L., (2016). Graphene oxide membranes with strong stability in aqueous solutions and controllable lamellar spacing. *Appl. Mater. Interfaces, 24*, 15557–15566. https://doi.org/10.1021/acsami.6b00928.
26. Xi, Y., Liu, Z., Ji, J., Wang, Y., Faraj, Y., Zhu, Y., & Xie, R., (2018). Graphene-based membranes with uniform 2D nanochannels for precise sieving of mono- /

multi-valent metal ions. *J. Memb. Sci., 550*, 208–218. https://doi.org/10.1016/j.memsci.2017.12.057.

27. Kang, L., Zhao, L., Yao, S., & Duan, C., (2019). Short communication A new architecture of super-hydrophilic β -SiAlON / graphene oxide ceramic membrane for enhanced anti-fouling and separation of water / oil emulsion. *Ceram. Int., 45*, 16717–16721. https://doi.org/10.1016/j.ceramint.2019.05.195.
28. Chu, K. H., Fathizadeh, M., Yu, M., Flora, J. R. V., Jang, M., Park, C. M., Yoo, S. S., et al., (2017a). *Evaluation of Removal Mechanisms in a Graphene Oxide Coated Ceramic Ultrafiltration Membrane for Retention of Natural Organic Matter, Pharmaceuticals, and Inorganic Salts*. https://doi.org/10.1021/acsami.7b14217.
29. Chu, K. H., Huang, Y., Yu, M., Heo, J., Flora, J. R. V., Jang, M., Jung, C., et al., (2017b). Evaluation of graphene oxide-coated ultrafiltration membranes for humic acid removal at different pH and conductivity conditions. *Sep. Purif. Technol.* https://doi.org/10.1016/j.seppur.2017.03.026.
30. Chu, K. H., Huang, Y., Yu, M., Her, N., Raymond, J., Flora, V., Park, C. M., et al., (2016). *Evaluation of Humic Acid and Tannic Acid Fouling in Graphene Oxide-Coated Ultrafiltration Membranes*. https://doi.org/10.1021/acsami.6b08020.
31. Kumar, A., Kumar, A., Sharma, G., Al-Muhtaseb, A. H., Naushad, M., Ghfar, A. A., & Stadler, F. J., (2018). Quaternary magnetic $BiOCl/g\text{-}C_3N_4/Cu_2O/Fe_3O_4$ nano-junction for visible light and solar powered degradation of sulfamethoxazole from aqueous environment. *Chem. Eng. J., 334*, 462–478. https://doi.org/10.1016/j.cej.2017.10.049.
32. Karimi-Maleh, H., Shafieizadeh, M., Taher, M. A., Opoku, F., Kiarii, E. M., Penny, G. P., Ranjbari, S., et al., (2020). The role of magnetite/graphene oxide nanocomposite as a high efficiency adsorbent for removal of phenazopyridine residues from water samples, an experimental/theoretical investigation. *J. Mol. Liq., 298*, 112040. https://doi.org/10.1016/j.molliq.2019.112040.
33. Nodeh, M. M., Radfard, M., Zardari, L. A., & Nodeh, H. R., (2018). Enhanced removal of naproxen from wastewater using silica magnetic nanoparticles decorated onto graphene oxide; parametric and equilibrium study. *Sep. Sci. Technol.,* 1–10. https://doi.org/10.1080/01496395.2018.1457054.
34. Sani, M. J., Hossein, A., Khavar, C., Mahjoub, A. R., Luque, R., Rodríguez-padrón, D., Satari, M., et al., (2019). Enhanced visible-light-driven photocatalytic degradation of emerging water contaminants by a modified zinc oxide-based photocatalyst – *in-vivo* and *in vitro*. *Sep. Purif. Technol.*, 116430. https://doi.org/10.1016/j.seppur.2019.116430.
35. Masud, A., Soria, N. G. C., Aga, D. S., & Aich, N., (2020). Adsorption and advanced oxidation of diverse pharmaceuticals and personal care products (PPCPs) from water using highly efficient rGO-nZVI nanohybrids. *Environ. Sci. Water Res. Technol., 6*, 2223–2238. https://doi.org/10.1039/D0EW00140F.
36. Motlagh, Y. P., Khataee, A., Hassani, A., & Sadeghi, T., (2020). ZnFe-LDH / GO nanocomposite coated on the glass support as a highly efficient catalyst for visible light photodegradation of an emerging pollutant. *J. Mol. Liq., 302*, 112532. https://doi.org/10.1016/j.molliq.2020.112532.

37. Shargh, A. Y., Sayadi, M. H., & Heidari, A., (2019). *Sonocatalyst Efficiency of Palladium-Graphene Nanocomposite for Ibuprofen Degradation from Aqueous Solution, 4*, 333–342. https://doi.org/10.22090/jwent.2019.04.007.
38. Heu, R., Ateia, M., & Yoshimura, C., (2020). Photocatalytic nanofiltration membrane using Zr-MOF / GO nanocomposite with high-flux and anti-fouling properties. *Catalysts, 10*, 711.
39. Li, C., Lu, Z., Ao, X., Sun, W., & Huang, X., (2022). Degradation kinetics and removal efficiencies of pharmaceuticals by photocatalytic ceramic membranes using ultraviolet light-emitting diodes. *Chem. Eng. J., 427*, 130828. https://doi.org/10.1016/j.cej.2021.130828.
40. Esmaeili, A., & Entezari, M. H., (2015). Cubic Ag / AgBr – graphene oxide nanocomposite: Sono-synthesis and use as a solar photocatalyst for pollutant. *RSC Adv., 5*, 97027–97035. https://doi.org/10.1039/C5RA16814G.
41. Banerjee, P., Barman, S. R., Mukhopadhyay, A., & Das, P., (2016a). Ultrasound-assisted mixed azo dye adsorption by chitosan- graphene oxide nanocomposite. *Chem. Eng. Res. Des*. https://doi.org/10.1016/j.cherd.2016.10.009.
42. Wang, X., Chen, Z., & Wang, X., (2015). *Graphene Oxides for Simultaneous Highly Efficient Removal of Trace Level Radionuclides from Aqueous Solutions, 58*, 1766–1773. https://doi.org/10.1007/s11426-015-5435-5.
43. Balasubramani, K., Sivarajasekar, N., Muthusaravanan, S., Ram, K., Naushad, M., Ahamad, T., & Sharma, G., (2020). Efficient removal of antidepressant flupentixol using graphene oxide/cellulose nanogel composite: Particle swarm algorithm-based artificial neural network modelling and optimization. *J. Mol. Liq., 319*, 114371. https://doi.org/10.1016/j.molliq.2020.114371.
44. Deng, Y., Li, Y., Nie, W., Gao, X., Zhang, L., & Yang, P., (2019). Fast removal of Propranolol from water by attapulgite / graphene oxide magnetic. *Materials (Basel), 12*, 924. https://doi.org/10.3390/ma12060924.
45. Kyzas, G. Z., Koltsakidou, A., Nanaki, S. G., Bikiaris, D. N., & Lambropoulou, D. A., (2015). Science of the total environment removal of beta-blockers from aqueous media by adsorption onto graphene oxide. *Sci. Total Environ., 537*, 411–420. https://doi.org/10.1016/j.scitotenv.2015.07.144.
46. Naskar, A., Khan, H., & Jana, S., (2018). One pot low temperature synthesis of graphene coupled Gd-doped $ZnFe_2O_4$ nanocomposite for effective removal of antibiotic levofloxacin drug. *J. Sol-Gel Sci. Technol*., 1–10. https://doi.org/10.1007/s10971-018-4668-0.
47. Tabrizian, P., Ma, W., Bakr, A., & Rahaman, S., (2018). pH-sensitive and magnetically separable Fe/Cu bimetallic nanoparticles supported by graphene oxide (GO) for high-efficiency removal of tetracyclines. *J. Colloid Interface Sci*. https://doi.org/10.1016/j.jcis.2018.09.034.
48. Sirés, I., Brillas, E., Oturan, M. A., Rodrigo, M. A., & Panizza, M., (2014). Electrochemical advanced oxidation processes: Today and tomorrow: A review. *Organic Compound*. https://doi.org/10.1007/s11356-014-2783-1.
49. Kanakaraju, D., Glass, B. D., & Oelgem, M., (2018). *Advanced Oxidation Process-Mediated Removal of Pharmaceuticals from Water: A Review,* 219. https://doi.org/10.1016/j.jenvman.2018.04.103.

50. Thakur, K., & Kandasubramanian, B., (2019). *Graphene and Graphene Oxide-Based Composites for Removal of Organic Pollutants: A Review*. https://doi.org/10.1021/acs.jced.8b01057.
51. Kong, Y., Zhuang, Y., & Shi, B., (2020). Tetracycline removal by double-metal-crosslinked alginate / graphene hydrogels through an enhanced Fenton reaction. *J. Hazard. Mater., 382*, 121060. https://doi.org/10.1016/j.jhazmat.2019.121060.
52. Yu, L., Chen, J., Liang, Z., Xu, W., Chen, L., & Ye, D., (2016). Degradation of phenol using $Fe_3O_4$-GO nanocomposite as a heterogeneous photo-Fenton catalyst degradation of phenol using $Fe_3O_4$ -GO nanocomposite as a heterogeneous photo-Fenton catalyst. *Sep. Purif. Technol., 171*, 80–87. https://doi.org/10.1016/j.seppur.2016.07.020.
53. Qi, K., Sun, Y., Duan, H., & Guo, X., (2015). A corrosion-protective coating based on a solution-processable polymer-grafted graphene oxide nanocomposite. *Corros. Sci., 98*, 500–506. https://doi.org/10.1016/j.corsci.2015.05.056.
54. Li, G., Lu, Y., Lu, C., Zhu, M., Zhai, C., Du, Y., & Yang, P., (2015). *Efficient Catalytic Ozonation of Bisphenol-A Over Reduced Graphene Oxide Modified Sea Urchin-Like α-$MnO_2$ Architectures*. Elsevier B.V. https://doi.org/10.1016/j.jhazmat.2015.03.045.
55. Checa, M., Figueredo, M., Aguinaco, A., & Beltrán, F. J., (2019). Graphene oxide / titania photocatalytic ozonation of primidone in a visible LED photoreactor. *J. Hazard. Mater., 369*, 70–78. https://doi.org/10.1016/j.jhazmat.2019.02.025.
56. Jothinathan, L., & Hu, J., (2018). Kinetic evaluation of graphene oxide based heterogeneous catalytic ozonation for the removal of Ibuprofen. *Water Res., 134*, 63–73. https://doi.org/10.1016/j.watres.2018.01.033.
57. Sheydaei, M., Reza, H., Shiadeh, K., Ayoubi-feiz, B., & Ezzati, R., (2018). Preparation of nano N-$TiO_2$ / graphene oxide / titan grid sheets for visible light assisted photocatalytic ozonation of cefixime. *Chem. Eng. J., 353*, 138–146. https://doi.org/10.1016/j.cej.2018.07.089.
58. Fu, K., Pan, Y., Ding, C., Shi, J., & Deng, H., (2020). Photocatalyst with enhanced photocatalytic performance for the degradation of naproxen under. *Catalysts, 10*, 710.
59. Akkari, M., Aranda, P., Belver, C., Bedia, J., Haj, A. B., & Ruiz-hitzky, E., (2018). Applied clay science ZnO / sepiolite heterostructured materials for solar photocatalytic degradation of pharmaceuticals in wastewater. *Appl. Clay Sci., 156*, 104–109. https://doi.org/10.1016/j.clay.2018.01.021.
60. Eleftheriadou, N. M., Ofrydopoulou, A., & Papageorgiou, M., (2020). *Development of Novel Polymer Supported Nanocomposite GO / $TiO_2$ Films, Based on Poly (L-Lactic Acid) for Photocatalytic Applications*.
61. Moztahida, M., Jang, J., Nawaz, M., Lim, S., & Sung, D., (2019). Science of the total environment effect of rGO loading on $Fe_3O_4$: A visible light assisted catalyst material for carbamazepine degradation. *Sci. Total Environ., 667*, 741–750. https://doi.org/10.1016/j.scitotenv.2019.02.376.
62. Banerjee, P., Bhattacharya, S., Das, R., & Das, P., (2021). *Microfiltration and Ultrafiltration Membranes for Water Purification*.
63. Li, B., Cui, Y., Japip, S., Thong, Z., & Chung, T., (2018). Graphene oxide (GO) laminar membranes for concentrating pharmaceuticals and food additives in organic solvents. *Carbon N. Y., 130*, 503–514. https://doi.org/10.1016/j.carbon.2018.01.040.

64. Wang, Y., Liu, Y., Yu, Y., & Huang, H., (2018). *Influence of CNT-rGO Composite Structures on Their Permeability and Selectivity for Membrane Water Treatment, 551*, 326–332. https://doi.org/10.1016/j.memsci.2018.01.031.
65. Nasseri, S., Ebrahimi, S., Abtahi, M., & Saeedi, R., (2018). Synthesis and characterization of polysulfone/graphene oxide nanocomposite membranes for removal of bisphenol A from water. *J. Environ. Manage., 205*, 174–182. https://doi.org/10.1016/j.jenvman.2017.09.074.
66. Banerjee, P., Roy, S., Sikdar, D., & Roy, U., (2017). *Enhanced Degradation of Ternary Dye Effluent by Developed Bacterial Consortium with RSM Optimization, ANN Modeling and Toxicity Evaluation, 72*, 20422. https://doi.org/10.5004/dwt.2017.20422.
67. Mukherjee, R., Chakraborty, R., & Dutta, A., (2019). Comparison of optimization approaches (response surface methodology and artificial neural network-genetic algorithm) for a novel mixed culture approach in soybean meal fermentation. *J. Food Process Eng., 42*, 1–10. https://doi.org/10.1111/jfpe.13124.
68. Sen, S., Nandi, S., & Dutta, S., (2018). Application of RSM and ANN for optimization and modeling of biosorption of chromium(VI) using cyanobacterial biomass. *Appl. Water Sci., 8*. https://doi.org/10.1007/s13201-018-0790-y.
69. Talib, N. S. R., Halmi, M. I. E., Ghani, S. S. A., Zaidan, U. H., & Shukor, M. Y. A., (2019). Artificial neural networks (ANNs) and response surface methodology (RSM) approach for modeling the optimization of chromium (VI) reduction by newly isolated *Acinetobacter radioresistens* strain NS-MIE from agricultural soil. *Biomed Res. Int., 2019*. https://doi.org/10.1155/2019/5785387.
70. Balasubramani, K., Sivarajasekar, N., & Naushad, M., (2020). Effective adsorption of antidiabetic pharmaceutical (Metformin) from aqueous medium using graphene oxide nanoparticles: Equilibrium and statistical modelling. *J. Mol. Liq., 301*, 112426. https://doi.org/10.1016/j.molliq.2019.112426.
71. Gupta, V. K., Agarwal, S., Asif, M., Fakhri, A., & Sadeghi, N., (2017). Application of response surface methodology to optimize the adsorption performance of a magnetic graphene oxide nanocomposite adsorbent for removal of methadone from the environment young researchers and elites club, science, and research branch, Islamic. *J. Colloid Interface Sci.* https://doi.org/10.1016/j.jcis.2017.03.006.

## CHAPTER 5

# Carbon-Based Nanomaterials for Energy Storage: A Review

SHAYERI DAS,[1,2*], PRABHAT RANJAN,[1] and TANMOY CHAKRABORTY[3]

[1]*Department of Mechatronics Engineering, Manipal University Jaipur, Dehmi Kalan – 303007, Rajasthan, India*

[2]*Department of Electrical Engineering, Ideal Institute of Engineering, Kalyani, Nadia, West Bengal – 741235, India*

[3]*Department of Chemistry and Biochemistry, School of Basic Sciences and Research, Sharda University, Greater Noida – 201310, Uttar Pradesh, India, E-mails: tanmoychem@gmail.com; tanmoy.chakraborty@sharda.ac.in*

**Corresponding author. E-mail: prabhat23887@gmail.com; Prabhat.ranjan@jaipur.manipal.edu*

## ABSTRACT

The energy supply chain has acknowledged energy storage as a vital component. Energy storage has been found to increase the efficiency of energy systems and even conserve traditional sources of fuel, along with having a huge environmental impact. Among various elements put into implementation for energy storage, carbon has been playing a crucial role in the enhancement of alternative energy sources and energy storage. Carbon being one of the most abundantly available materials, only adds to the attraction of thoroughly using carbon and its forms to maximum usage. This review chapter summarizes the various sources as well as applications

Carbon Composites: Composites with Nanotubes, Nanomaterials, and Graphene Oxide.
Eduardo A. Castro, Ann Rose Abraham, & A. K. Haghi (Eds.)

of different forms of carbon in energy storage. The integration of carbon along with its allotropes in the traditional as well as upcoming fields of energy storage has been reviewed.

## 5.1 INTRODUCTION

Any form of energy is one of the most essential entities in the world. Energy plays a key role in the global development. It is one of the most consumed commodities worldwide. Innovation and development of science has given rise to various forms of energy, but broadly we classify them into primary and secondary forms of energy [1]. The sources which comprise extraction or capture, with or without separation from contiguous material, cleaning, or grading, before the energy contained in it can be transformed into heat or mechanical work [1]. These sources are naturally found. They comprise all energy forms which have not been subjected to any conversion or transformation process. Classic examples are crude oil, coal, biomass, wind, solar, tidal, natural uranium, geothermal, falling, and flowing water, natural gas, etc. Globally, energy consumption has been increasing exponentially over the years. Although there is an increasing trend in the global energy supply, the percentage share of fossil fuel has been decreasing gradually due to the penetration of renewable energy systems. For example, approximately 82% of the primary energy supply in 2012 came from fossil fuels compared to 87% in 1973 [2]. Nevertheless, this decline in fossil fuels share in the primary energy supply does not portray in actual terms a reduction in $CO_2$ emission. For example, fossil fuels contributed about 31,734 Mt of $CO_2$ emissions in 2012 compared to 16,633 Mt in 1973 [2, 3].

Carbon dioxide ($CO_2$) emissions from fossil fuels have been identified as a foremost global environmental threat due to its contribution to global warming. For the last few years, many efforts have been made to diminish $CO_2$ emission in order to alleviate the environmental impact. These vary from generating new and innovative energy conversion technologies to refining the efficiency of existing energy conversion technologies. Additionally, reducing energy wastage from a variety of industries, both domestic and commercial, by storing them for future use has a very significant impact in reducing toxic emissions. The need to mitigate the mismatch between energy supplied to the grid and the energy actually used from the grid by storing the excess energy is equally imperative to accomplishing a low carbon economy. It is against this framework that energy storage is considered to be essential in the modern energy supply chain. Consequently, energy storage has recently fascinated the attention of governments, stakeholders, researchers, and investors as it may be used to mend the performance of the energy supply chain.

Carbon is one of the amplest elements on the planet, and has an acute role in the living entities and their surrounding ecosystems. For decades, carbon has also been a source of energy, and history as we know it today, is closely associated with the struggle to extract and utilize power from carbon materials. Any modern technology that we might name completely is dependent on pure carbon- or carbon-based materials. Both aspects have been supported by the endless supply of energy. Over the past several decades, noteworthy progress has been achieved in evolving alternative technologies to harvest and use clean, sustainable energy, including solar energy, wind power, biofuels, and hydrogen, as well as clean energy technologies, such as fuel cells and lithium-ion batteries. Even though these forms of energy have played a marginal role in the past, new technology is moving ahead impressively to make alternative energy more practical and cheaper than fossil fuels. It is expected that the coming decades will usher in a long-expected alteration away from fossil fuel as our primary fuel.

Carbon-based materials have a substantial role in advance of alternative clean, and sustainable energy. For instance, fullerene-containing p-type semiconducting polymers are one of the key fundamentals in rapidly advancing organic photovoltaics [4–7]. Furthermore, carbon nanotubes (CNTs) and graphenes are emergent as areas of carbon materials, and are being explored as critical additives for the next generation of optically transparent films for solar cells [8–10]. CNTs and graphenes are also considered for the development of batteries, supercapacitors, and fuel cells [11–18]. Graphite and graphitic carbons, on the other hand, are excellent electrical conductors. Graphene, CNTs, and fullerenes are all derivatives of graphite and have attracted significant attention lately in the research community, partly due to their unique and technically important physical properties.

Carbon has been widely used as catalyst supports, filters, sorbents, scaffolds, and matrices in various technically significant fields. Water purification [19–21], artificial livers or kidneys [22–25], and catalyst support [26–29], are few of the applications. Highly porous carbon can be allocated into two groups: (i) derived from naturally occurring carbonaceous precursors such as coal, wood, coconut shells, fruit stones, and other agricultural byproducts [30–34]; and (ii) synthetic porous carbon [35–37]. There are a few ways to synthesize porous carbon, including sol-gel processing [38–40], imprint of metal carbides [41–43], and templated carbon [44–47]. In this review, we have summarized the recent progress in carbon nanomaterials for energy storage.

## 5.2 CARBON AND ITS SOURCES

Carbon materials are generally classified based on the carbon precursor: graphitic, diamond-like, and amorphous. Graphitic carbons have been expansively pursued and remarkable efforts put into the development of materials with the graphite structure due to its high electrical conductivity. Graphite is molded by 2D sheets of $sp^2$ hybridized carbon atoms with hexagonal lattice. Each graphene sheet is oriented to its adjacent sheets by 180° rotations generating in an ideal structure, a repeating ABA stacking sequence. The structure is known as the 3D crystal of graphite. The edges of its planes have terminations with carbon atoms arranged with either zigzag or armchair configurations. Graphite is found in nature in coal mines and also produced from various sources. For the fabrication of carbon with graphitic, the sources can be categorized as soft and as hard carbons. Soft carbons are referred to as graphitizable carbons due to the ease of graphitization, typically at temperatures below 2,000°C. Typical soft carbon precursors are pitch-based chemicals (coal tar, petrol pitch and organic polymers). Hard carbons, however, are non-graphitizable carbon materials, even after thermal treatments at temperatures as high as 3,000°C. Phenolic resins, carbohydrates, and biomass (fruit stones, trees, etc.), are some of the examples of hard carbon precursors. Soft and hard carbons have domains of microcrystalline graphite that are randomly stacked with a turbostratic structure. The latter largely resembles the graphite structure, but with large numbers of defect sites such as broken and dangling bonds, unpaired electrons, and oxidized carbon sites. Most importantly, the term turbostratic relates to the lack of 3D stacking ordering found in graphite.

Expansion and control over the porosity of a material is not completely dependent on the type of templates used, but also the carbon precursor selected. Significant for energy storage applications, the selection of the carbon precursors also hinges on the precursor availability, cost, on the presence of heteroatoms such as nitrogen, boron, and phosphorous that could dope the final carbons and on the extent to which it can be graphitized. The search for new, available, and economically viable carbon sources that result in a porous carbon material has intensified, with a focus on synthetic carbon sources that can combine all aforesaid properties. The most common chemical precursors to carbon materials have biomass and oil origins, including alcohols [48, 49], carbohydrates, polysaccharides, lignocellulose sources [50–53], pitches [54], phenolic resins [55, 56], and organic polymers [56]. As more carbon sources are explored, a greater understanding of the surface chemistry involved will allow for the design of enhanced materials that will permit scientists to overcome existing barriers set by the present energy storage and conversion technologies.

### *5.2.1 BIOMASS SOURCES*

The biomass-based sources have gained huge importance due to their sufficient availability as well as environmental and financial benefits in comparison to fossil fuel-based resources. Steam is a widely used activating process used for biomass materials due to the inexpensive cost and lack of post-activation work-up. Microporosity is predominant in the carbon materials; yet, a broad circulation of mesopores can also be obtained. Other forms of carbon, such as fruit stones and coconut hulls are also used to produce activated carbon (AC) [52, 53]. In this case, the biomass is often mixed with phosphoric acid prior to carbonization or it is treated with steam at a greater temperature [52, 53]. Sucrose, xylose, maltose, amylopectin, starch, and biomass derivatives such as hydroxymethyl furfural (HMF) and furfural can also be hydrothermally dehydrated to form micrometer-sized carbon spheres with diameters determined by the reaction conditions, the use of stabilizers and precursor concentration [51]. Polysaccharides like starch have received much attention lately due to the likelihood of obtaining carbons with huge mesopores devoid of templating methods and with pore volumes comparable to hard-templated carbons [50]. The appeal of sucrose and starch is the renewable source aspect coupled with reasonable pioneer materials.

### 5.2.2 PITCH-BASED SOURCES

The materials obtained from these sources are soft carbon in nature and can be easily graphitized in comparison to hard carbon [54]. Pitch contains aromatic hydrocarbons and can also be obtained from nature other than coal or petrol. Generally, the preparation of carbon fibers may be delayed by the softening point of pitches. Stabilization in air at temperatures below 300°C is usually required prior to carbonization and graphitization [54, 57]. By blending pitches from tar and coal with polymers such as polyacrylonitrile (PAN), the structures of final fibers are custom-made in order to increase either the conductivity, or the mechanical strength of the fibers by changing the molecular orientations with the PAN addition [58].

## 5.3 APPLICATION IN ENERGY STORAGE

### 5.3.1 CARBON FOR SUPERCAPACITORS

The hybrid vehicles mostly lack power and the models that accelerate fast use large internal combustion engines. This considerably degrades the fuel efficiency, just about making the extra cost worthwhile in terms of fuel efficiency [59]. The batteries provided in these vehicles have the capacity to store large quantities of energy, but they cannot neither be charged nor discharged quickly. This lack of power density necessitates the battery packs to be oversized, resulting in increased vehicle weight and reduced efficiency. As with the poor discharge rate, battery charging is limited by the same kinetics, thus reducing efficiency gains through full regenerative braking. Additionally, the peak power demands that are put on the battery packs reduce the life of the battery, reducing the overall longevity of the vehicle [60]. An ideal electrical energy storage device should have high cycle life as well as high energy and power density when measured in terms of weight, volume, and cost.

According to the charge storage mechanism, there are two types of electrochemical capacitors (ECs). One is the electrochemical double layer capacitors (EDLCs) based on ACs with capacitance proportional to the electrode surface area. The other, known as pseudo-capacitors or redox supercapacitors, uses transition metal oxides or electrical conducting polymers as electrode materials, with the charge storage depending on the fast Faradaic redox reactions. Obviously, the combination of high surface area and small charge separation is necessary for an extremely high capacitance. EDLCs, also known as supercapacitors, operate by adsorbing/desorbing charged ions from an electrolyte onto their highly porous high surface area electrodes. While traditional capacitors rely on a dielectric material to store a charge, EDLCs rely on the charge of the adsorbed double layer.

CNTs and graphenes are also been used for making of batteries and supercapacitors because of their high conductivity [61, 62]. Carbon materials-based supercapacitors have drawn interest towards reviews [63]. There are several approaches to improve charge storage in carbon supercapacitors. A higher capacitance can be accomplished by thermal, chemical, or electrochemical treatment to enhance the accessible surface area and surface functional groups, or by extending the operating voltage range beyond the limit of an aqueous electrolyte solution. Several critical factors add to a high capacitance.

Interconnected mesoscale porosity has an important role in confirming that charged ions can freely access all the surfaces (2–50 nm). Hence, various researchers investigated surfactant-templated mesoporous carbon having controllable pore sizes [64, 65]. However, the contemporary study reported the effect of pore sizes on the charge storage properties and furnished fresh information on the relative role of the mesoscale and microscale porosity [66]. The charge storage in carbide-derived carbon by high-temperature chlorination [66]. This material possesses precise control of the pore sizes down to less than 1 nm. Three regions have been observed. In the first segment, where the mesopore dominates, the capacitance increases with the pore sizes due to enhanced pore accessibility and less overlapping of the double-layer structure. However, as the pore size becomes smaller in the second segment, the capacitance starts to increase. In this zone, the capacitance increases sharply with a decrease in pore sizes. The effect of the ultra-small pore on the capacitance is attributed to the distortion of the double layers in the small pores and decrease in the double layer thickness.

CNTs with larger surface area have drawn interest in the field of supercapacitors having specific capacitances [67–72]. Capacitance is remarkably affected by the surface areas and the pore size of the nanotubes [73]. Niu et al. [69] fabricated multi-walled nanotubes ECs with a specific surface area of around 430 $m^2$/gm, which exhibited capacitance of 113 F/gm. An et al. [72] found a maximum specific capacitance of 180 F/g with a large power density of 20 kW/kg by heating CNTs to improve their surface and pore distribution. Researchers have used ammonia with the nanotubes in order to enhance the electrochemical properties of the electrodes [74–76]. Yoon et al. [74] increased the capacitance of the CNTs electrode from 38.7 to 207.3 F/g through surface treatment using ammonia plasma. Yet, some scientists are of the opinion that the oxygen groups might lead to capacitor instability with an increased resistance and deterioration of capacitance [77, 78]. Additionally, the introduction of surface oxygen groups to the carbon nanotube electrode would not work with organic electrolytes. There are various ways to improve the energy storage properties:

- Current collectors when deposited with CNTs as electrodes tend to have a reduction in their contact resistance [69, 72, 74, 79–81]. The electrodes form a binder-free film which enhances the electrochemical performance due to lack of impurities. Researchers have designed even lighter electrodes with dense CNTs as current collector as well as active electrode in the ECs [82].
- Hybrid composites which combine CNTs with transition metals oxides or other polymers can gain a much larger capacitance by dual storage mechanism [83]. These systems exhibit large specific capacitance [84–86], in which CNTs act as a backbone as well as conductor during long cycling.
- Incorporating the advantages of ECs with Li-ion batteries, new hybrid capacitors have been fabricated [87].

Huge research attention has been grabbed by graphene-based supercapacitors. These supercapacitors possess a capacitance of above 100 F/g [88–90]. Graphene needs to be doped in order to enhance the electrochemical properties. In order to avoid the graphene sheets from agglomeration, a gas-solid reduction process has been used [91], wherein aqueous electrolytes have better capacitance than carbon nanotube-based capacitors [67–71]. Liu et al. [92] fabricated supercapacitors using curved graphene and acquired the highest energy density (85.6 Wh/kg at 1 A/g at room temperature (RT) or 136 Wh/kg at 80°C) reported so far till date. Combining graphene with other nanomaterials is also one of the explored methods of enhancing storage capacity [93–104]. Numerous hybrid composites have been manufactured, such as graphene (or grapheme oxide) – polyaniline composites [93–97], graphene nanosheet–CNTs–PANI composites [98], graphene–CNTs composites [99], graphene–metal oxide composites [100, 101, 105], graphene-metal hydroxide composites [102, 103], and graphene–$Sn_3S_4$ composites [104]. For example, the composite electrode of PANI nanowire arrays aligned vertically on graphene oxide (GO) nanosheets showed a synergistic effect of PANI and GO, with higher electrochemical capacitance and better stability than each individual component [93]. Research interest in fields like photocatalysis [106], direct methanol fuel cells [107], electrochemical cells [108], electrochemical biosensing [109–113], and even providing platforms for sensing TNT [114], studying charge transport [115], and analyzing small molecules by matrix-assisted laser desorption/ionization time-of-flight mass spectrometry [116] have also grown for graphene nanosheets.

### *5.3.2 CARBON FOR BATTERIES*

Battery systems of the past have been replaced by lithium-ion batteries in almost all devices. This transition has come to effect due to their high energy density and long life cycle [117–119]. Till date, proper application of Li-ion batteries has not been possible in high specific power and energy storage applications, such as power tools, electric vehicles, and efficient use of renewable energies [120]. Carbon has a predominantly important part in the development of Li-ion batteries. Earlier to the discovery of graphite anode materials, Li metal had been the main contender for Li-ion batteries, but had a severe drawback of dendrite formation, which triggered a short circuit resulting in a safety issue. The use of graphite in rechargeable batteries has been anticipated a long time ago [121] and advanced as the anode material for intercalating Li ions [122]. As of today, graphite is still the main anode material used in commercial Li-ion batteries [123]. Graphite is appealing due to its high in-plane electron conductivity due to the p-bond and weak interaction with Li ions, resulting in high Li-ion storage capacity and fast Li ion diffusion. Regrettably, the intercalation capacity of Li ions in graphite is still within bounds (372 mAh/g, $LiC_6$). Huge efforts have been made to intensify this capacity by alteration of the carbon structures, enhancing lattice disordering, creating pore spaces, and magnifying surfaces areas. The excessive Li-ion storage capacity could be obtained from bulk storage for the formation of different species, such as the proposed formation of $Li_2$ molecules in polymer-derived carbon. The higher capacity might also result from storage in microcavities or nanopores [124]. A different important mechanism is Li storage on the surfaces and interfaces of microcrystalline or nanocrystalline graphite or stacked graphene sheets [125–128]. On the surfaces and interfaces the Li storage capacity can be much greater than 372 mAh/g. The fundamental to accomplishing the high capacity is to regulate the starting materials and the processing conditions.

CNTs have been expansively investigated as anode materials for Li-ion batteries due to the mesoporous character, high chemical stability, low resistance, strong mechanical strength, and high activated surfaces [117, 130–132]. However, the nature of carbon materials limits their capacity, although some researchers have fabricated particular microstructures to expand the electrochemical properties of CNTs. For instance, 1D highly aligned carbon nanotube arrays have been fabricated by the chemical-vapor-deposition (CVD) method [133, 134]. The Fisher group fabricated a tube-in-tube structure with $Li^+$ intercalation capacity two times higher than that of the template synthesized CNTs, as the inner tubules delivered more electrochemical active sites for intercalation of Li ions [135]. The capacity of the carbon anodes can be improved by fabricating composite electrodes of CNTs with other materials. Several researchers have established that in some hybrid systems, the CNTs function as an effective confining buffer of mechanical stress created by volumetric changes in charging and discharging reactions, whereas the other nanomaterials provide a high capacity. By developing different systems such as metal (Sn, Sb, Bi, etc.)–C, metal oxide (SnO, $SnO_2$, $MnO_2$, $Fe_2O_3$, $Fe_3O_4$, CuO, etc.)–C, Si–C, and alloy (SnSb, SnCo, SnMn, SnFe, AgFeSn, etc.)–C electrodes exhibited charge capacities and good resilience [107, 136–150]. Reddy et al. [151] fused tubular coaxial $MnO_2$ with carbon nanotube array electrodes with a distinctive amalgamation of high porosity and low internal resistance [151]. The Li group fabricated coaxial CNTs with $MoS_2$ nanomaterials in which electrochemical measurements exposed the improvement of lithium storage by $MoS_2$ sheath resulting in an increase of release properties at the nanoscale over a distinctive synergy [152]. Smarter, reduced, lighter, and superior energy density would be the main requirements for portable battery technology. The flexibility, porosity, and conductivity of carbon nanotube membranes provided researchers a comprehension on fabricating "paper electrodes" [153–156]. Single-wall carbon nanotubes have been used to fabricate free-standing electrode without polymer binder or a metal substrate, but the capacity was inadequate [152]. Carbon nanotube networks and aligned CNTs with conducting polymer have been then subjugated for paper electrodes, and the capacity of the latter was 50% higher than that of single-wall carbon nanotubes paper electrodes [155, 156]. In particular, the Cui group developed Si-doped carbon nanotube film free-standing electrodes with high specific charge storage capacity and better cycling performance [153]. Sometime after that, vertically aligned Si-doped carbon nanotube arrays created by the Kumta group also had a capacity of 2,000 mAh/gm [159]. Hybrid paper electrodes tend to show a favorable research prospect.

In recent times, graphene, which is composed of monolayers of carbon atoms organized in a honeycombed network, has arisen tremendously, and drawn huge attention in the fields of materials science and condensed-matter physics [160–162]. As the tiniest carbon materials, graphene and graphene-based materials have high potential applications in energy-related electrochemical devices, such as Li-ion batteries, ECs, fuel cells, and solar cells [163–165]. The materials have superior electrical conductivities compared to graphitic carbon, higher surface area than CNTs, and a wide-ranging electrochemical range which would be more beneficial in energy storage. Consequently, a sequence of research work on Li-ion batteries and ECs based on graphene or GO have been conducted intensively with the analogous means for Li-ion batteries. Some researchers implemented graphene sheets directly as an anode material for lithium-ion batteries and established that those had enhanced electrochemical properties. In spite of the assumption that graphene nanostructures have substantial disorder and defects, which might lower their electrical conductivity, some reports verified highly disordered graphene nanosheets as electrodes with high reversible capacities (794–1,054 mAh/gm) and good cyclic stability [166]. The surface oxidation of carbon materials might enhance the electrochemical properties [167], and Bhardwaj et al. established that oxidized graphene nanoribbons (ox-GNRs) outperformed multi-wall carbon nanotubes and GNRs, offering a first charge capacity of 1,400 mAh/gm and a reversible capacity of 800 mAh/gm [168]. In particular, 3D graphene-based hybrid structures have been created to enhance the storage capacity of Li-ion batteries by increasing specific surface area and more suitable layer spacing of graphene sheets. Some metal oxides, CNTs, fullerenes ($C_{60}$), carbon nanofibers (CNFs), and even organic agents might be presented to fabricate 3D-structured graphene [16, 169–171]. Yin et al. [169] created honeycomb-like electrode materials with hierarchical graphene nanoarchitectures modified by the organic agent DODA with electrostatic interaction. This novel structure simultaneously optimized ion transport and capacity, leading to a high performance of reversible capacity.

Mesoporous carbon possessing larger surface area, and good conductivity has come up as an ideal conductive material for energy storage. Capacity, cycling stability, and rate capability are improved with the trapping of active materials inside the pores. Li–S batteries possess high theoretical specific capacity and energy density, but suffer from the poor electrical conductivity of sulfur and the fast capacity degradation from polysulfide dissolution into the electrolyte [172–175]. In recent times use of mesoporous carbon composites has improved electrical conductivity [129, 157, 158, 172]. This composite improves the electrical conductivity of the sulfur cathode and improves the fast capacity degradation from polysulfide dissolution into the electrolyte [129]. For achieving high energy density in Li–S batteries, mesoporous carbon with large pore volume and full sulfur filling are desirable. Nevertheless, there is a predicament between sulfur utilization, cycling stability, and sulfur loading. High sulfur loading and full sulfur filling will result in low sulfur utilization and fast capacity fading. The performances of a series of mesoporous carbon composite electrodes were investigated. It has been found that at full sulfur-filling conditions, the mesoporous carbon pore structure has no influence on battery performance other than increasing the maximum sulfur loading for increased pore volume [157]. Various other porous materials have also been researched for the same application, which have brought into light graphene-sandwiched structures that theoretically provide new strategies for electrode design in energy storage.

## 5.4 CONCLUSION

During the last decade, innovative structures and forms of carbon have been developed. The best specimens include the finding of fullerenes, and the associated single-wall and multi-wall carbon nanotubes. Even though atomic and electron level regulation of CNTs is still a large scientific and technological challenge, the perspective of different varieties of CNTs and their counterparts for energy storage and conversion has been well established. Furthermore, the long-term stability of the materials and devices, as well as the charge of manufacturing both materials and devices, might become the main concern.

In the last few years, another allotrope of carbon, molecularly dispersed graphite, graphene, has attracted huge attention. Graphene possesses unique physical and chemical properties but also provides an abundant prospect to comprehend the fundamental chemical and physical processes involved. Nonetheless, it is imperative to keep in mind that although new carbon materials such as CNTs and graphene may exhibit great potential, they still need to compete with traditional carbon and graphite in terms of cost and performance.

## KEYWORDS

- **carbon nanotubes**
- **chemical-vapor-deposition**
- **electrochemical capacitors**
- **electrochemical double-layer capacitors**
- **hydroxymethyl furfural**
- **oxidized graphene nanoribbons**

## REFERENCES

1. Øvergaard, S., (2008). Issue paper: Definition of primary and secondary energy. In: *Standard International Energy Classification (SIEC) in the International Recommendation on Energy Statistics (IRES)* (pp. 1–7). Oslo Group on Energy Statistics.
2. IEA, (2014). *Key World Energy Statistics*. International Energy Agency.
3. IEA, (2011). *$CO_2$ Emissions from Fuel Combustion: Highlights*. International Energy Agency.
4. Thompson, B. C., & Frechet, J. M. J., (2008). *Angewandte Chemie— International Edition, 47*, 58.
5. Park, S. H., et al., (2009). *Nature Photonics, 3*, 297.
6. Spanggaard, H., & Krebs, F. C., (2004). *Solar Energy Materials and Solar Cells, 83*, 125.
7. Dennler, G., Scharber, M. C., & Brabec, C. J., (2009). *Advanced Materials, 21*, 1323.

8. Pasquier, A. D., Unalan, H. E., Kanwal, A., Miller, S., & Chhowalla, M., (2005). *Applied Physics Letters, 87.*
9. Kim, Y. K., & Min, D. H., (2009). *Langmuir, 25,* 11302.
10. Hecht, D. S., Hu, L. B., & Irvin, G., (2011). *Advanced Materials, 23,* 1482.
11. Sun, Y. Q., Wu, Q. O., & Shi, G. Q., (2011). *Energy & Environmental Science, 4,* 1113.
12. Zhang, L. L., Zhou, R., & Zhao, X. S., (2010). *Journal of Materials Chemistry, 20,* 5983.
13. Vivekchand, S. R. C., Rout, C. S., Subrahmanyam, K. S., Govindaraj, A., & Rao, C. N. R., (2008). *Journal of Chemical Sciences, 120,* 9.
14. Subrahmanyam, K. S., Vivekchand, S. R. C., Govindaraj, A., & Rao, C. N. R., (2008). *Journal of Materials Chemistry, 18,* 1517.
15. Frackowiak, E., & Beguin, F., (2002). *Carbon, 40,* 1775.
16. Yoo, E., et al., (2008). *Nano Letters, 8,* 2277.
17. Wu, Y. P., Rahm, E., & Holze, R., (2003). *Journal of Power Sources, 114,* 228.
18. Landi, B. J., Ganter, M. J., Cress, C. D., DiLeo, R. A., & Raffaelle, R. P., (2009). *Energy & Environmental Science, 2,* 638.
19. Otowa, T., Nojima, Y., & Miyazaki, T., (1997). *Carbon, 35,* 1315.
20. Lu, C. Y., & Chiu, H. S., (2006). *Chemical Engineering Science, 61,* 1138.
21. Gomez-Serrano, V., Macias-Garcia, A., Espinosa-Mansilla, A., & Valenzuela-Calahorro, C., (1998). *Water Research, 32,* 1.
22. Annesini, M. C., Di Carlo, C., Piemonte, V., & Turchetti, L., (2008). *Biochemical Engineering Journal, 40*(205).
23. Malik, D. J., Warwick, G. L., Mathieson, I., Hoenich, N. A., & Streat, M., (2005). *Carbon, 43,* 2317.
24. Bigsby, R. J. A., Rider, R. J., & Blount, G. N., (1998). Proceedings of the institution of mechanical engineers; Part H. *Journal of Engineering in Medicine, 212,* 373.
25. Ye, C., Gong, Q. M., Lu, F. P., & Liang, J., (2007). *Separation and Purification Technology, 58,* 2.
26. Joo, S. H., et al., (2001). *Nature, 412,* 169.
27. Auer, E., Freund, A., Pietsch, J., & Tacke, T., (1998). *Applied Catalysis A— General, 173,* 259.
28. Park, G. G., Yang, T. H., Yoon, Y. G., Lee, W. Y., & Kim, C. S., (2003). *International Journal of Hydrogen Energy, 28,* 645.
29. Guha, A., Lu, W. J., Zawodzinski, T. A., & Schiraldi, D. A., (2007). *Carbon, 45,* 1506.
30. Aygun, A., Yenisoy-Karakas, S., & Duman, I., (2003). *Microporous and Mesoporous Materials, 66,* 189.
31. Ahmadpour, A., & Do, D. D., (1996). *Carbon, 34,* 471.
32. Jagtoyen, M., & Derbyshire, F., (1998). *Carbon, 36,* 1085.
33. Hu, Z. H., & Srinivasan, M. P., (1999). *Microporous and Mesoporous Materials, 27,* 11.
34. Ahmedna, M., Marshall, W. E., & Rao, R. M., (2000). *Bioresource Technology, 71,* 113.
35. Wu, D. C., Fu, R. W., Zhang, S. T., Dresselhaus, M. S., & Dresselhaus, G., (2004). *Journal of Non-Crystalline Solids, 336,* 26.
36. Pekala, R. W., (1989). *Journal of Materials Science, 24,* 3221.

37. Tamon, H., Ishizaka, H., Yamamoto, T., & Suzuki, T., (1999). *Carbon, 37*, 2049.
38. Kawashima, D., Aihara, T., Kobayashi, Y., Kyotani, T., & Tomita, A., (2000). *Chemistry of Materials, 12*, 3397.
39. Job, N., Pirard, R., Marien, J., & Pirard, J. P., (2004). *Carbon, 42*, 619.
40. Lin, C., & Ritter, J. A., (1997). *Carbon, 35*, 1271.
41. Gogotsi, Y., et al., (2003). *Nature Materials, 2*, 591.
42. Dash, R., et al., (2006). *Carbon, 44*, 2489.
43. Kockrick, E., et al., (2010). *Carbon, 48*, 1707.
44. Beck, J. S., et al., (1992). *Journal of the American Chemical Society, 114,* 10834.
45. Lee, J., Kim, J., & Hyeon, T., (2006). *Advanced Materials, 18*, 2073.
46. Ryoo, R., Joo, S. H., & Jun, S., (1999). *Journal of Physical Chemistry B, 103*, 7743.
47. Lee, J., Yoon, S., Hyeon, T., Oh, S. M., & Kim, K. B., (1999). *Chemical Communications*, 2177.
48. Fitzer, E., Schaefer, W., & Yamada, S., (1969). *Carbon, 7*, 643.
49. Wang, H. T., & Yao, J. F., (2006). *Ind. Eng. Chem. Res., 45*, 6393.
50. White, R. J., Budarin, V., Luque, R., Clark, J. H., & Macquarrie, D. J., (2009). *Chem. Soc. Rev., 38*, 3401.
51. Titirici, M. M., & Antonietti, M., (2010). *Chem. Soc. Rev., 39*, 103.
52. Ioannidou, O., & Zabaniotou, A., (2007). *Renewable Sustainable Energy Rev., 11,* 1966.
53. Mohamed, A. R., Mohammadi, M., & Darzi, G. N., (2010). *Renewable Sustainable Energy Rev., 14*, 1591.
54. Otani, S., (1981). *Mol. Cryst. Liq. Cryst., 63*, 249.
55. Inagaki, M., Park, C. R., Skowronski, J. M., & Morawski, A. W., (2008). *Adsorpt. Sci. Technol., 26,* 735;.
56. Saufi, S. M., & Ismail, A. F., (2004). *Carbon, 42*, 241.
57. Mochida, I., Toshima, H., Korai, Y., & Matsumoto, T., (1988). *J. Mater. Sci., 23*, 670.
58. Edie, D. D., (1998). *Carbon, 36*, 345.
59. Hybrid Cars, (2011). http://www.hybridcars.com/make/honda-mpgS (accessed on 06 October 2022).
60. AFS Trinity Power Corporation, (2010). http://afstrinity.com (accessed on 06 October 2022).
61. Obreja, V. V. N., (2008). *Physica E-Low-Dimensional Systems & Nanostructures, 40*, 2596.
62. Feng, W., & Bin, X., (2006). *New Carbon Materials, 21*, 176.
63. Frackowiak, E., (2007). *Physical Chemistry Chemical Physics, 9*, 1774.
64. Liang, C., Li, Z., & Dai, S., (2008). *Angewandte Chemie—International Edition, 47*, 3696.
65. Shi, Y., Wan, Y., & Zhao, D., (2011). *Chemical Society Reviews, 40*, 3854.
66. Chmiola, J., et al., (2006). *Science, 313*, 1760.
67. Frackowiak, E., Metenier, K., Bertagna, V., & Beguin, F., (2000). *Applied Physics Letters, 77*, 2421.
68. Chen, J. H., et al., (2002). *Carbon, 40*, 1193.
69. Niu, C., Sichel, E. K., Hoch, R., Moy, D., & Tennent, H., (1997). *Applied Physics Letters, 70*, 1480.
70. Ma, R. Z., et al., (1999). *Journal of Power Sources, 84*, 126.

71. Kim, J. H., Nam, K. W., Ma, S. B., & Kim, K. B., (2006). *Carbon, 44*, 1963.
72. An, K. H., et al., (2001). *Advanced Materials, 13*, 497.
73. Vix-Guterl, C., et al., (2005). *Carbon, 43*, 1293.
74. Yoon, B. J., et al., (2004). *Chemical Physics Letters, 388*, 170.
75. Huang, Z. P., et al., (1998). *Applied Physics Letters, 73*, 3845.
76. Frackowiak, E., & Beguin, F., (2001). *Carbon, 39*, 937.
77. Pandolfo, A. G., & Hollenkamp, A. F., (2006). *Journal of Power Sources, 157*, 11.
78. Aza¨ıs, P., et al., (2007). *Journal of Power Sources, 171*, 1046.
79. Du, C., Yeh, J., & Pan, N., (2005). *Nanotechnology, 16*, 350.
80. Gao, L., et al., (2008). *Solid State Communications, 146*, 380.
81. Emmenegger, C., et al., (2003). *Carbon, 41*, 539.
82. Kaempgen, M., Ma, J., Gruner, G., Wee, G., & Mhaisalkar, S. G., (2007). *Applied Physics Letters, 90*, 264104.
83. Liu, H. T., He, P., Li, Z. Y., Liu, Y., & Li, J. H., (2006). *Electrochimica Acta, 51*, 1925.
84. Gupta, V., & Miura, N., (2006). *Materials Letters, 60*, 1466.
85. Fan, L. Z., & Maier, J., (2006). *Electrochemical Communications, 8*, 937.
86. Yana, J., et al., (2009). *Journal of Power Sources, 194*, 1202.
87. Wang, Q., Wen, Z. H., & Li, J. H., (2006). *Advanced Functional Materials, 16*, 2141.
88. Meryl, S. D., Sungjin, P., Yanwu, Z., Jinho, A., & Rodney, R. S., (2008). *Nano Letters, 8*, 3498.
89. Vivekchand, S. R. C., Rout, C. S., Subrahmanyam, K. S., Govindaraj, A., & Rao, C. N. R., (2008). *Journal of Chemical Science, 120*, 9.
90. Yu, A., Roes, I., Davies, A., & Chen, Z., (2010). *Applied Physics Letters, 96*, 253105.
91. Wang, Y., et al., (2009). *Journal of Physical Chemistry C, 113*, 13103.
92. Liu, C., Yu, Z., Neff, D., Zhamu, A., & Jang, B. Z., (2010). *Nano Letters, 10*, 4863.
93. Xu, J., Wang, K., Zu, S. Z., Han, B. H., & Wei, Z., (2010). *ACS Nano, 4*, 5019.
94. Zhang, L. L., Zhao, S., Tian, X. N., & Zhao, X. S., (2010). *Langmuir, 26*, 17624.
95. Zhang, K., Zhang, L. L., Zhao, X. S., & Wu, J., (2010). *Chemical Materials, 22*, 1392.
96. Wang, D. W., et al., (2009). *ACS Nano, 3*, 1745.
97. Gomeza, H., et al., (2011). *Journal of Power Sources, 196*, 4102.
98. Yan, J., et al., (2010). *Journal of Power Sources, 195*, 3041.
99. Yu, D. S., & Dai, Y. M., (2010). *Journal of Physical Chemistry Letters, 1*, 467.
100. Wang, S., Jiang, S. P., & Wang, X., (2011). *Electrochimica Acta, 56*, 3338.
101. Zhang, L. L., Wei, T., Wang, W., & Zhao, X. S., (2009). *Microporous and Mesoporous Materials, 123*, 260.
102. Jiang, J., et al., (2011). *Applied Materials Interfaces, 3*, 99.
103. Wang, H., Casalongue, H. S., Liang, Y., & Dai, H., (2010). *Journal of American Chemical Society, 132*, 7472.
104. Yuan, C., et al., (2011). *Materials Letters, 65*, 374.
105. Yan, J., et al., (2010). *Carbon, 48*, 3825.
106. Zhang, H., Lv, X., Li, Y., Wang, Y., & Li, J. H., (2009). *ACS Nano, 4*, 380.
107. Li, Y. M., Tang, L. H., & Li, J. H., (2009). *Electrochemical Communications, 11*, 846.
108. Chang, H. X., Lv, X., Zhang, H., & Li, J. H., (2010). *Electrochemical Communications, 12*, 483.
109. Wang, Y., Shao, Y., Matson, D. W., Li, J. H., & Lin, Y. H., (2010). *ACS Nano, 4*, 1790.

110. Wang, Y., Li, Y., Tang, L., Lu, J., & Li, J. H., (2009). *Electrochemical Communications, 11*, 889.
111. Wang, Y., Lu, J., Tang, L. H., Chang, H. X., & Li, J. H., (2009). *Analytical Chemistry*, 81.
112. Chang, H. X., Tang, L. H., Wang, Y., Jiang, J. H., & Li, J. H., (2010). *Analytical Chemistry, 82,* 2341.
113. Dong, X. L., Cheng, J. S., Li, J. H., & Wang, Y. S., (2010). *Analytical Chemistry, 82*, 6208.
114. Tang, L. H., Feng, H. B., Cheng, J. S., & Li, J. H., (2010). *Chemical Communications, 46*, 5882.
115. Chen, F., Qing, Q., Xia, J., Li, J. H., & Tao, N. J., (2009). *Journal of American Chemical Society, 131*, 9908.
116. Wang, Y., et al., (2010). *Journal of American Chemical Society, 132*, 9274.
117. Tarascon, J. M., & Armand, M., (2001). *Nature, 414*, 359.
118. Hassoun, J., Reale, P., & Scrosati, B., (2007). *Journal of Materials Chemistry, 17*, 3668.
119. Armand, M., & Tarascon, J. M., (2008). *Nature, 451*, 652.
120. Taberna, L., Mitra, S., Poizot, P., Simon, P., & Tarascon, J. M., (2006). *Nature Materials, 5*, 567.
121. Rudorff, W., & Hofmann, U., (1938). *Journal of Inorganic and General Chemistry, 238*, 1.
122. Armand, M., & Touzain, P., (1977). *Materials Science and Engineering, 31*, 319.
123. Kaskhedikar, N. A., & Maier, J., (2009). *Advanced Materials, 21*, 2664.
124. Sato, K., Noguchi, M., Demachi, A., Oki, N., & Endo, M., (1994). *Science, 264*, 556.
125. Zheng, T., Reimers, J. N., & Dahn, J. R., (1995). *Physical Review B, 51*, 734.
126. Dahn, J. R., Zheng, T., Liu, Y. H., & Xue, J. S., (1995). S*cience, 270*, 590.
127. Zheng, T., et al., (1995). *Journal of the Electrochemical Society, 142,* 2581.
128. Zheng, T., Zhong, Q., & Dahn, J. R., (1995). *Journal of the Electrochemical Society, 142*, L211.
129. Ji, X., Lee, K. T., & Nazar, L. F., (2009). *Nature Materials, 8*, 500.
130. Claye, A. S., Fischer, J. E., Huffman, C. B., Rinzler, A. G., & Smalley, R. E., (2000). *Journal of Electrochemical Society, 147*, 2845.
131. Barisci, J. N., Wallace, G. C., & Baughman, R. H., (2000). *Electrochimica Acta, 46*, 509.
132. Zhao, J., Buldum, A., Han, J., & Lu, J. P., (2000). *Physics Review Letters, 85*, 1706.
133. Zhao, J., Gao, Q. Y., Gu, C., & Yang, Y., (2002). *Chemical Physics Letters, 358*, 77.
134. Zhang, H., et al., (2010). *Electrochimica Acta, 55*, 2873.
135. Che, G. L., Lakshmi, B. B., Fisher, E. R., & Martin, C. R., (1998). *Nature, 393*, 346.
136. Chen, W. X., Lee, J. Y., & Liu, Z. L., (2003). *Carbon, 41*, 959.
137. Wang, W., & Kumta, P. N., (2010). *ACS Nano, 4*, 2233.
138. Li, Y. M., Lv, X. J., Lu, J., & Li, J. H., (2010). *Journal of Physical Chemistry C, 114*, 21770.
139. Cui, L., Yang, Y., Hsu, C., & Cui, Y., (2009). *Nano Letters, 9*, 3370.
140. Zhao, Y., Li, J. X., Ding, Y. H., & Guan, L. H., (2011). *Chemical Communications, 47*, 7416.
141. Xia, H., Lai, M. O., & Lu, L., (2010). *Journal of Materials Chemistry, 20*, 6896.

142. Nuli, Y., Yang, J., & Jiang, M. S., (2008). *Materials Letters, 62*, 2092.
143. Yin, J. T., et al., (2005). *Journal of Electrochemical Society, 152*, A1341.
144. Beaulieu, L. Y., & Dahn, J. R., (2000). *Journal of Electrochemical Society, 147*, 3237.
145. Mao, O., & Dahn, J. R., (1999). *Journal of Electrochemical Society, 146*, 423.
146. He, Y. S., et al., (2010). *Electrochemical Communications, 12*, 570.
147. Zheng, S. F., et al., (2008). *Chemical Materials, 20*, 3617.
148. Li, H., Wang, Q., Shi, L., Chen, L., & Huang, X., (2002). *Chemical Materials, 14*, 103.
149. Wang, Z., Chen, G., & Xia, D., (2008). *Journal of Power Sources, 184*, 432.
150. Kumar, T. P., Ramesh, R., Lin, Y. Y., & Fey, G. T., (2004). *Electrochemical Communications, 6*, 520.
151. Reddy, A. L. M., Shaijumon, M. M., Gowda, S. R., & Ajayan, P. M., (2009). *Nano Letters, 9*, 1002.
152. wang, Q., & Li, J. H., (2007). *Journal of Physical Chemistry C, 111*, 1675.
153. Cui, L. F., Hu, L., Choi, J. W., & Cui, Y., (2010). *ACS Nano, 4*, 3671.
154. Ng, S. H., et al., (2005). *Electrochimica Acta, 51*, 23.
155. Kiebele, A., & Gruner, G., (2007). *Applied Physics Letters, 91*, 144104.
156. Chen, J., et al., (2007). *Chemical Materials, 19.*
157. Jayaprakash, N., Shen, J., Moganty, S. S., Corona, A., & Archer, L. A., (2011). *Angewandte Chemie—International Edition, 50*, 5904.
158. Liang, C., Dudney, N. J., & Howe, J. Y., (2009). *Chemistry of Materials, 21*, 4724.
159. Wang, W., Epur, R., & Kumta, P. N., (2011). *Electrochemical Communications, 13*, 429.
160. Novoselov, K. S., et al., (2004). *Science, 306*, 666.
161. Zhang, Y. B., Tan, Y. W., Stomer, H. L., & Kim, P., (2005). *Nature, 438*, 201.
162. Altland, A., (2006). *Physics Review Letters, 97*, 236802.
163. Chen, D., Tang, L. H., & Li, J. H., (2010). *Chem. Soc. Rev., 39*, 3157.
164. Brownson, D. A. C., & Kampouris, D. K., (2011). *Journal of Power Sources, 196*, 4873.
165. Pumera, M., (2011). *Energy Environmental Science, 4*, 668.
166. Pan, D., et al., (2009). *Chemical Materials, 21*, 3136.
167. Fu, L. J., et al., (2006). *Solid State Science, 8*, 113.
168. Bhardwaj, T., Antic, A., Pavan, B., Barone, V., & Fahlman, B. D., (2010). *Journal of American Chemical Society, 132*, 12556.
169. Yin, S., et al., (2011). *ACS Nano, 5*, 3831.
170. Fan, Z. J., et al., (2011). *ACS Nano, 5*, 2787.
171. Wang, J. Z., et al., (2011). *Chemical European Journal, 17*, 661.
172. Kolosnitsyn, V. S., & Skundin, A. M., (2008). *Russian Journal of Electrochemistry, 44*, 505.
173. Rauh, R. D., Abraham, K. M., Pearson, G. F., Surprenant, J. K., & Brummer, S. B., (1979). *Journal of the Electrochemical Society, 126*, 523.
174. Shim, J., Striebel, K. A., & Cairns, E. J., (2002). *Journal of the Electrochemical Society, 149*, A1321.
175. Peramunage, D., & Licht, S., (1993). *Science, 261*, 1029.

## CHAPTER 6

# Carbon Nanotubes and Their Biotechnological and Biomedical Applications

T. R. ANILKUMAR

*Inter-University Center for Evolutionary and Integrative Biology, University of Kerala, Karyavattom, Trivandrum, Kerala, India, E-mail: anilkumartr97@gmail.com*

## ABSTRACT

Carbon nanotubes (CNTs) are nanomaterials composed solely of carbon atoms arranged in benzene rings forming graphene sheets rolled up to form cylinders. There are two main types of CNTs, single-walled (SWCNTs) and multi-walled carbon nanotubes (MWCNTs). These CNTs have wide application in biomedicine and biotechnology owing to their chemical, physical, and biological characteristics. This chapter describes the biotechnological and biomedical application of CNTs which includes bio-sensing, gene therapy, drug delivery, tissue engineering, etc. A major drawback of CNTs is their non-compatibility with biological systems. The organic functionalization approaches of CNTs have been shown to enhance chemical reactivity to reagents and lead to increased biocompatibility (Figure 6.1).

Carbon Composites: Composites with Nanotubes, Nanomaterials, and Graphene Oxide.
Eduardo A. Castro, Ann Rose Abraham, & A. K. Haghi (Eds.)

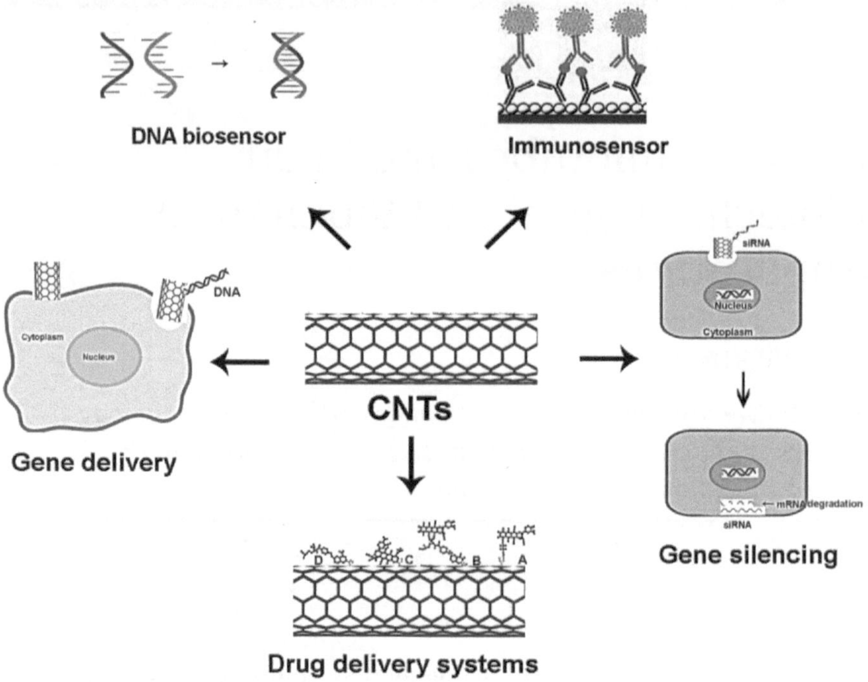


**FIGURE 6.1** Graphical abstract.

## 6.1 INTRODUCTION

The growth of nanotechnology over these years has been unprecedented, and nanomaterials are known to play an immense role in the field of biomedicine and biotechnology, such as bio-sensing, gene therapy, and drug delivery systems. The versatility of the application of Nanomaterials is because of its unique properties [1, 2]. Carbon nanotubes (CNTs) are a class of nanomaterials, chemically allotropes of carbon formed of a graphene sheet rolled into a tube with capping at both ends by one half of a fullerene-like molecule [3]. These CNTs are classified into Single-walled carbon nanotubes (SWCNTs) and multi-walled carbon nanotubes (MWCNTs) based on the number of layers of graphene sheets. SWCNTs consist of a single graphene sheet rolled into a cylinder of length in several microns and diameter ranges in 0.4–2 nm, while MWCNTs consist of two or more multiple layers of concentrically arranged graphene sheets which are separated by approximately 0.34 nm and diameter ranges between 2 nm and 100 nm [4] (Figure 6.2). Other types of carbon-based nanomaterials include fullerenes, carbon nanoporous, carbon nanohorns, carbon nanopeapods, and carbon nanobuds [5]. CNTs are mainly synthesized employing three different techniques: the carbon arc-discharge technique [6, 7], laser-ablation technique and chemical vapor deposition (CVD). Even though the above-mentioned methods were used for the synthesis of nanomaterials, catalysts became part of CVD methods [8]. One of the limitations of CNTs in biomedical application is its lack of solubility. Functionalization is an approach by which this can be overcome. There are three main approaches for CNT modification which include (a) covalent modification of π-conjugated skeleton of CNT using chemical reactions and by adding various chemical groups, (b) noncovalent modification by adsorbing various biomolecules, and (c) endohedral filling of inner cavity [9]. 1,3-dipolar cycloaddition of azomethine ylides was one of the methods to solubilize CNTs.

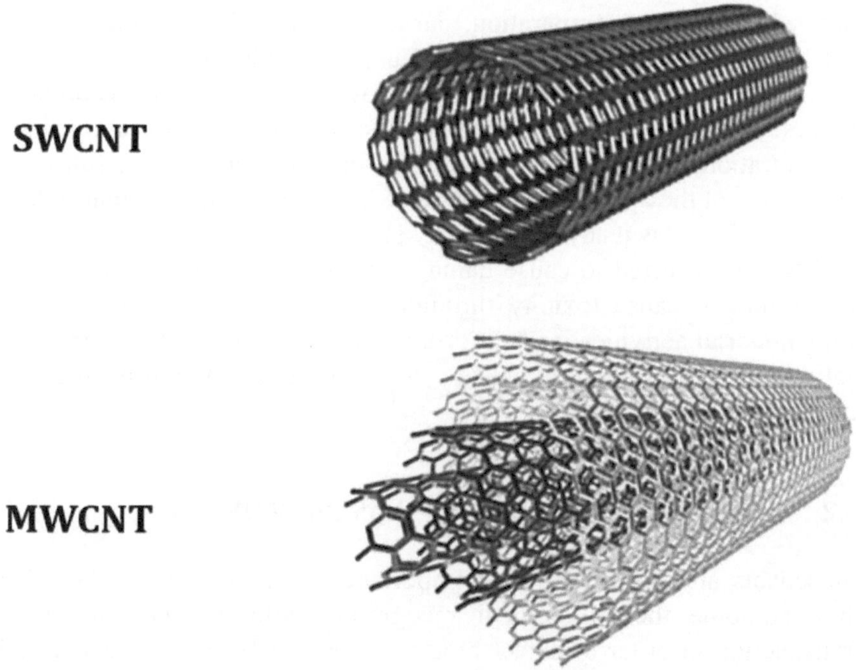


**FIGURE 6.2** Schematic representation of the structure of single-walled carbon nanotubes and muti-walled carbon nanotubes.

Modern medicine has been benefitted a lot with the merging of Nanotechnology with medicine which leads to the development of many novel materials useful in clinical diagnosis and treatment of various diseases. It has also played a role in boosting the economy as the production of these materials in industrial scales will be the driving force for the emerging economies [10]. CNTs play an active role among nanomaterials in revolutionizing biomedical research owing to their chemical, thermal, mechanical, electrical, and structural properties. They are also known for their other properties like elasticity and electron transport properties such as metallic, semiconducting, and superconducting [11]. There has been a growing demand for CNTs in the past decade owing to its application in the biomedical field, which includes biosensors, probes [6], diagnostic tools and devices in radiation oncology, biopharmaceutical applications such as drug delivery and drug discovery [12, 13]. The most essential part of CNTs in clinical settings is its safety feature with biological, environmental, and safety profiles should have been characterized. CNTs vary significantly in size, structure, morphology, and purity and it mainly depends on synthesis part, which includes preparation, purification, and functionalization [14, 15]. This biocompatibility of CNTs is unpredictable since, in most cases, CNTs exhibit different levels of toxicity which mainly depends on many factors like mode of manufacturing, surface-to-volume ratio, shape, concentration, aspect ratio, extent of oxidation, composition, functional group(s), and the applied dosage [16–18]. One of the factors that reduces biocompatibility is its hydrophobicity [19].

CNTs is reported to cause damage in DNA and cell membrane. It is also found to cause toxicity through oxidative stress, modification of mitochondrial activities, protein synthesis, and altered intracellular metabolic routes. It is known to cause cytotoxicity through both necrosis and apoptosis [13].

## 6.2 CARBON NANOTUBES (CNTS) AS BIOSENSORS

Biosensors are one of the most important applications of CNTs, where they combine their biological recognition with a chemical/physical transduction in order to detect biomolecules. CNTs are the most widely used nanomaterials as electrochemical (bio) sensors owing to their high surface-to-volume ratio, electronic properties, and edge-plane-like defects

[11, 20]. The first CNTs-based electrochemical sensor was reported by Britto et al. [21]. Since the development of this CNTs has gained enormous attention as electrochemical sensors in the subsequent years [22]. CNTs are the most prominent nanomaterials for biomedical applications because of their unique structure, which contributes to its mechanical, electrical, and optical properties [23]. Despite these advantageous features, in order to utilize CNTs as electrochemical sensors, they must be biocompatible to use in vitro and *in vivo* systems. There should be efforts to minimize the van der Waals and π π stacking interactions by functionalizing it covalently or non-covalently [24–26]. The National Institute of Health has coined a definition for Biomarkers as "a characteristic that is objectively measured and evaluated as an indicator of normal biological/pathogenic processes or pharmacological responses to a therapeutic intervention" (2001). Biomarkers are known to be used for diagnosis, prognosis, evaluation of therapy effectiveness, and risk assessment. Electrochemical (bio)sensors must be of low cost, high sensitivity, specificity, and portability [27–29].

Ultrasensitive biosensing properties of CNTs are mainly due to their larger surface area-to-volume ratio, and this enables CNTs to detect biological molecules at low concentrations. CNTs have a fast response time due to high electron-transfer time which are measured employing NADH and hydrogen peroxide ($H_2O_2$) reaction CNTs are stable due to the less surface fouling and low redox potential [30]. CNT-based biosensors are superior in quality than material-based sensors such as silicon sensors with regard to sensitivity, high surface-to-volume ratio, and hollow tubular structure. CNTs are also used as a substrate for immobilizing enzymes [31]. Tyrosinase biosensor is a glassy carbon electrode functionalized with MWCNT, 1-butyl-3-methylimidazolium chloride (IL), and tyrosinase (Tyr) within a dihexadecyl phosphate (DHP) film. During its synthesis, MWCNT, IL, and Tyr were efficiently immobilized in the film using 1-ethyl-3-(3-dimethylaminopropyl) carbodiimide/*N*-hydroxysuccinimide (EDC/NHS) as crosslinking agents. This was characterized by cyclic voltammetry (CV) in the presence of substrate catechol wherein IL-MWCNT nanocomposite (NC) showed good biocompatibility and conductivity and it showed the biocatalytic activity in the oxidation of catechol to *o*-quinone This was then electrochemically reduced to catechol at a potential of 0.04 V [32]. Tyrosinase sensors are found to have a better response signal since it was designed to combine the electrocatalytic activity of MWCNTs with the conductivity and biocompatibility of 1-butyl-3-methylimidazolium chloride (IL).

Biosensor for detecting androsterone in human serum samples is yet another example for CNTs-based sensors, which are proven to be so fast in detection, highly sensitive and stable. It was designed for the electrochemical detection of NADH generated in the dehydrogenation of androsterone by the enzyme 3α-hydroxysteroid dehydrogenase. Here the enzyme is immobilized onto a composite electrode platform of MWCNTs, octylpyridinium hexafluorophosphate (OPPF(6)) ionic liquid and NAD(+) cofactor [33]. CNTs-based biosensors to measure glucose levels has been gaining momentum in medical diagnostics and food industries. Here in glucose biosensors, glucose molecule gets catalytically oxidized into gluconic acid and $H_2O_2$ in the presence of oxygen. A single-wall carbon nanotube has been designed and analytically modeled. Here the glucose level detected with the concentration is measured as a function of gate voltage. The simulated data demonstrate that the analytical model can be employed with an electrochemical glucose sensor to predict the behavior of the sensing mechanism in biosensors [34].

In DNA sensors, the main sensing element is either single-stranded or double-stranded (dsDNA). It is also found that single-stranded DNA is highly adsorptive to CNTs than dsDNA. It has been widely exploited in the identification of new strains and organisms by DNA sequencing. The detection of genosensors relies on DNA hybridization on the surface of a physical transducer. This has got wide applications like DNA chips for diagnosis and molecular diagnosis [35]. DNA biosensors are simple in setup and chemistry when compared to other biosensors, such as electrochemical and optical DNA biosensors [36]. DNA biosensor has been found to be modified with MWCNTs, polydopamine (PDA), and gold nanoparticles and used as DNA sequencing detection.

### *6.2.1 CNT-BASED BIOSENSORS FOR PROTEIN BIOMARKERS*

CNTs-based electrochemical (bio)sensors have been extensively used for the detection of protein biomarkers.

1. **Carcinoembryonic Antigen (CEA):** These are cell-surface glycoproteins [37] with a molecular weight of 200 kDa, produced by cells of the gastrointestinal tract during embryonic development [38]. It is known that the concentration of CEA decreases after birth and the normal range of CEA in the blood in healthy

adults is lower than 2.5 ng $mL^{-1}$) [39]. Elevated levels of CEA are associated with ovarian carcinoma, lung, and breast cancer [40], and especially colorectal adenocarcinoma [41]. CEA is the most widely used tumor biomarkers to evaluate tumor progression, success of a tumor surgery, effectiveness of a cancer therapy. There are different ways for the quantification of CEA [42] in which anti-CEA primary antibody has been immobilized on various bioanalytical platforms and transduces the bioaffinity event.

2. **Prostate Specific Antigen (PSA):** It is reported to be a biomarker for prostate cancer. PSA is a 34 kDa protein that consists of 240 amino acids. PSA values above 1.5 ng $mL^{-1}$ predict the risk of malignancy [43]. CNTs-based biosensors with biorecognition platform made by the incorporation of carboxylated MWC-NTs (cMWCNTs), poly(dimethyl diallyl ammonium chloride) (PDDA), and ceria mesoporous nanospheres ($CeO_2$NSs) and the covalent immobilization of anti-PSA primary antibody (Ab1). Biorecognition was transduced by DPV through the decrease in the oxidation current of o-phenylenediamine (OPD) catalyzed by $CeO_2$NSs [44]. Schematic representations of an immunosensor and DNA sensor are shown in Figure 6.3.

3. **Alpha-Fetoprotein (AFP):** It is a glycoprotein of molecular weight 70 kDa found in plasma secreted by liver, yolk sac and gastrointestinal tract of human fetus [45, 46]. It is structurally similar to albumin in adult and shows similar physicochemical properties [47]. The normal level of AFP in healthy adults is around 3.4 ng mL, while higher levels were noted in newborns up to one year. AFP is a biomarker of neural tube and ventral wall defects in pregnant women. It is also reported as a biomarker of malignancies such as hepatocellular carcinoma (HCC) and yolk sac-derived germ cell tumors, benign liver diseases (hepatitis and cirrhosis), and less frequently in patients with other tumors [48–50].

   Many electrochemical biosensors have been designed to detect AFP and early detection of these biomarkers will enable to analyze the progress of treatment and offer many advantages [37, 45]. A label-free glycobiosensor was developed by the covalent immobilization of wheat-germ agglutinin (WGA) lectin at SPE

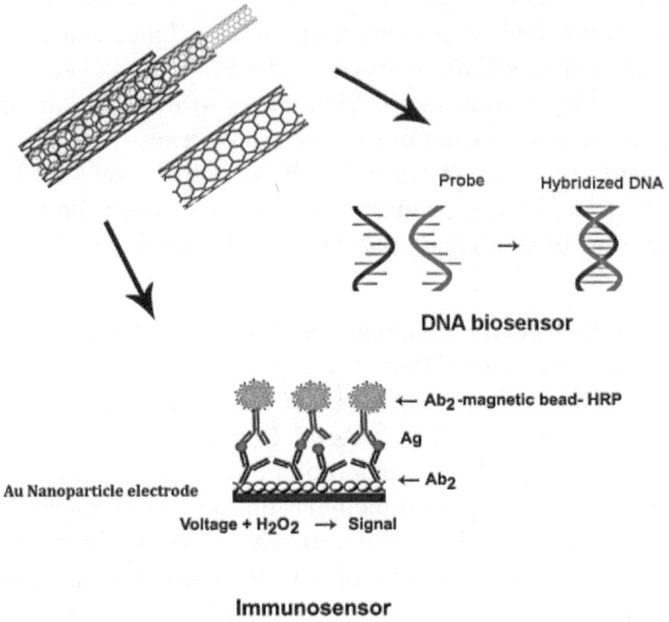


**FIGURE 6.3** Schematic representation of an immunosensor and DNA sensor. Ag–Ab binding and $H_2O_2$ conversion using horse radish peroxide generates a signal.

modified with carboxylated SWCNTs (cSWCNTs) in which an increase in RCT (randomized controlled trials) was observed with the concentration of AFP is the analytical signal [51]. This lectin-based biosensor is also used to evaluate glycan expression of AFP (AFP N-glycan) in serum samples which will enable to distinguish between healthy and cancer patients. Another biosensor employed with sensitive label-free detection of AFP was prepared by attaching cSWCNTs with anti-AFP primary antibody (Ab1) mesoporous silica (MSP) previously grafted with aminopropylethoxysilane [52]. A biosensor which is label-free AFP based immunosensor Prussian blue (PB) film-modified GCE coated with SWCNTs functionalized with polylysine (PLL-SWCNTs has been designed in which antibody against AFP conjugated with HRP was immobilized. Here the concentration of AFP was measured as an electric current of $H_2O_2$ obtained by DPV, which is linearly proportional to the concentration. Here the measurement range falls in the range of 0.05–10.0 ng $mL^{-1}$ with a detection limit of 0.011 ng $mL^{-1}$.

4. **Cytokines:** These are proteins involved in cell signaling and immune modulation their quantification plays an important role in early prognosis and diagnosis of many diseases [53]. There are many CNTs-based electrochemical biosensors developed either in the primary biorecognition platform or as part of the label conjugate. Tumor necrosis factor-alpha (TNF-α) is a well-known cytokine associated with many disorders such as diabetes, graft rejection, and many infectious and inflammatory diseases such as rheumatoid arthritis, HIV infection, neonatal listeriosis, systemic erythema nodosum leprosum, endotoxic shock, and severe meningococcemia [54]. There are label-free immunosensors designed based on SPE, modified with MWCNTs, and functionalized with fullerene and the ionic liquid 1-butyl-3-methylimidazolium-bis(trifluoromethylsulfonyl)imide ([C4mimi][NTf2]) (MWCNTs-$C_{60}$-[C4mimi][NTf2]) [55] and bimetallic Ag@Pt core-shell NPs and Chit (MWCNTs-Ag@Pt-Chit) [55]. The signals obtained from the decrease in the oxidation current of catechol are due to the surface blockage produced by TNF. Linear ranges from 5 to 75 pg $mL^{-1}$ and 6 to 60 pg $mL^{-1}$, and detection limits of 2 and 1.6 pg $mL^{-1}$, were obtained for SPE/MWCNTs-$C_{60}$-[C4mimi][NTf2]/anti-TNF-_ and SPE/MWCNTs-Ag@Pt.

5. **Troponin:** Cardiovascular diseases (CVDs) are blood circulation disorders of the heart,which include coronary heart disease, congenital heart disease, and stroke, which causes 46.2% of deaths among non-communicable diseases as per World Health Organization (WHO) statistics [56]. Early and quick diagnosis plays an important role in the successful prognosis of any disease. CVDs are characterized by the increase in the level of biochemical markers, especially Cardiac markers in blood caused by the leak out of the damaged myocardial cells [57, 58]. There are four biomarkers known to be associated with the diagnosis of myocardial infarction, which includes cardiac troponin I (cTnI) and T (cTnT), MB isoform of creatine kinase (CK-MB), and myoglobin (Mb) [59]. Mb is a small protein cardiac biomarker ranges in size in the range 17.8 kDa, increases in the serum in high quantity after cardiac

injury. This is one of the markers released as early as 1–3 h upon onset of symptoms reaching maximum at 6–12 h [60]. Normal range of Mb is 50 to 200 ng $mL^{-1}$ and its level increase up to ~600 ng $mL^{-1}$ at the onset of symptoms. Another biomarker is muscle isoenzyme CK-MB specific for cardiac injury and its level shoots up in serum within 4–6 h after the onset of Myocardial Infarction peaks to 39–185 ng $mL^{-1}$ at 18–24 h [61, 62]. cTnI and T (cTnT are more sensitive and specific than Mb and CK-MB [63]. Cardiac troponins I and T are released from dead cells after 2–4 h and 3–4 h, respectively, after Myocardial infarction. These markers remain in the bloodstream for more than 10 days with the concentration peaks at 1–2 days after Myocardial infarction. Normal range of cardiac troponin is 0.001 _g $L^{-1}$ and it shoot ups to 100 _g $L^{-1}$ during MI patients [64] and an even a low concentration of 0.01 _g $L^{-1}$ is associated with heart failure.

Detection of cardiac biomarkers was done by CNT-based electrochemical biosensors in which antitroponin antibodies are bound to CNTs at the electrode surface [65]. Redox probes [Fe(CN)6]3–/4 – and $H_2O_2$ are suitable for the detection of cTnT. Another CNT-based immunosensor for cardiac Troponin T is a conductive polymer film in which carboxylated CNTs are covalently bound to the electrode surface through polyethyleneimine. This enables the detection of cardiac Troponin using anti-cTnT monoclonal antibodies. This sensor has a low limit of detection around 0.033 ng $mL^{-1}$ and a linear range between 0.1 ng $mL^{-1}$ and 10 ng $mL^{-1}$ [66].

There are MWCNTs embedded SU-8 electrospun nanofibers have been designed for the ultrasensitive detection of cardiac biomarkers using electrochemical impedance spectroscopy (EIS). MWCNTs confer better electrical and transduction properties to composite nanofibers while SU-8 enables the functionalization and biocompatibility. These synthesized Nanofibers have been tested for the detection of all MI biomarkers, such as Mb, cTnI, and creatine kinase MB (CK-MB) and were functionalized with the antibodies of the biomarkers. The detection of these biomarkers was done by using EIS and a minimum detection limit of nano-gram/ml is achievable in this [67].

### *6.2.2 CNT-BASED BIOSENSORS FOR NUCLEIC ACIDS BIOMARKERS*

#### *6.2.2.1 MICRORNA*

microRNAs (miRNAs) are non-coding ribonucleic acid (RNA) molecules with length ranges between 19 and 25 nucleotides that regulate the expression of genes by repressing the translation of mRNA [68]. miRNA is found to regulate 60% of protein-coding genes in humans [69], and a single miRNA molecule is known to regulate hundreds of mRNAs [70, 71]. Several pathologies have been associated with abnormal miRNAs, including cardiovascular [72, 73] neurodegenerative diseases and central nervous system injury [74, 75], liver [76, 77], kidney [78] diseases, and even immune dysfunction [79, 80]. There are many reports on several miRNAs reported as reliable markers for diagnosis and prognosis due to their occurrence in different body fluids (blood, saliva, urine) in stable form which enable their extraction and analysis [81]. Thus, the quantification of a set of miRNAs holds a clinical significance and impact on health care [82].

## 6.3 CARBON NANOTUBES (CNTS) IN BIOMEDICAL IMAGING

Biomedical imaging is a field known to have received inputs from different disciplines and different fields of science. It is one of the most essential and emerging tools which enables to capture high-resolution images of cells, tissues, and organs. Imaging will help to study the behavior of cells, tissues, and organs. It is known that CNTs are versatile and can be manipulated in different ways to use in various biomedical imaging to analyze and improve functionalities [83]. SWNTs exhibit quasi-1-D nature and the optical properties of SWNTs enable their use as optical probes. They also exhibit high optical absorption, strong resonance Raman scattering, and photoluminescence in the NIR range enable its use in biological systems *in vitro* and *in vivo* [84].

### *6.3.1 IN VITRO PHOTOLUMINESCENCE IMAGING*

Individual semiconducting SWNTs with band gaps of the order of ~1 eV, dependent on the diameter and chirality, exhibits photoluminescence near-infrared (900–1,600 nm) enable biological imaging. This is due to the high optical transparency of biological tissue near 800–1,000 nm and the inherently low autofluorescence from tissue in the NIR range [85]. Another advantage of SWNTs is the large separation between the excitation (550–850 nm) and emission bands (900–1,600 nm) which will reduce background from autofluorescence and Raman scattering. It is known that NIR photoluminescence from micelle-encapsulated SWNTs, yields a quantum efficiency as low as $10^{-3}$ [86]. Cell surface receptors can be probed by bio-inert PEGylated SWNTs conjugated with antibodies to these receptors as NIR fluorescent tags [87].

### *6.3.2 IN VITRO RAMAN IMAGING*

SWNTs exhibit strong resonance Raman scattering due to the quasi-1-D nature. SWNTs also possess several Raman scattering features, such as radial breathing mode (RBM) and tangential mode (G-band) [88]. Raman microscopy is used to image SWNTs in liver cells, as well as tissue slices, using either the RBM peak or G-band peak of SWNTs [89–92]. This also helps in the imaging of Cancer cells by labeling with three isotopically unique formulations of "colored" SWNTs, conjugated with various targeting ligands, including Herceptin (anti-Her2), Erbitux (anti-Her1), and RGD peptide [89, 90]. The SWNT Raman excitation and scattering photons are in the NIR region, which is the most transparent optical window for biological systems *in vitro* and *in vivo*.

### *6.3.3 SWNTS FOR IN VIVO ANIMAL IMAGING*

SWNT-based biomedical imaging has been performed in a living animal with the first such an imaging was performed by Weisman group in 2007 [93]. In this work, Drosophila larvae were fed by food containing SWNTs and imaged by NIR fluorescence microscopy. *In vivo* tumor imaging of RGD-conjugated PEGylated SWNTs was done in live mice [94, 95].

CNTs with fluorescence emission near-infrared region enables the high-speed label-free detection in live cells. Single-walled carbon nanotubes (SWNTs) and MWCNTs exhibit fluorescence near-infrared II (NIR-II) region with a good tissue penetration and resolution. Raman imaging employing Raman scattering of SWNTs by various groups enable *in vitro* and *in vivo* imaging. The strong absorbance of CNTs in the NIR-II region is used for photoacoustic imaging, which can be enhanced by adding organic dyes, or coating with gold shells [96]. There are many reports of Raman [89, 97, 98] photoacoustic [99] and near-infrared photoluminescence imaging [100, 101] have been used to visualize nanotubes in biological environments [102].

Photoacoustic imaging allows imaging of deeper tissues with high contrast and spatial resolution. Tomographic imaging of skin and other superficial organs can be done using laser-induced photoacoustic microscopy. It also enables early detection of breast cancers by near-infrared light or radio-frequency–wave-induced photoacoustic imaging [103]. It is also known that photoacoustic signal can be enhanced by using SWCNTs conjugated with cyclic Arg-Gly-Asp (RGD) peptides as contrast agents for photoacoustic imaging of tumors and intravenous administration of these nanotubes showed eight times efficacy in determining tumor in mice than non-targeted nanotubes [99]. Low-temperature scanning gate microscopy (SGM) technique employing two coupled single-wall carbon nanotube quantum dots in a multiple quantum dot system was performed at a temperature of 170 mK. Localization of single-wall carbon nanotube quantum dots was achieved by conductance images contacted by two metallic electrodes. The single electron transport has been observed by varying the position or voltage bias of conductive atomic force microscopy [104].

There are also reports on Inorganic nanoparticles being used as biomedical imaging agents, which impart high sensitivity, high spatial, and temporal resolution. A major hurdle in the use of these nanomaterials is the toxicity of inorganic nanomaterials, which causes the release of ROS and chemical instability [105]. Modification of CNTs and the addition of elements to their structure can bring new perspectives of analysis of their behavior. One technology is assembling gold nanostructures, which has proven to enhance fluorescence intensity and has achieved ferritin receptor-mediated targeting and biomedical imaging both *in vitro* and *in vivo*, proving itself as an important biomedical imaging agent. Effective multifunctional platforms can be made out of the nanotubes, thus enhancing their properties and the width of their application, as is for the effectiveness of the imaging [104].

## 6.4 CARBON NANOTUBES (CNTS) IN DRUG AND GENE DELIVERY

CNTs have been extensively used as drug-delivery systems for the treatment of a variety of diseases. CNT-based anticancer drugs have gained much attention recently due to targeted delivery and controlled release of the drugs [106–108]. Such systems are known to improve the efficacy of the drug while minimizing the hazardous effects of the systemic toxicity of the drug in the whole body.

Single-wall carbon nanotubes (SWCNTs) have attracted considerable interest in this regard, as they offer potential advantages over the more widely studied metal nanoparticle systems, including their ability to carry a high cargo loading, their intrinsic stability and structural flexibility, which could prolong the circulation time and hence the bioavailability of the carried drug molecules. SWCNTs modified with asparagine-glycine-arginine (NGR) developed by noncovalent approach has been shown to have loaded with anticancer drug tamoxifen (TAM). This TAM-loaded NGR-modified SWCNTs has been shown to retain optical properties of SWCNTs and cytotoxicity of TAM and accumulate in tumors, facilitates the combination of chemotherapy with photothermal therapy [109]. Another single-wall carbon nanotubes (SWCNTs) based targeted drug delivery system triggers the release of the drug with the change in pH. Here single-wall carbon nanotubes (SWCNTs) are derivatized with carboxylate groups and coated with a polysaccharide material, and loaded with the anticancer drug doxorubicin. The drug binds at physiological pH (pH 7.4) and is only released at a lower pH, a characteristic of the tumor environments. Loading efficiency and release rate of the associated doxorubicin can be controlled by modifying the polysaccharide coating of the nanotubes. Folic acid (FA) can be additionally tethered to the SWCNTs as an additional agent to selectively deliver doxorubicin into the lysosomes of HeLa cells with much higher efficiency than free doxorubicin [110].

CNTs deliver drugs to specific tumor sites, which minimizes the systemic toxicity and undesirable side effects of anticancer drugs with a CNT-based anticancer drug is also known to overcome the problem of multidrug-resistant cancer cells and effective in sensitizing cancer cells without affecting cell proliferation and cell cycle [111]. CNTs can be used as carrier of immunization against some antigens [18]. There are instances of immunizations against tumors through CNT-based vaccination [112]. SWCNTs can also be used to transport acetylcholine (Ach) to the brain by traditional ways, and this could be a treatment strategy for Alzheimer's disease where neurons are unable to synthesize Ach [113] (Figure 6.4).

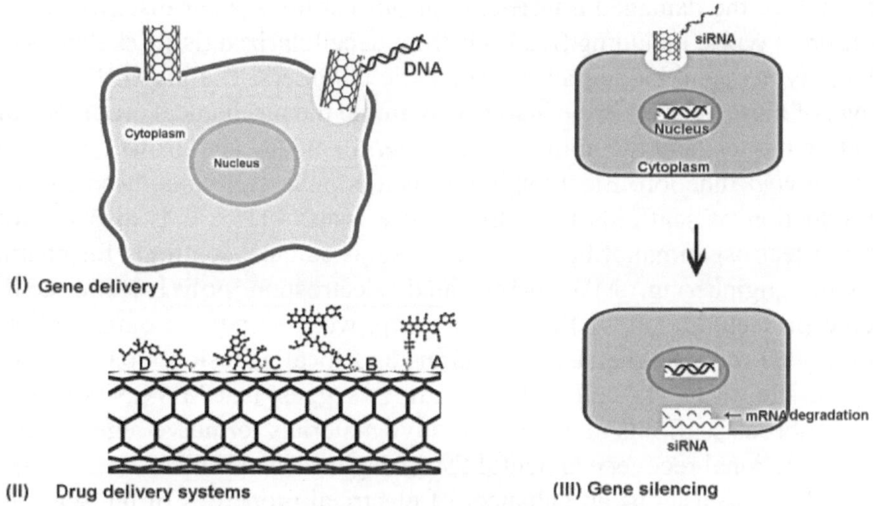


**FIGURE 6.4** Schematic representation of (I) gene delivery systems; (II) drug delivery systems with: (A) covalent modification of carbon nanotubes with DOX by PEGylation of carboxylic acids; (B) attachment of DOX to polysaccharide coated carbon nanotubes; (C) π-π stacking of acid-treated carbon nanotubes with epirubicin; (D) methotrexate attachment to amino-functionalized carbon nanotubes. (iii) Gene silencing employing transfection of carbon nanotubes conjugated with small interfering RNAs.

## 6.5 TISSUE ENGINEERING AND REGENERATIVE MEDICINE

Tissue engineering and regenerative medicine are the modern approaches in medicine which mainly deals with the development of engineered artificial tissues which has wide applications as replacement grafts and in drug delivery. It is known that delivering pro-angiogenic genes such as vascular endothelial growth factor-165 (VEGF) is a good therapeutic interruption in cardiovascular diseases. In an approach, it was found to be delivered along with a nanocomplex of GO employing an injectable and biocompatible hydrogel. In this study, polyethylenimine (PEI) functionalized GO (fGO) nanosheets complexed with $DNA_{VEGF}$ was formulated and incorporated in the low-modulus methacrylated gelatin (GelMA) hydrogel to promote controlled and localized gene therapy. This was found to be effective for transfection into myocardial tissues and induce favorable therapeutic effects without invoking cytotoxic effects [114]. In these approaches, cells are seeded or encapsulated in a suitable biomaterial for growing engineered tissues. Major tissue engineering vascular grafts (TEVGs) designed to replace the damaged arteries of various cardiovascular diseases are of different types, including Scaffolds from decellularized tissue skeletons to biopolymers and biodegradable synthetic polymers. The major disadvantage of these TEVGs is the inability to mimic the mechanical properties of native tissues, and the ability for long-term patency and growth required for *in vivo* function. Electrospinning is a popular technique used for the production of scaffolds to address these issues [115]. It is also known that Electrospun nanofibers are also used as suitable scaffolds for neural tissue engineering. MWCNTs-coated electrospun poly(L-lactic acid-co-caprolactone) (PLCL) nanofibers improved the neurite outgrowth of rat dorsal root ganglia (DRG) neurons and focal adhesion kinase (FAK) expression of PC-12 cells. These findings suggest that MWCNTs-coated nanofibrous scaffolds may be alternative materials for nerve regeneration and functional recovery in neural tissue engineering [116].

CNTs are used as an enhancer of electrical properties of the scaffold for neural and cardiac tissue growth, while functionalization with groups attracting calcium cations would help in the enhancement of bone growth [117]. There are reports on the functionalization of MWCNTs with fibroblast growth factors used in scaffolds that enable bone formation [118].

## 6.6 CONCLUSION

In this chapter, we have discussed about the biotechnological and biomedical applications of CNTs which includes biosensors, delivery of therapeutic drugs and genes, biomedical imaging, and use as scaffolds in tissue engineering, etc. CNTs are amenable to functionalization using various groups or biomolecules, either covalently or noncovalently to increase biocompatibility and decrease cytotoxicity. The functionalization is found to be essential for the delivery of drugs and gene delivery systems. However, cytotoxicity has still remained as the limiting factor for the use of CNTs in biological systems. These cytotoxicities of various CNTs depend upon the parameters such as functionalization, method of preparation, and the doses of CNTs.

There are conflicting data concerning the safety and biocompatibility of CNTs. There are reports on various metal impurities used in the synthesis of CNTs could have an impact on toxicity. It is known that different types of cell-viable indicator dyes used in combination with CNTs contribute to cytotoxicity profiles. The most common indicator dyes in use are coomassie blue, Alamar blue, neutral red, MTT (3-(4,5-dimethylthiazol-2-yl)-2,5-diphenyltetrazolium bromide), and WST-1 (a water-soluble tetrazolium salt) [119]. Cytotoxicity of CNTs is mostly contributed by the concentration and type of metal impurities, length, and type of CNTs, presence of functionalization groups, etc. [120]. In another study, even though CNTs were able to enter macrophages and influence cell physiology and function, they did not show any toxicity on cell viability as measured by the quantitative analysis of inflammatory mediators such as NO, TNF-alpha, and IL-8. It is also found that the metal particles associated with commercial nanotubes are responsible for the biological effects [121]. Several MWCNT preparations were assessed ranging from 97% purity with or without acid treatment; in comparison to MWCNTs of 99% purity, all CNT preparations showed no development of oxidative stress during prolonged cell culture. In another study conducted to investigate the physicochemical features of MWCNTs on toxicity and biocompatibility, it was found that an increase in the concentration of CNTs affects cell viability, and a concentration of 5–10 mug/mL MWCNTs is ideal for the design and development of artificial MWCNT nano vectors for gene and drug therapy against cancer [122]. Many studies show that the contaminated tubes with impurities cause immunological toxicity and localized alopecia, whereas extremely pure implanted tubes showed good biocompatibility. However, many studies for toxicological characterization of CNTs in biological systems, detailed understanding of the uptake CNTs by the cells, internalization, and altered gene expression associated with CNT toxicity has been elusive. This understanding will enable for the future development of biocompatible CNTs for biomedical applications.

## KEYWORDS

- **biosensors**
- **carbon nanotubes**
- **drug delivery**
- **gene therapy**
- **microbial fuel cells**
- **multi-walled carbon nanotubes**

## REFERENCES


1. Zhang, A., & Lieber, C. M., (2016). Nano-bioelectronics. *Chem Rev., 116*(1), 215–257. doi: 10.1021/acs.chemrev.5b00608.
2. Zhu, X., Li, J., He, H., Huang, M., Zhang, X., & Wang, S., (2015). Application of nanomaterials in the bioanalytical detection of disease-related genes. *Biosens. Bioelectron., 74*, 113–133. doi: 10.1016/j.bios.2015.04.069.
3. Dresselhaus, M. S., Dresselhaus, G., & Jorio, A., (2004). Unusual properties and structure of carbon nanotubes. *Annual Review of Materials Research, 34*(1), 247–278. doi: 10.1146/annurev.matsci.34.040203.114607.
4. Georgakilas, V., Perman, A. J., Tucek, J., & Zboril, R., (2015). Broad family of carbon nanoallotropes: Classification, chemistry, and applications of fullerenes, carbon dots, nanotubes, graphene, nanodiamonds, and combined superstructures. *Chem. Rev., 115*(11), 4744–4822. doi: 10.1021/cr500304f.
5. Mehra, N. K., Keerti, J., & Narendra, K. J., (2015). Pharmaceutical and biomedical applications of surface engineered carbon nanotubes. *Drug Discovery Today, 20*(6), 750–759. doi: https://doi.org/10.1016/j.drudis.2015.01.006.
6. Xue, Y., (2017). Chapter 11 – Carbon nanotubes for biomedical applications. In: Huisheng, P., Qingwen, L., & Tao, C., (eds.), *Industrial Applications of Carbon Nanotubes* (pp. 323–346). Boston: Elsevier.
7. Iijima, S., (1991). Helical microtubules of graphitic carbon. *Nature, 354*(6348), 56–58. doi: 10.1038/354056a0.
8. José-Yacamán, M., Miki-Yoshida, M., Rendón, L., & Santiesteban, J. G., (1993). Catalytic growth of carbon microtubules with fullerene structure. *Applied Physics Letters, 62*(2), 202–204. doi: 10.1063/1.109315.
9. Tasis, D., Nikos, T., Alberto, B., & Maurizio, P., (2006). Chemistry of carbon nanotubes. *Chemical Reviews, 106*(3), 1105–1136. doi: 10.1021/cr050569o.
10. Alshehri, R., Asad, M. I., Anwarul, H., Adnan, A., Farid, A., & Adnan, M., (2016). Carbon nanotubes in biomedical applications: Factors, mechanisms, and remedies of toxicity. *Journal of Medicinal Chemistry, 59*(18), 8149–8167. doi: 10.1021/acs.jmedchem.5b01770.
11. Pumera, M., (2009). The electrochemistry of carbon nanotubes: Fundamentals and applications. *Chemistry, 15*(20), 4970–4978. doi: 10.1002/chem.200900421.
12. Beg, S., Rizwan, M., Sheikh, M. A., Hasnain, S. M., Anwer, K., & Kohli, K., (2011). Advancement in carbon nanotubes: Basics, biomedical applications and toxicity. *J. Pharm. Pharmacol., 63*(2), 141–163. doi: 10.1111/j.2042-7158.2010.01167.x.
13. Vardharajula, S., Ali, Z. S., Tiwari, M. P., Eroğlu, E., Vig, K., Dennis, A. V., & Singh, S. R., (2012). Functionalized carbon nanotubes: Biomedical applications. *Int. J. Nanomedicine, 7*, 5361–5374. doi: 10.2147/ijn.s35832.
14. Karousis, N., Rigini, P. M., Argiris, S., Paraskevi, V., Petros, G., Yiannis, S., Georgios, S., et al., (2009). Peptidomimetic–functionalized carbon nanotubes with antitrypsin activity. *Carbon, 47*(15), 3550–3558. doi: https://doi.org/10.1016/j.carbon.2009.08.025.

15. Pagona, G., & Tagmatarchis, N., (2006). Carbon nanotubes: Materials for medicinal chemistry and biotechnological applications. *Curr. Med. Chem., 13*(15), 1789–1798. doi: 10.2174/092986706777452524.
16. Lanone, S., & Boczkowski, J., (2006). Biomedical applications and potential health risks of nanomaterials: Molecular mechanisms. *Curr Mol Med., 6*(6), 651–663. doi: 10.2174/156652406778195026.
17. Bhirde, A. A., Patel, S., Sousa, A. A., Patel, V., Molinolo, A. A., Ji, Y., Leapman, D. R., et al., (2010). Distribution and clearance of PEG-single-walled carbon nanotube cancer drug delivery vehicles in mice. *Nanomedicine (Lond), 5*(10), 1535–1546. doi: 10.2217/nnm.10.90.
18. Foldvari, M., & Bagonluri, M., (2008). Carbon nanotubes as functional excipients for nanomedicines: II. Drug delivery and biocompatibility issues. *Nanomedicine, 4*(3), 183–200. doi: 10.1016/j.nano.2008.04.003.
19. Yan, L., Feng, Z., Shoujian, L., Zhongbo, H., & Yuliang, Z., (2011). Low-toxic and safe nanomaterials by surface-chemical design, carbon nanotubes, fullerenes, metallofullerenes, and graphenes. *Nanoscale, 3*(2), 362–382. doi: 10.1039/C0NR00647E.
20. Li, Z., Zheng, L., Haiyan, S., & Chao, G., (2015). Superstructure assembly of nanocarbons: Fullerenes, nanotubes, and graphene. *Chemical Reviews, 115*(15), 7046–7117. doi: 10.1021/acs.chemrev.5b00102.
21. Britto, J. P., Santhanam, V. S. K., & Ajayan, M. P., (1996). Carbon nanotube electrode for oxidation of dopamine. *Bioelectrochemistry and Bioenergetics, 41*(1), 121–125. doi: https://doi.org/10.1016/0302-4598(96)05078-7.
22. Tiwari, N. J., Vij, V., Kemp, C. K., & Kim, K. S., (2016). Engineered carbon-nanomaterial-based electrochemical sensors for biomolecules. *ACS Nano, 10*(1), 46–80. doi: 10.1021/acsnano.5b05690.
23. Kunzmann, A., Britta, A., Tina, T., Harald, K., Annika, S., & Bengt, F., (2011). Toxicology of engineered nanomaterials: Focus on biocompatibility, biodistribution and biodegradation. *Biochimica et Biophysica Acta (BBA) - General Subjects., 1810*(3), 361–373. doi: https://doi.org/10.1016/j.bbagen.2010.04.007.
24. Gao, C., Zheng, G., Jin-Huai, L., & Xing-Jiu, H., (2012). The new age of carbon nanotubes: An updated review of functionalized carbon nanotubes in electrochemical sensors. *Nanoscale, 4*(6), 1948–1963. doi: 10.1039/C2NR11757F.
25. Kim, S. W., Taehoon, K., Yern, S. K., Hong, S. C., Hyeong, J. L., Seung, J. Y., & Chong, R. P., (2012). Surface modifications for the effective dispersion of carbon nanotubes in solvents and polymers. *Carbon, 50*(1), 3–33. doi: https://doi.org/10.1016/j.carbon.2011.08.011.
26. Primo, N. E., Gutierrez, A. F., Luque, L. G., Dalmasso, R. P., Gasnier, A., Jalit, Y., Moreno, M., et al., (2013). Comparative study of the electrochemical behavior and analytical applications of (bio)sensing platforms based on the use of multi-walled carbon nanotubes dispersed in different polymers. *Anal. Chim. Acta, 805*, 19–35. doi: 10.1016/j.aca.2013.10.039.
27. Sánchez, S., Martin, P., Enric, C., & Esteve, F., (2007). Carbon nanotube/polysulfone composite screen-printed electrochemical enzyme biosensors. *Analyst., 132*(2), 142–147. doi: 10.1039/B609137G.

28. Ronkainen, J. N., & Okon, S. L., (2014). Nanomaterial-based electrochemical immunosensors for clinically significant biomarkers. *Materials (Basel), 7*(6), 4669–4709. doi: 10.3390/ma7064669.
29. Campuzano, S., Yáñez-Sedeño, P., & Pingarrón, M. J., (2017). Electrochemical genosensing of circulating biomarkers. *Sensors (Basel), 17*(4). doi: 10.3390/s17040866.
30. Tîlmaciu, M. C., & Morris, M. C., (2015). Carbon nanotube biosensors. *Front Chem., 3*, 59. doi: 10.3389/fchem.2015.00059.
31. Jacobs, C. B., Jennifer, P. M., & Jill, V. B., (2010). Review: Carbon nanotube based electrochemical sensors for biomolecules. *Analytica Chimica Acta, 662*(2), 105–127. doi: https://doi.org/10.1016/j.aca.2010.01.009.
32. Vicentini, F. C., Bruno, J. C., Christopher, B., A, M., & Fatibello-Filho, O., (2013). Tyrosinase biosensor based on a glassy carbon electrode modified with multi-walled carbon nanotubes and 1-butyl-3-methylimidazolium chloride within a dihexadecylphosphate film. *Sensors and Actuators B: Chemical, 188*, 1101–1108. doi: https://doi.org/10.1016/j.snb.2013.07.109.
33. Mundaca, A. R., Moreno-Guzmán, M., Eguílaz, M., Yáñez-Sedeño, P., & Pingarrón, J. M., (2012). Enzyme biosensor for androsterone based on 3α-hydroxysteroid dehydrogenase immobilized onto a carbon nanotubes/ionic liquid/NAD+ composite electrode. *Talanta., 99*, 697–702. doi: 10.1016/j.talanta.2012.07.008.
34. Pourasl, H. A., Ahmadi, T. M., Rahmani, M., Chin, C. H., Lim, S. C., Ismail, R., & Tan, M. L., (2014). Analytical modeling of glucose biosensors based on carbon nanotubes. *Nanoscale Res. Lett., 9*(1), 33. doi: 10.1186/1556-276x-9-33.
35. Teles, R, R. F., & Fonseca, L. P., (2008). Trends in DNA biosensors. *Talanta., 77*(2), 606–623. doi: 10.1016/j.talanta.2008.07.024.
36. Huang, Y., Shi, M., Zhao, L., Zhao, S., Hu, K., Chen, F. Z., Chen, J., & Liang, H., (2014). Carbon nanotube signal amplification for ultrasensitive fluorescence polarization detection of DNA methyltransferase activity and inhibition. *Biosens Bioelectron., 54*, 285–291. doi: 10.1016/j.bios.2013.10.065.
37. Wang, H., Li, H., Zhang, Y., Wei, Q., Ma, H., Wu, D., Li, Y., et al., (2014). Label-free immunosensor based on Pd nanoplates for amperometric immunoassay of alpha-fetoprotein. *Biosens. Bioelectron., 53*, 305–309. doi: 10.1016/j.bios.2013.10.010.
38. Hammarström, S., Engvall, E., Johansson, G. B., Svensson, S., Sundblad, G., & Goldstein, I. J., (1975). Nature of the tumor-associated determinant(s) of carcinoembryonic antigen. *Proc. Natl. Acad. Sci U S A., 72*(4), 1528–1532. doi: 10.1073/pnas.72.4.1528.
39. Fakih, G. M., & Padmanabhan, A., (2006). CEA monitoring in colorectal cancer. What you should know. *Oncology (Williston Park), 20*(6), 579–587. Discussion 588, 594, 596 passim.
40. Ballesta, M. A., Molina, R., Filella, X., Jo, J., & Giménez, N., (1995). Carcinoembryonic antigen in staging and follow-up of patients with solid tumors. *Tumor Biol., 16*(1), 32–41. doi: 10.1159/000217926.
41. Naghibalhossaini, F., & Ebadi, P., (2006). Evidence for CEA release from human colon cancer cells by an endogenous GPI-PLD enzyme. *Cancer Lett., 234*(2), 158–167. doi: 10.1016/j.canlet.2005.03.028.

42. Hasanzadeh, M., & Nasrin, S., (2017). Advanced nanomaterials for use in electrochemical and optical immunoassays of carcinoembryonic antigen: A review. *Microchimica Acta, 184*(2), 389–414. doi: 10.1007/s00604-016-2066-2.
43. Ross, T., Ahmed, K., Raison, N., Challacombe, B., & Dasgupta, P., (2016). Clarifying the PSA grey zone: The management of patients with a borderline PSA. *Int. J. Clin. Pract., 70*(11), 950–959. doi: 10.1111/ijcp.12883.
44. Kavosi, B., Abdollah, S., Rahman, H., & Kamal, A., (2014). A highly sensitive prostate-specific antigen immunosensor based on gold nanoparticles/PAMAM dendrimer loaded on MWCNTS/chitosan/ionic liquid nanocomposite. *Biosensors and Bioelectronics, 52*, 20–28. doi: https://doi.org/10.1016/j.bios.2013.08.012.
45. Li, N., Ma, H., Cao, W., Wu, D., Yan, T., Du, B., & Wei, Q., (2015). Highly sensitive electrochemical immunosensor for the detection of alpha-fetoprotein based on PdNi nanoparticles and N-doped graphene nanoribbons. *Biosens. Bioelectron., 74*, 786–791. doi: 10.1016/j.bios.2015.07.049.
46. Gao, Q., Han, J., & Ma, Z., (2013). Polyamidoamine dendrimers-capped carbon dots/Au nanocrystal nanocomposites and its application for electrochemical immunosensor. *Biosens. Bioelectron., 49*, 323–328. doi: 10.1016/j.bios.2013.05.048.
47. Justino, C. L. I., Armando, D. C., & Rocha-Santos, P, A. T., (2016b). Critical overview on the application of sensors and biosensors for clinical analysis. *TrAC Trends in Analytical Chemistry, 85*, 36–60. doi: https://doi.org/10.1016/j.trac.2016.04.004.
48. Justino, L. I. C., Duarte, C. A., & Rocha-Santos, P. A. T., (2016a). Critical overview on the application of sensors and biosensors for clinical analysis. *Trends Analyt Chem., 85*, 36–60. doi: 10.1016/j.trac.2016.04.004.
49. Guo, J., Han, X., Wang, J., Zhao, J., Guo, Z., & Zhang, Y., (2015). Horseradish peroxidase functionalized gold nanorods as a label for sensitive electrochemical detection of alpha-fetoprotein antigen. *Anal. Biochem., 491*, 58–64. doi: 10.1016/j.ab.2015.09.006.
50. Wang, Y., Li, X., Cao, W., Li, Y., Li, H., Du, B., & Wei, Q., (2014). Ultrasensitive sandwich-type electrochemical immunosensor based on a novel signal amplification strategy using highly loaded toluidine blue/gold nanoparticles decorated KIT-6/carboxymethyl chitosan/ionic liquids as signal labels. *Biosens. Bioelectron., 61*, 618–624. doi: 10.1016/j.bios.2014.05.059.
51. Yang, H., Zhejian, L., Xiaomin, W., Ru, H., Honglan, Q., Qiang, G., Chenzhong, L., & Chengxiao, Z., (2013). Detection and discrimination of alpha-fetoprotein with a label-free electrochemical impedance spectroscopy biosensor array based on lectin functionalized carbon nanotubes. *Talanta, 111*, 62–68. doi: https://doi.org/10.1016/j.talanta.2013.01.060.
52. Lin, J., Zhijing, W., Huihui, Z., & Meijia, S., (2013). Sensitive immunosensor for the label-free determination of tumor marker based on carbon nanotubes/mesoporous silica and graphene modified electrode. *Biosensors and Bioelectronics, 41*, 342–347. doi: https://doi.org/10.1016/j.bios.2012.08.051.
53. Raphael, I., Nalawade, S., Eagar, N. T., & Forsthuber, T. G., (2015). T cell subsets and their signature cytokines in autoimmune and inflammatory diseases. *Cytokine, 74*(1), 5–17. doi: 10.1016/j.cyto.2014.09.011.
54. Blandizzi, C., Gionchetti, P., Armuzzi, A., Caporali, R., Chimenti, S., Cimaz, R., Cimino, L., et al., (2014). The role of tumor necrosis factor in the pathogenesis of

immune-mediated diseases. *Int. J. Immunopathol Pharmacol., 27*(1), 1–10. doi: 10.1177/03946320140270s101.

55. Mazloum-Ardakani, M., & Laleh, H., (2015). Highly-sensitive label-free immunosensor for tumor necrosis factor α based on Ag@Pt core-shell nanoparticles supported on MWCNTs as an efficient electrocatalyst nanocomposite. *RSC Advances, 5*(87), 70781–70786. doi: 10.1039/C5RA08016A.
56. Roger, L. V., Go, S. A., Lloyd-Jones, M. D., Benjamin, J. E., Berry, D. J., Borden, B. W., Bravata, M. D., et al., (2012). Heart disease and stroke statistics--2012 update: A report from the American heart association. *Circulation, 125*(1), e2-e220. doi: 10.1161/CIR.0b013e31823ac046.
57. Yang, Z., & Min, Z. D., (2006). Cardiac markers and their point-of-care testing for diagnosis of acute myocardial infarction. *Clin. Biochem., 39*(8), 771–780. doi: 10.1016/j.clinbiochem.2006.05.011.
58. Hasanzadeh, M., Nasrin, S., Morteza, E., Miguel De La, G., & Eskandar, O., (2013). Electrochemical nano-immunosensing of effective cardiac biomarkers for acute myocardial infarction. *Trends in Analytical Chemistry, 49*, 20–30.
59. Aldous, J. S., (2013). Cardiac biomarkers in acute myocardial infarction. *Int. J. Cardiol., 164*(3), 282–294. doi: 10.1016/j.ijcard.2012.01.081.
60. Singh, V., Martinezclark, P., Pascual, M., Shaw, S. E., & O'Neill, W. W., (2010). Cardiac biomarkers - the old and the new: A review. *Coron. Artery Dis., 21*(4), 244–256. doi: 10.1097/MCA.0b013e328338cd1f.
61. Gibler, B. W., Young, P. G., Hedges, R. J., Lewis, M. L., Smith, S. M., Carleton, C. S., Aghababian, V. R., Jorden, O. R., Allison, J. E. Jr., Otten, J. E., et al., (1992). Acute myocardial infarction in chest pain patients with nondiagnostic ECGs: Serial CK-MB sampling in the emergency department. The emergency medicine cardiac research group. *Ann. Emerg. Med., 21*(5), 504–512. doi: 10.1016/s0196-0644(05)82515-6.
62. Moreira, T. F., Dutra, A. R., Noronha, P. J., & Sales, M. G., (2014). Novel sensory surface for creatine kinase electrochemical detection. *Biosens Bioelectron., 56*, 217–222. doi: 10.1016/j.bios.2013.12.052.
63. Garg, P., Morris, P., Fazlanie, L. A., Vijayan, S., Dancso, B., Dastidar, G. A., Plein, S., et al., (2017). Cardiac biomarkers of acute coronary syndrome: From history to high-sensitivity cardiac troponin. *Intern. Emerg. Med., 12*(2), 147–155. doi: 10.1007/s11739-017-1612-1.
64. Agewall, S., Giannitsis, E., Jernberg, T., & Katus, H., (2011). Troponin elevation in coronary vs. non-coronary disease. *Eur. Heart J., 32*(4), 404–411. doi: 10.1093/eurheartj/ehq456.
65. Silva, V. B., Cavalcanti, T. I., Silva, M. M., & Dutra, R. F., (2013). A carbon nanotube screen-printed electrode for label-free detection of the human cardiac troponin T. *Talanta., 117*, 431–437. doi: 10.1016/j.talanta.2013.08.059.
66. Gomes-Filho, S. L. R., Dias, A. C. M. S., Silva, M. M. S., Silva, B. V. M., & Dutra, R. F., (2013). A carbon nanotube-based electrochemical immunosensor for cardiac troponin T. *Microchemical Journal, 109*, 10–15. doi: https://doi.org/10.1016/j.microc.2012.05.033.
67. Prakash, D. M., Singh, G. S., Sharma, S. C., & Siva, R. K. V., (2017). Electrochemical detection of cardiac biomarkers utilizing electrospun multi-walled carbon nanotubes

embedded SU-8 nanofibers. *Electroanalysis, 29*(2), 380–386. doi: https://doi.org/10.1002/elan.201501163.

68. Jayaraj, G. G., Smita, N., & Souvik, M., (2015). Nonconventional chemical inhibitors of microRNA: Therapeutic scope. *Chemical Communications, 51*(5), 820–831. doi: 10.1039/C4CC04514A.
69. Friedman, C. R., Farh, K. K., Burge, B. C., & Bartel, P. D., (2009). Most mammalian mRNAs are conserved targets of microRNAs. *Genome Res., 19*(1), 92–105. doi: 10.1101/gr.082701.108.
70. Pritchard, C. C., Cheng, H. H., & Tewari, M., (2012). MicroRNA profiling: Approaches and considerations. *Nat. Rev. Genet., 13*(5), 358–369. doi: 10.1038/nrg3198.
71. Bartel, D. P., (2009). MicroRNAs: Target recognition and regulatory functions. *Cell, 136*(2), 215–233. doi: 10.1016/j.cell.2009.01.002.
72. Wang, H., & Cai, J., (2017). The role of microRNAs in heart failure. *Biochim Biophys Acta Mol. Basis Dis., 1863*(8), 2019–2030. doi: 10.1016/j.bbadis.2016.11.034.
73. Vegter, L. E., Ovchinnikova, S. E., Van V. J. D., Jaarsma, T., Berezikov, E., Van, D. M. P., & Voors, A. A., (2017). Low circulating microRNA levels in heart failure patients are associated with atherosclerotic disease and cardiovascular-related rehospitalizations. *Clin. Res. Cardiol., 106*(8), 598–609. doi: 10.1007/s00392-017-1096-z.
74. Stoicea, N., Du, A., Lakis, C. D., Tipton, C., Arias-Morales, E. C., & Bergese, S. D., (2016). The MiRNA journey from theory to practice as a CNS biomarker. *Front Genet., 7*, 11. doi: 10.3389/fgene.2016.00011.
75. Faravelli, I., & Corti, S., (2018). *MicroRNA-Directed Neuronal Reprogramming as a Therapeutic Strategy for Neurological Diseases, 55*(5), 4428–4436. doi: 10.1007/s12035-017-0671-7.
76. Wang, S., Wang, Q. J., & Lv, X. W., (2017a). Exosomal miRNAs as biomarkers in the diagnosis of liver disease. *Biomark Med., 11*(6), 491–501. doi: 10.2217/bmm-2017-0011.
77. Long, R. X., Zhang, J. Y., Zhang, Y. M., Chen, K., Zheng, S, F. X., & Wang, H. Y., (2017). Identification of an 88-microRNA signature in whole blood for diagnosis of hepatocellular carcinoma and other chronic liver diseases. *Aging (Albany NY)., 9*(6), 1565–1584. doi: 10.18632/aging.101253.
78. Kashani, K., Cheungpasitporn, W., & Ronco, C., (2017). Biomarkers of acute kidney injury: The pathway from discovery to clinical adoption. *Clin. Chem. Lab Med., 55*(8), 1074–1089. doi: 10.1515/cclm-2016-0973.
79. Jin, C., Cheng, L., Höxtermann, S., Xie, T., Lu, X., Wu, H., Skaletz-Rorowski, A., et al., (2017). MicroRNA-155 is a biomarker of T-cell activation and immune dysfunction in HIV-1-infected patients. *HIV Med., 18*(5), 354–362. doi: 10.1111/hiv.12470.
80. Le, X., Yu, X., & Shen, N., (2017). Novel insights of microRNAs in the development of systemic lupus erythematosus. *Curr. Opin. Rheumatol., 29*(5), 450–457. doi: 10.1097/bor.0000000000000420.
81. Wang, S., Jian-Qing, W., & Xiong-Wen, L., (2017b). Exosomal miRNAs as biomarkers in the diagnosis of liver disease. *Biomarkers in Medicine, 11*(6), 491–501. doi: 10.2217/bmm-2017-0011.

82. Mahasneh, A., Al-Shaheri, F., & Jamal, E., (2017). Molecular biomarkers for an early diagnosis, effective treatment and prognosis of colorectal cancer: Current updates. *Exp. Mol. Pathol., 102*(3), 475–483. doi: 10.1016/j.yexmp.2017.05.005.
83. Wu, Y., Phillips, A. J., Liu, H., Yang, R., & Tan, W., (2008). Carbon nanotubes protect DNA strands during cellular delivery. *ACS Nano, 2*(10), 2023–2028. doi: 10.1021/nn800325a.
84. Liu, Z., Tabakman, S., Welsher, K., & Dai, H., (2009a). Carbon nanotubes in biology and medicine: *In vitro* and *in vivo* detection, imaging and drug delivery. *Nano Res., 2*(2), 85–120. doi: 10.1007/s12274-009-9009-8.
85. Aubin, J. E., (1979). Autofluorescence of viable cultured mammalian cells. *J. Histochem. Cytochem., 27*(1), 36–43. doi: 10.1177/27.1.220325.
86. O'Connell, J. M., Bachilo, M. S., Huffman, B. C., Moore, C. V., Strano, S. M., Haroz, H. E., Rialon, L. K., et al., (2002). Band gap fluorescence from individual single-walled carbon nanotubes. *Science, 297*(5581), 593–596. doi: 10.1126/science.1072631.
87. Welsher, K., Liu, Z., Daranciang, D., & Dai, H., (2008). Selective probing and imaging of cells with single walled carbon nanotubes as near-infrared fluorescent molecules. *Nano Lett., 8*(2), 586–590. doi: 10.1021/nl072949q.
88. Rao, M. A., Richter, E., Bandow, S., Chase, B., Eklund, C. P., Williams, A. K., Fang, S., et al., (1997). Diameter-selective Raman scattering from vibrational modes in carbon nanotubes. *Science, 275*(5297), 187–191. doi: 10.1126/science.275.5297.187.
89. Heller, D. A., Baik, S., Eurell, T. E., & Strano, M. S., (2005). Single-walled carbon nanotube spectroscopy in live cells: Towards long-term labels and optical sensors. *Advanced Materials, 17*(23), 2793–2799. doi: https://doi.org/10.1002/adma.200500477.
90. Liu, Z., Davis, C., Cai, W., He, L., Chen, X., & Dai, H., (2008). Circulation and long-term fate of functionalized, biocompatible single-walled carbon nanotubes in mice probed by Raman spectroscopy. *Proc. Natl. Acad. Sci. U S A., 105*(5), 1410–1415. doi: 10.1073/pnas.0707654105.
91. Liu, Z., Winters, M., Holodniy, M., & Dai, H., (2007). siRNA delivery into human T cells and primary cells with carbon-nanotube transporters. *Angew Chem. Int. Ed. Engl., 46*(12), 2023–2037. doi: 10.1002/anie.200604295.
92. Schipper, L. M., Nakayama-Ratchford, N., Davis, R. C., Kam, W. N., Chu, P., Liu, Z., Sun, X., et al., (2008). A pilot toxicology study of single-walled carbon nanotubes in a small sample of mice. *Nat. Nanotechnol., 3*(4), 216–221. doi: 10.1038/nnano.2008.68.
93. Leeuw, K. T., Reith, M. R., Simonette, A. R., Harden, E. M., Cherukuri, P., Tsyboulski, A. D., Beckingham, M. K., & Weisman, R. B., (2007). Single-walled carbon nanotubes in the intact organism: Near-IR imaging and biocompatibility studies in drosophila. *Nano Lett., 7*(9), 2650–2654. doi: 10.1021/nl0710452.
94. Keren, S., Zavaleta, C., Cheng, Z., De La Zerda, A., Gheysens, O., & Gambhir, S. S., (2008). Noninvasive molecular imaging of small living subjects using Raman spectroscopy. *Proc. Natl. Acad. Sci. U S A, 105*(15), 5844–5849. doi: 10.1073/pnas.0710575105.
95. Zavaleta, C., De La Zerda, A., Liu, Z., Keren, S., Cheng, Z., Schipper, M., Chen, X., et al., (2008b). Noninvasive Raman spectroscopy in living mice for evaluation of

tumor targeting with carbon nanotubes. *Nano Lett., 8*(9), 2800–2805. doi: 10.1021/nl801362a.

96. Gong, H., Peng, R., & Liu, Z., (2013). Carbon nanotubes for biomedical imaging: The recent advances. *Adv Drug Deliv Rev., 65*(15), 1951–1963. doi: 10.1016/j.addr.2013.10.002.
97. Liu, Z., Xiaolin, L., Scott, T. M., Kaili, J., Shoushan, F., & Hongjie, D., (2008). Multiplexed multicolor Raman imaging of live cells with isotopically modified single walled carbon nanotubes. *Journal of the American Chemical Society., 130*(41), 13540–13541. doi: 10.1021/ja806242t.
98. Zavaleta, C., De La Zerda, A., Liu, Z., Keren, S., Cheng, Z., Schipper, M., Chen, X., et al., (2008a). Noninvasive Raman spectroscopy in living mice for evaluation of tumor targeting with carbon nanotubes. *Nano Letters, 8*(9), 2800–2805. doi: 10.1021/nl801362a.
99. De La Zerda, A., Cristina, Z., Shay, K., Srikant, V., Sunil, B., Zhuang, L., Jelena, L., et al., (2008). Carbon nanotubes as photoacoustic molecular imaging agents in living mice. *Nature Nanotechnology, 3*(9), 557–562. doi: 10.1038/nnano.2008.231.
100. Cherukuri, P., Sergei, B. M., Silvio, L. H., & Bruce, W. R., (2004). Near-infrared fluorescence microscopy of single-walled carbon nanotubes in phagocytic cells. *Journal of the American Chemical Society, 126*(48), 15638–15639. doi: 10.1021/ja0466311.
101. Jin, H., Daniel, H. A., & Michael, S. S., (2008). Single-particle tracking of endocytosis and exocytosis of single-walled carbon nanotubes in NIH-3T3 cells. *Nano Letters, 8*(6), 1577–1585. doi: 10.1021/nl072969s.
102. Tong, L., Yuxiang, L., Bridget, D. D., Yookyung, J., Mikhail, S. N., Donald, B. E., & Ji-Xin, C., (2012). Label-free imaging of semiconducting and metallic carbon nanotubes in cells and mice using transient absorption microscopy. *Nature Nanotechnology, 7*(1), 56–61. doi: 10.1038/nnano.2011.210.
103. Xu, M., & Lihong, W. V., (2006). Photoacoustic imaging in biomedicine. *Review of Scientific Instruments, 77*(4), 041101. doi: 10.1063/1.2195024.
104. Zhou, X., James, H., Yoichi, M., Grutter, P., & Koji, I., (2014). Scanning gate imaging of two coupled quantum dots in single-walled carbon nanotubes. *Nanotechnology, 25*, 495703. doi: 10.1088/0957–4484/25/49/495703.
105. Li, J., Chang, X., Chen, X., Gu, Z., Zhao, F., Chai, Z., & Zhao, Y., (2014). Toxicity of inorganic nanomaterials in biomedical imaging. *Biotechnol Adv., 32*(4), 727–743. doi: 10.1016/j.biotechadv.2013.12.009.
106. Zhang, X., Meng, L., Lu, Q., Fei, Z., & Dyson, P. J., (2009a). Targeted delivery and controlled release of doxorubicin to cancer cells using modified single wall carbon nanotubes. *Biomaterials, 30*(30), 6041–6047. doi: 10.1016/j.biomaterials.2009.07.025.
107. Huang, H., Yuan, Q., Shah, S. J., & Misra, R. D., (2011). A new family of folate-decorated and carbon nanotube-mediated drug delivery system: Synthesis and drug delivery response. *Adv. Drug Deliv. Rev., 63*(14, 15), 1332–1339. doi: 10.1016/j.addr.2011.04.001.
108. Chen, C., Xie, X. X., Zhou, Q., Zhang, Y. F., Wang, L. Q., Liu, Q. Y., Zou, Y., et al., (2012). EGF-functionalized single-walled carbon nanotubes for targeting delivery of etoposide. *Nanotechnology, 23*(4), 045104. doi: 10.1088/0957–4484/23/4/045104.

109. Chen, C., Hou, L., Zhang, H., Zhu, L., Zhang, H., Zhang, C., Shi, J., et al., (2013). Single-walled carbon nanotubes mediated targeted tamoxifen delivery system using aspargine-glycine-arginine peptide. *J. Drug Target., 21*(9), 809–821. doi: 10.3109/1061186x.2013.829071.
110. Zhang, X., Lingjie, M., Qinghua, L., Zhaofu, F., & Paul, D. J., (2009b). Targeted delivery and controlled release of doxorubicin to cancer cells using modified single wall carbon nanotubes. *Biomaterials, 30*(30), 6041–6047. doi: https://doi.org/10.1016/j.biomaterials.2009.07.025.
111. Cheng, J., Meziani, J. M., Sun, P. Y., & Cheng, H. S., (2011). Poly(ethylene glycol)-conjugated multi-walled carbon nanotubes as an efficient drug carrier for overcoming multidrug resistance. *Toxicol. Appl. Pharmacol., 250*(2), 184–193. doi: 10.1016/j.taap.2010.10.012.
112. Villa, H. C., Dao, T., Ahearn, I., Fehrenbacher, N., Casey, E., Rey, A. D., Korontsvit, T., et al., (2011). Single-walled carbon nanotubes deliver peptide antigens into dendritic cells and enhance IgG responses to tumor-associated antigens. *ACS Nano, 5*(7), 5300–5311. doi: 10.1021/nn200182x.
113. Yang, Z., Zhang, Y., Yang, Y., Sun, L., Han, D., Li, H., & Wang, C., (2010). Pharmacological and toxicological target organelles and safe use of single-walled carbon nanotubes as drug carriers in treating Alzheimer's disease. *Nanomedicine, 6*(3), 427–441. doi: 10.1016/j.nano.2009.11.007.
114. Paul, A., Anwarul, H., Hamood Al, K., Akhilesh, G. K., Vijayaraghava, R. S. T., Mehdi, N., Su Ryon, S., et al., (2014). Injectable graphene oxide/hydrogel-based angiogenic gene delivery system for vasculogenesis and cardiac repair. *ACS Nano, 8*(8), 8050–8062. doi: 10.1021/nn5020787.
115. Hasan, A., Adnan, M., Nasim, A., Monowar, H., Arghya, P., Mehmet, D. R., Fariba, D., and Ali. K., (2014). Electrospun scaffolds for tissue engineering of vascular grafts. *Acta Biomaterialia., 10*(1), 11–25. doi: https://doi.org/10.1016/j.actbio.2013.08.022.
116. Jin, Guang-Zhen, Meeju, K., Ueon, S. S., & Hae-Won, K., (2011). Neurite outgrowth of dorsal root ganglia neurons is enhanced on aligned nanofibrous biopolymer scaffold with carbon nanotube coating. *Neuroscience Letters, 501*(1), 10–14. doi: https://doi.org/10.1016/j.neulet.2011.06.023.
117. Zhao, B., Hui, H., Swadhin, M. K., & Robert, H. C., (2005). A bone mimic based on the self-assembly of hydroxyapatite on chemically functionalized single-walled carbon nanotubes. *Chemistry of Materials, 17*(12), 3235–3241. doi: 10.1021/cm0500399.
118. Hirata, E., Ménard-Moyon, C., Venturelli, E., Takita, H., Watari, F., Bianco, A., & Yokoyama, A., (2013). Carbon nanotubes functionalized with fibroblast growth factor accelerate proliferation of bone marrow-derived stromal cells and bone formation. *Nanotechnology, 24*(43), 435101. doi: 10.1088/0957-4484/24/43/435101.
119. Cui, Hui-Fang, Sandeep, K. V., Al-Rubeaan, K., John, L. T. H., & Fwu-Shan, S., (2010). Interfacing carbon nanotubes with living mammalian cells and cytotoxicity issues. *Chemical Research in Toxicology, 23*(7), 1131–1147. doi: 10.1021/tx100050h.
120. Liu, Z., Scott, T., Kevin, W., & Hongjie, D., (2009b). Carbon nanotubes in biology and medicine: *In vitro* and *in vivo* detection, imaging and drug delivery. *Nano Research, 2*(2), 85–120. doi: 10.1007/s12274-009-9009-8.

121. Pulskamp, K., Diabaté, S., & Krug, H. F., (2007). Carbon nanotubes show no sign of acute toxicity but induce intracellular reactive oxygen species in dependence on contaminants. *Toxicol. Lett., 168*(1), 58–74. doi: 10.1016/j.toxlet.2006.11.001.
122. Vittorio, O., Raffa, V., & Cuschieri, A., (2009). Influence of purity and surface oxidation on cytotoxicity of multi-walled carbon nanotubes with human neuroblastoma cells. *Nanomedicine, 5*(4), 424–431. doi: 10.1016/j.nano.2009.02.006.
123. Biomarkers and Surrogate Endpoints: Preferred Definitions and Conceptual Framework, (2001). *Clin Pharmacol Ther., 69*(3), 89–95. doi: 10.1067/mcp.2001.113989.
124. Mazloum-Ardakani, M., Laleh, H., & Alireza, K., (2015). Label-free electrochemical immunosensor for detection of tumor necrosis factor α based on fullerene-functionalized carbon nanotubes/ionic liquid. *Journal of Electroanalytical Chemistry, 757*, 58–64. doi: https://doi.org/10.1016/j.jelechem.2015.09.006.

## CHAPTER 7

# Carbon Nanomaterials for Hydrogen Gas Sensing Applications

KEERTHI G. NAIR[1,2] and P. BIJI[2]

[1]*Department of Science and Humanities, Federal Institute of Science and Technology, Angamaly, Ernakulam, Kerala, India*

[2]*Nano-Sensor Laboratory, PSG Institute of Advanced Studies, Coimbatore, Tamil Nadu, India*

**Corresponding author. E-mail: drkeerthignair@gmail.com*

## ABSTRACT

Hydrogen gas ($H_2$) has been regarded as one of the promising next-generation energy sources as it is profuse in nature and its combustion reaction only produces water ($H_2O$) as the byproduct. Therefore, with the depletion of fossil fuels, the field of $H_2$-powered fuel cells has emerged rapidly. However, $H_2$ is not only a flammable gas with a lower flammability limit of 4% but also colorless, odorless, and buoyant in the air. These properties demand quick detection of any leakage of $H_2$ to ensure safety. Therefore, $H_2$ sensors are needed in various fields, such as hydrogen fuel cells, $H_2$ storage systems, and infrastructure/industry having or using $H_2$. Emerging materials, such as carbon nanotubes (CNTs) [19], graphenes [20], carbon nanofibers (CNFs) [21], and so on, also exhibit an interesting and efficient $H_2$ sensing performance at low operating temperatures. These intriguing approaches have solved some issues in $H_2$ sensors and have demonstrated that chemi-resistive $H_2$ sensors are one of the most promising $H_2$ sensing

Carbon Composites: Composites with Nanotubes, Nanomaterials, and Graphene Oxide.
Eduardo A. Castro, Ann Rose Abraham, & A. K. Haghi (Eds.)

platforms for practical applications. One of the most studied and currently used materials in the nanotechnology field is carbon-based materials due to their remarkable properties. Carbonaceous structures possess numerous advantages compared to other conventionally employed materials, especially on their extraordinary physicochemical properties. Manufacturing processes of carbon-based sensing materials can be simple, yielding an adequate quantity of material with low densification defects. In this scenario, carbon-based materials can be considered an alternative to currently used expensive electronic compounds, presenting excellent performance, and can be considered an eco-friendly material. Therefore, carbon nanostructures have been investigated to be used in powerful sensor devices, since they possess superior physical and chemical parameters yielding high-quality sensing properties [23].

## 7.1 NEED FOR $H_2$ SENSORS

Hydrogen ($H_2$) is an assured clean energy carrier, which can be generated from various renewable energy sources, helping to solve serious problems such as climate changes due to greenhouse gas emissions, depletion of fossil fuel resources, and pollution [1, 2]. Also, hydrogen is one of the most important reducing gases and is used in many chemical processes and various industries, including aerospace, medical, petrochemical, transportation, and energy. The usage of $H_2$ in fuel cells for power generation applications and transportation has renewed attention recently, which can direct to pollution-less vehicles, where the emissions are only water. Likewise, hydrogen finds extensive application as a cryogenic fuel in rockets and as a lift gas in weather balloons. Hydrogen batteries and fuel cells are used in satellites. In power plants, gaseous hydrogen is used for removing friction-heat in turbines. Hydrogen is involved as a product in many chemical reactions, including the charging of acid electrolyte batteries [3]. An overview of production, usage, and applications of hydrogen are shown in Figure 7.1. Hydrogen is a colorless, odorless gas which, when mixed with air, is flammable/explosive in the concentration range of 4–75% (by volume). The minimum energy for $H_2$ gas detonation in air at ambient pressure was 0.02 mJ, and fled $H_2$ can be very easily ignited as the ignition temperature in air is 520–580°C. $H_2$ gas has a relative density of 0.07, also called the lightest element and is enormously floating and will gather close to the roof of the room. Therefore, hydrogen production, storage, and transport can be hazardous since hydrogen is flammable or explosive if not handled properly [4–6].

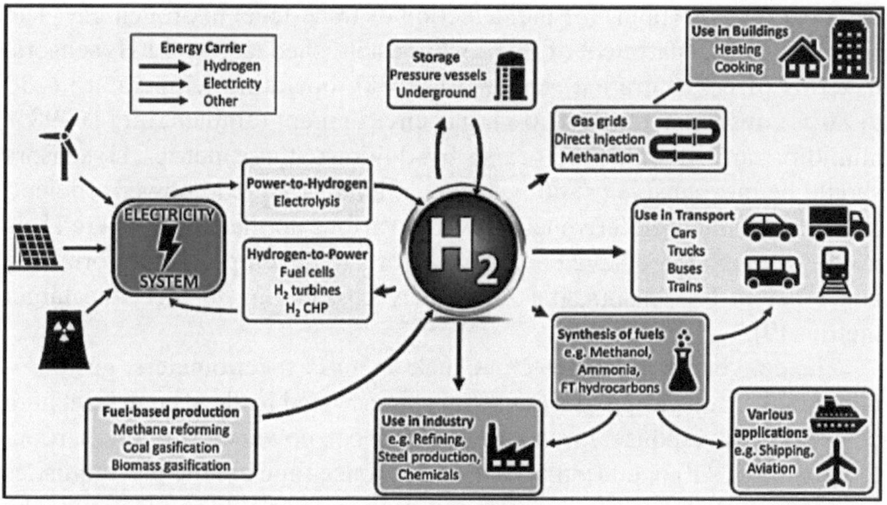


**FIGURE 7.1** An overview of hydrogen production and usage and applications [7].
*Source:* Reproduced from Ref. [7] with permission from the Royal Society of Chemistry.

With the recent market introduction of the first series-type hydrogen fuel cell cars and the prospect of a hydrogen economy at hand, development of safe and reliable hydrogen sensors is critical. Hence, a rapidly increasing need for accurate hydrogen sensors optimized to operate under different environments can be anticipated. The wide flammability limits of hydrogen in air (4–75 vol.%) in combination with low ignition energies make fast and accurate detection of hydrogen leakages, a vital part of any hydrogen-driven vehicles and also in indoor places where such vehicles are parked and surrounding necessary hydrogen infrastructure [8]. Moreover, at high concentrations, hydrogen excludes an adequate supply of oxygen, causing asphyxiation. Hydrogen must be monitored for many reasons. For process control/manufacturing applications, $H_2$ monitoring is very important to provide control of reactions. Measurement of diffusible hydrogen is important in the certification of welds. Accumulation of $H_2$ can be dangerous and therefore, safety monitoring of confined spaces is important. Transportation applications further increase the demand for cheaper, simple, reliable, and low-cost hydrogen gas sensors to guard against possible accidents. The emergence of a safety sensor is vital for widespread use and public acceptance of hydrogen as a fuel. The required sensors for hydrogen need to have sensitivity that can range in the order of parts per million (ppm) for the detection of trace-level hydrogen gas. The United States Department of Energy has established targets for $H_2$ sensors, in terms of concentration range (0.1–10%), operating temperature (–30 to 80°C), response time (<1.0 s), gas environment (ambient air; 10–98% humidity), and lifetime (>10 years). In addition to these metrics, $H_2$ sensors should be inexpensive (<$40 per unit), miniaturized, and power-efficient, to maximize their effectiveness towards various applications, where $H_2$ is used. Although there have been huge efforts to develop high-performance and efficient $H_2$ sensors, it is extremely challenging to meet the target metrics [9].

The conventional $H_2$ detectors such as mass spectrometers, gas chromatographs, and gas pressure sensors are restricted by their huge size, high cost and slow response. Lower production cost, power consumption, room temperature (RT) operation and faster response time are the pre-requisites for extensive use, such as *in-situ* monitoring and portable applications. The rapid development of the hydrogen economy also has endorsed research on new types of $H_2$ gas sensors with swifter and more precise $H_2$ sensing with near RT operation and compatibility with microelectronic ICs.

At present, there are various $H_2$ sensors are accessible commercially, namely electrochemical [10], semiconductor [11], thermoelectric [12], metallic, optical [13] and acoustic ones [14], etc. Among them, both electrochemical and chemi-resistive $H_2$ sensors are commercially available and regarded as a current state-of-the-art $H_2$ sensor technology [3]. Electrochemical $H_2$ sensors are operated by the transduction of redox reactions of $H_2$ at the sensor interface into electrical signals. Electrochemical sensors display sensitive and selective $H_2$ sensing properties with low detection limits, but they are highly susceptible to interference issues under ambient conditions (oxygen and humidity levels) and have rather complicated systems that require high fabrication costs. On the other hand, chemi-resistive sensors, which are operated by the transduction of chemical reactions into electrical signals (resistances or conductance), offer several benefits, such as efficient sensing performances, low-cost fabrication, and portable applications. In particular, the detection of resistance (or conductance) changes in sensors can be efficiently interfaced with electronic readouts with minimal processing, while other transduction mechanisms require more complex signal processing. Because of these advantages, chemi-resistive $H_2$ sensor technology as summarized in Figure 7.2, and performance have dramatically advanced in recent years.

Palladium (Pd)-based resistor-type sensors are widely regarded as conventional $H_2$ sensing systems due to their simplicity, efficient sensing properties with high selectivity, and room-temperature operation. Pd can react with $H_2$, even at RT, resulting in the formation of palladium hydride ($PdH_x$) with a higher resistivity than Pd. Thus, $H_2$ can be easily and efficiently sensed in real-time by tracking the resistances of Pd-based sensors at RT [9, 15, 16]. However, the $H_2$ reactions with Pd are impeded by oxygen molecules ($O_2$) in the air. Thus, Pd-based sensors usually show degraded performance with reduced sensitivities and sluggish sensing speeds in ambient air.

Whereas, semiconductor sensors exhibit high sensitivity, fast response, long-term stability, and potential for integration in hydrogen sensing performance. However, this type of hydrogen sensors still suffers from high operation temperatures, which results in high power consumption and potential safety hazards [17]. Moreover, cross selectivity to other combustible or reducing gases is another critical issue, which should be restricted to enhance the sensing accuracy.

Recently carbon nanomaterials, like graphene, CNTs, CNFs, two-dimensional (2D) transition dichalcogenides (TMDs) [22], exhibits effective $H_2$ sensing performance at low operating temperatures. These nanomaterials have unraveled a few problems in $H_2$ sensors, which leads to the conclusion that chemi-resistive $H_2$ sensors can be the most promising $H_2$ sensing platforms for practical applications. Recently carbon-based materials are discussed in nanotechnology due to their remarkable properties. Fabrication of high-yield carbon-based gas sensors nanomaterials can be simple, with low densification defects. Hence, carbon-based materials are considered to be an eco-friendly material and also a substitute to presently used expensive electronic compounds, with exceptional performance. Thus, carbon nanomaterials have been considered as a material for powerful sensor devices, since they hold superior physical and chemical properties resulting high-quality sensing properties [23].

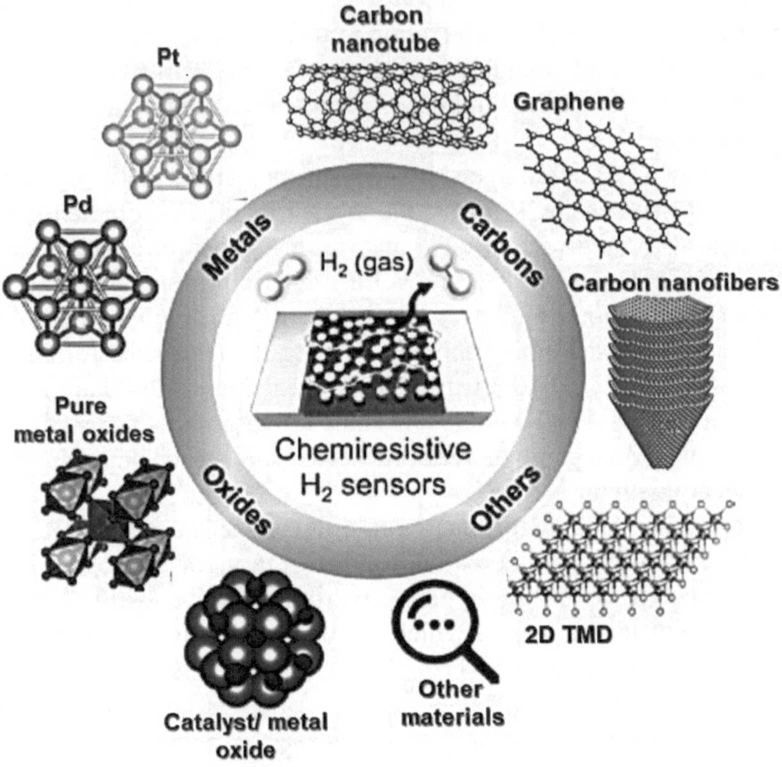


**FIGURE 7.2** Schematic illustration of representative materials for chemi-resistive $H_2$ sensors. The representative materials are categorized into metals, metal oxides, carbons, and other materials (2D TMDs: two-dimensional transition metal chalcogenides) [18].

*Source:* Reprinted with permission from Ref. [18]; Copyright (2020), American Chemical Society.

## 7.2 METAL-BASED $H_2$ SENSORS

Among various metals, palladium (Pd) as a catalytic metal appeared as the most frequently used active sensing material, because it can selectively and reversibly adsorb $H_2$ molecules, forming interstitial hydrogen atoms in the Pd hydride ($PdH_x$), at RT. The resistance of the sensors was increased in response to $H_2$ exposures, resulting in sigmoidal signal responses as shown in Figure 7.3. This type of Pd-based $H_2$ sensor enables simple and selective detection of $H_2$ even at RT [24]. However, low sensitivity, slow response and recovery speed remain as unmet challenges.

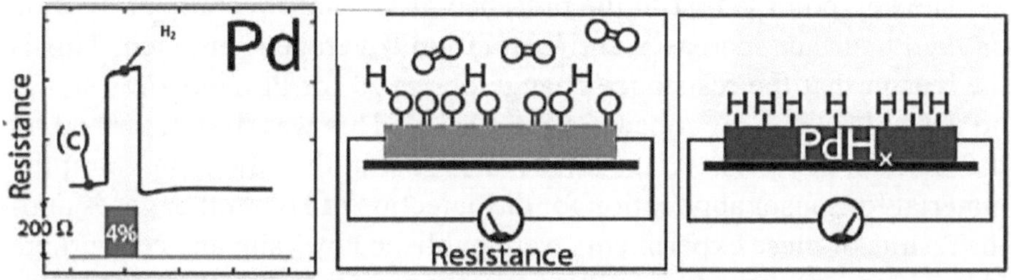


**FIGURE 7.3** R versus time for the exposure of a Pd nanowire to 4% $H_2$ showing the increased resistance in $H_2$. The chemical state of the Pd nanowire surface in air and in flowing $H_2$ [16].

*Source:* Reprinted with permission from Ref. [16]; Copyright (2012), American Chemical Society.

In addition to Pd, other metallic elements, such as Pt, have been recently demonstrated its $H_2$ sensing ability by using the surface scattering model that is attributed to the modulation of electron scattering by adsorbed gas molecules. Unlike Pd, platinum (Pt) does not absorb $H_2$, but it is able to chemisorb $H_2$ molecules, forming a surface hydride [16]. The response of Pt nanowire sensors to the same sequence of $H_2$ pulses in air (Figure 7.4(a)) shows that Pt nanowires exhibit an inverted response compared to Pd nanowires; the resistance decreases in the presence of $H_2$ instead of increasing.

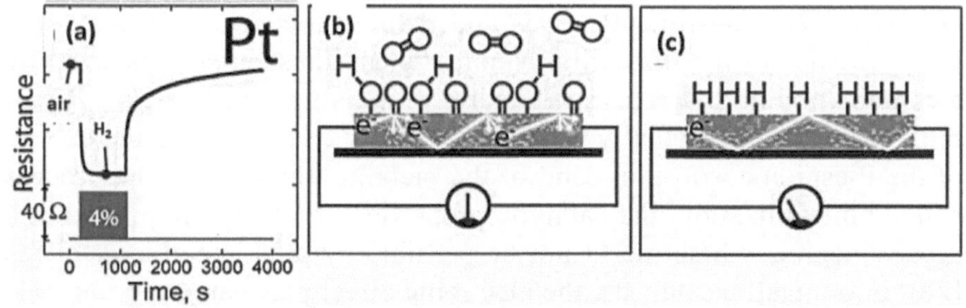


**FIGURE 7.4** (a) R versus time for the exposure of a Pt nanowire to 4% $H_2$. The surface chemical composition of the nanowire is similar to Pd, but in the absence of hydride formation, R is reduced in $H_2$ because diffuse electron scattering at the H-terminated wire surface is less prominent than at the oxygen-terminated surface that exists in dry air. The chemical state of the Pd nanowire surface in (b) air; and in (c) flowing $H_2$ [16].

*Source:* Reprinted with permission from Ref. [16]; Copyright (2012), American Chemical Society.

The key point is that in the presence of $H_2$, the coverage of hydride on the Pt surface increases until saturation coverage is attained. This is the reason that the resistance change observed for Pt nanowires shows no dependence on $H_2$ concentration [9, 16]. This was first described by the German physicist, Klaus Fuchs in 1938, but they did not exploit the material for sensor application for the detection of $H_2$ or other gases until the results of these experiments were published by Yang and co-workers [16]. $H_2$ sensors based upon Pt nanowires are unlikely to be very useful because they cannot discriminate different $H_2$ concentrations. However, in order to address these issues of Pt or Pd-based monometallic nanoparticles-based $H_2$ sensors, some interesting approaches using bimetals have been recently explored as discussed in the forthcoming sections.

## 7.3 BIMETALS-BASED $H_2$ SENSORS

Bimetallic nanoparticles are fascinating resolution to overcome the challenges of monometallic nanoparticles, as they are identified for their greater activity for surface reactions with respect to monometallic NPs [25]. In such cases, the synergistic effects are achieved by combinations of elements, which exceed the intrinsic properties of each individual element. Specifically, bimetallic nanoparticles; comprised of two dissimilar metals have become of greater interest rather than the monometallic NPs [26, 27]. Constituent metals and their nanometric size determine the properties of the bimetallic nanoparticles. These are synthesized by a combination of different architectures of metallic nanoparticles. Bimetallic nanoparticles actually offer the tendency of optimizing the energy of the Plasmon absorption band of the metallic mixture. By the process called bimetallization, the catalytic properties of NPs are improved to a greater degree, which might not be possible by monometallic catalysts [28]. In bimetallic catalysts, the electronic effect plays an important role which describes the charge transfer. Alloying of the constituent elements can cause structural changes of the bimetallic NPs [29]. According to Ferrando et al., in a bimetallic material, there are four possible types of mixing patterns between the two metals: (i) core-shell structures where a core of metal atoms is surrounded by a shell of metal atoms of a different type (Figure 7.5(a)), and a number of examples exist in the literature for materials with this kind of mixing pattern. (ii) Sub-cluster segregated nanoalloys (Figure 7.5(b)) in which, the two metal components have a pseudo-planar interface. Although these patterns have been proposed, physical examples of these kinds of alloys are rare. (iii) Homogeneously mixed alloys, where the two metals are intimately mixed in either an atomically ordered or a statistically random manner (termed as "ordered nanoalloys" and "random nanoalloys" respectively) (Figure 7.5(c)). Amongst these two sub-categories, examples of random alloys are more commonly found than ordered ones. Finally, multi-shell nanoalloys (Figure 7.5(d)) have more than one concentric shell covering the core metal, to form an "onion like" structure [30].

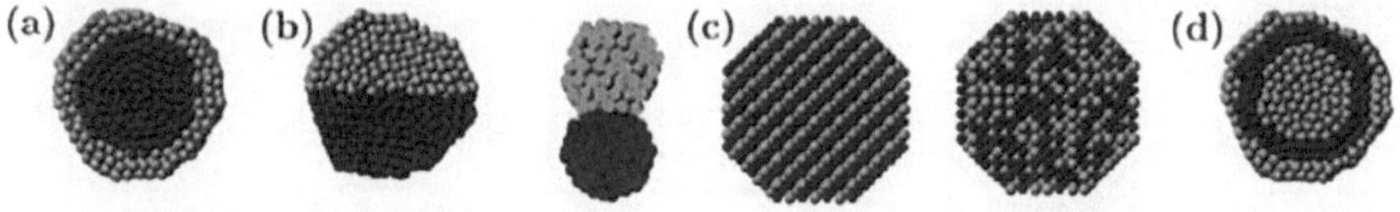


**FIGURE 7.5** Schematic representation of some possible mixing patterns in bimetallic systems: (a) core-shell alloys; (b) sub-cluster segregated alloys; (c) ordered and random homogeneous alloys; and (d) multishell alloys. The diagrams show cross-sectional representations of the clusters [30].

*Source:* Reprinted with permission from Ref. [30]; Copyright (2008), American Chemical Society.

One recent advance in Pd-based bimetals as $H_2$ gas sensor is the demonstration of platinum (Pt)-modified Pd nanowires (Pd@Pt nanowires) reported by Li et al. Inspired by the fact that Pt is a better catalyst for catalytic water formation than Pd [31]. In addition, various Pd-based multimetallic nanostructures can improve $H_2$ sensing performances by reducing the adsorption and dissociation energy of $H_2$ due to their synergistic effect in geometry and chemical/physical properties. In this regard, Jung and co-workers [32] developed Pd/Pt and Pd/Au bimetallic nanopattern arrays, which were prepared by the nanolithography via plasma ion bombardment for $H_2$ sensors.

Koo et al. [33] assumed that the Pd with Pt might be balancing for $H_2$ sensing in air owing to the fact that, Pt as well, can detect $H_2$ by reactions which eradicate surface-adsorbed oxygen. When the size of bimetallic NPs, kept below a few nanometers, it will exhibit very high catalytic activity due to the unsaturated surface states [18]. The capability to exactly alter the size and composition of bimetallic NPs are critically important to the sensory performance.

This is performed by the usage of a porous ion-exchange polymer, which permits homogeneous diffusion and immobilization of ionic metal precursors of complementary charge into the polymer matrix as shown in Figure 7.6(a). The porous polymer matrix too permits for gas accessibility to the bimetallic NPs and established the wireless sensing system, for $H_2$ detection in a head space sealed by the combination of RFID tag with ITPES-PdPt NPs. Here, the $H_2$ sensing mechanism of PdPt NPs in air was substantiated by two procedures: (i) the elimination of surface-adsorbed oxygen (resistance decreased) at Pd sites and the interfaces of Pd/Pt atoms and (ii) Hydrogen adsorption (resistance increased) within the PdPt NPs (Figure 7.6(b)) [18].

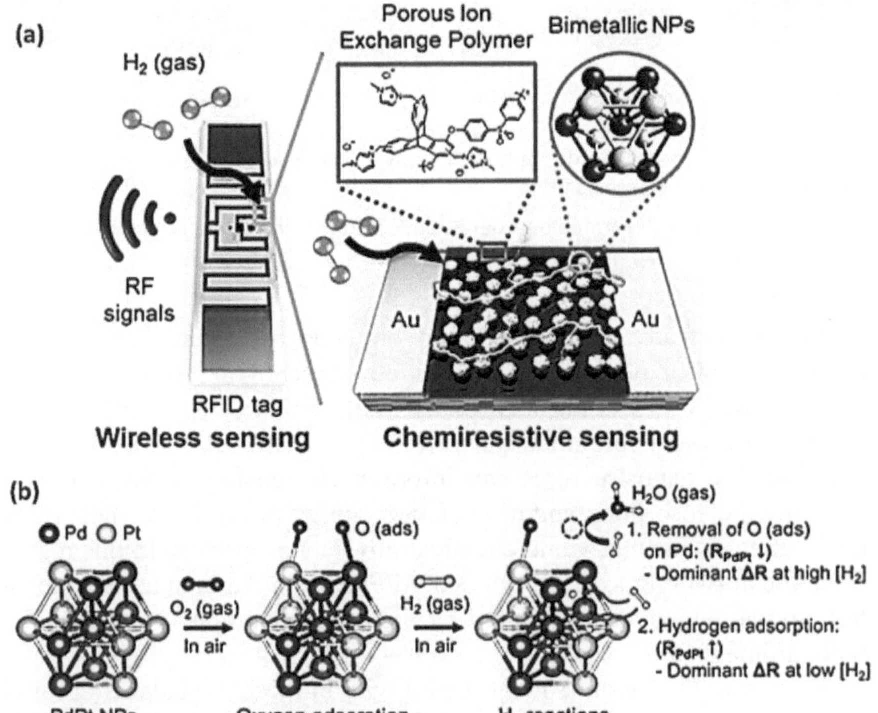


**FIGURE 7.6** Schematic illustration of (a) ITPES-PdPt NPs-based chemiresistors; (b) $H_2$ sensing mechanisms of ITPES-PdPt NPs [33].

*Source:* Reprinted from Ref. [33]; Copyright (2020), with permission from Elsevier.

## 7.4 CARBON-BASED $H_2$ GAS SENSORS

Carbon-based nanostructured materials, such as carbon nanotubes (CNTs) [19, 34], graphene [35, 36], and carbon fibers [21, 37] are shown in Figure 7.7, have attracted significant attention as a sensing element in chemical gas sensors. The unique structure of CNTs and graphene affords high sensitivity and very low detection limits to the surrounding gas species even at low temperatures. However, challenges remain in terms of selectivity, reversibility, recovery, and scalability. To achieve these properties, the active sensing spot must ideally pursue a high surface area/volume ratio to increase the probability of adsorption of molecules and increase the sensitivity. The selectivity of the sensor is mostly achieved by functionalization with specific materials that have a high affinity to the analytes of interest. Carbon nanostructures pursue all the ideal properties such as high surface area and can be easily functionalized. Currently, their relatively cheap and simple production steps as well as advanced manipulation methods of a wide variety of carbon nanostructures makes them perfect candidates to be used in different types of sensors [18, 23, 38].

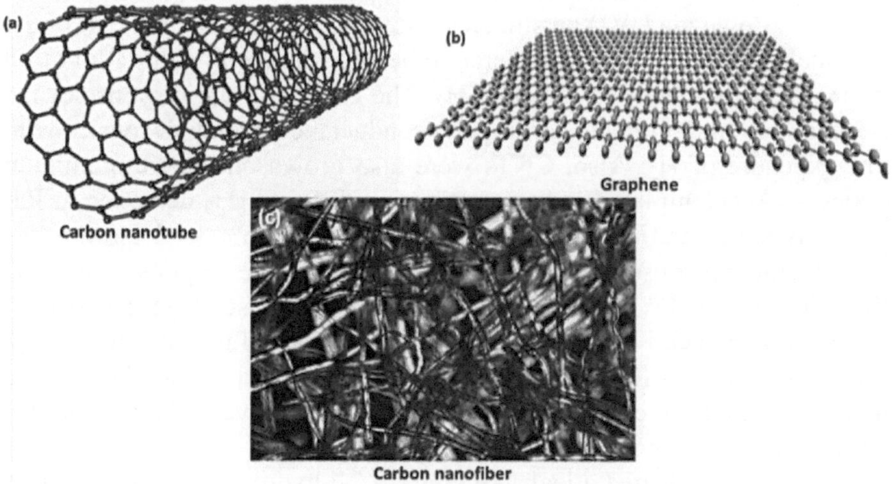


**FIGURE 7.7** (a) Carbon nanotube; (b) graphene; and (c) carbon nanofibers.

### 7.4.1 CNT-BASED $H_2$ GAS SENSORS

CNTs, since first discovered by Iijima in 1991, have drawn the most research interest because of their unique geometry, morphology, and properties [39]. Their preparation, properties (such as electronic, mechanical, thermal, and optical properties), and applications in various fields are all studied intensely. CNTs in the sensor field are extremely promising due to their enhanced sensitivity toward substances owing to the large nanotube electronic transport and thermo-power at RT. Further qualities of these nanomaterials are in the reduced number of required elements for a response owing to their excellent electron transfer, large surface area, and capability to be deployed with other conductive materials and polymers to form chemically active sites for application purposes. Like other gas-sensing materials, the response time and properties of CNTs are dissimilar for different target gases. The response property varies in physisorption and chemisorption. When utilized as sensing materials, on the one hand, CNTs are often decorated with other materials in order to enhance their sensitivity and selectivity [19, 34, 39, 40].

Hydrogen ($H_2$) has been one of the primary gases on which the CNT-based gas sensors have been employed upon. One of the pioneer works involving the detection of $H_2$ was based on the use of SWCNTs that was grown via CVD on silicon substrates. Semiconducting part of the conductive material was considered to be deposited with a thin film of Pd to increase its sensitivity towards $H_2$. The change in the response of the sensors was measured concerning the conductance, which decreased with the exposure of $H_2$. Also, CNTs were also grown on anodic aluminum oxide (AAO) templates and were mixed with Pd for $H_2$ detection at RT. The gas concentration was varied from 0.1% to 1.5% to determine the steady-state response of the sensors. The steady-state response increased from 0.17% to 0.48% when the concentration of $H_2$ risen from 100 ppm to 1,000 ppm, respectively. At average concentrations of the gas, the steady-state response was found out to be between 1.5% and 4.2%. The average response time for these sensors was 3–4 min with a reversible recovery for $H_2$ concentrations below 1.5% [41].

Lin et al. fabricated a leaf-like Pd/CNT/Ni hydrogen sensor operating at RT by the NC plating technique as shown in Figure 7.8. The Pd/CNT/Ni sensor displays a high response (7.3%), a fast response time (312 s), and a short recovery time (173 s) for low hydrogen concentrations (200 ppm).

Here, the Pd NPs modified on the CNT/Ni composite film significantly improved sensor response [40].

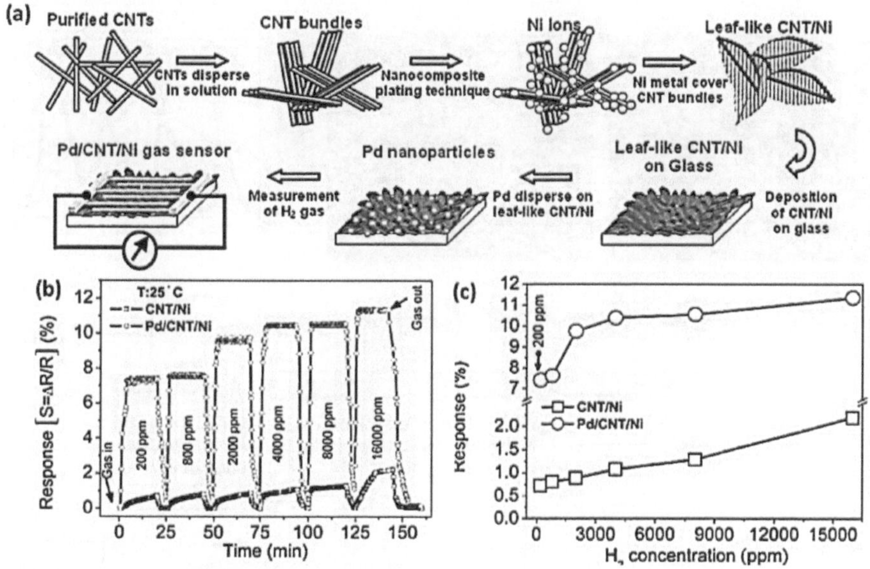


**FIGURE 7.8** (a) Schematic illustration of the major processes involved in the fabrication of hydrogen sensors on glass substrates with the use of the Pd nanoparticles functionalized CNT/Ni composite nanostructure. Comparisons of the hydrogen response for the two types of gas sensors. (b) Response vs. time with $H_2$ concentrations ranging from 200 ppm to 16,000 ppm. (c) Response as a function of hydrogen concentration [40].

*Source:* Reprinted from Ref. [40]; Copyright (2012), with permission from Elsevier.

An aerogel sheet of CNTs decorated with Pd nanoparticles reported by Kim et al. was suitable for mass production, showed $H_2$ sensitivity of 7% towards 1,000 ppm of $H_2$. Although the functionalization of carbon nanostructures with Pd nanocatalysts has shown a promising response towards $H_2$ due to its excellent hydrogen solubility at RT, it can undergo structural deformation owing to $PdH_x$ formation [38].

Kumar et al. have investigated the Pt functionalized MWCNTs RT sensor for the detection of 4% $H_2$ in air. It was found that Pt functionalized MWCNTs showed better $H_2$ sensitivity, but prolonged response and recovery times were found to limit its applications [43]. Dhall et al. sputtered Pt nanoparticles over pristine MWCNTs (Pt/P-MWCNTs) and functionalized MWCNTs (Pt/F-MWCNTs) and found that the Pt/F-MWCNTs showed enhanced sensor response (~3%) towards 1% of $H_2$ at RT with improved response and recovery time shown in Figure 7.9 [42].

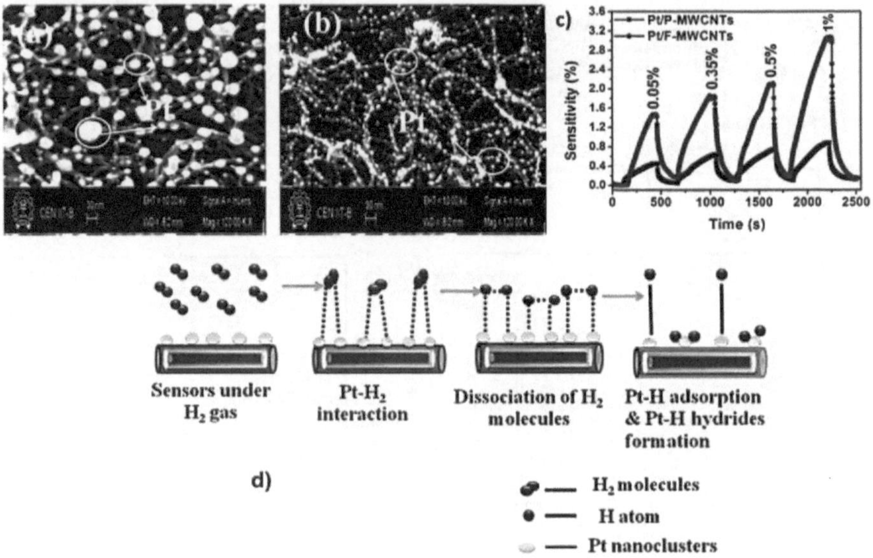


**FIGURE 7.9** SEM image of (a) P-MWCNTs/Pt; (b) F-MWCNTs/Pt; (c) $H_2$ sensing response of P-MWCNTs/Pt and F-MWCNTs/Pt; and (d) schematic representation of sensing mechanism [42].

*Source:* Reprinted with permission from Ref. [42]; Copyright (2006), American Chemical Society.

Recently, the effect of loading and size distribution of Pt NPs dispersed over wrinkled graphene-like carbon-wrapped CNTs, which showed sensitivity of 42.8% towards 4% of $H_2$ at RT was discussed by Baro and co-workers. It was revealed that, with high Pt loading on support material, the large crystallite size decreased the activity of the Pt nanocatalyst and, thereby, the sensor performance. Though Pt, which is also a well-known material for $H_2$ sensing, aids in the dissociation of $H_2$ molecules at the surface forming Pt-H, unlike Pd, it stays on the surface without diffusing into the bulk. Hence, Pt nanocatalysts supported on carbon nanostructures showed an improved recovery of the sensor than Pd-based sensing materials [44]. Table 7.1 shows a summary of CNTs-based sensors for $H_2$ detection.

**TABLE 7.1** A Summary of Carbon Nanotubes-based $H_2$ Sensors

| Sensing Material | Temperature (°C) | Sensitivity (%) | Concentration | Response Time | Recovery Time | References |
|---|---|---|---|---|---|---|
| Pt-CNT | RT | 16 | 23% | 110s | – | [45] |
| MWCNTs/ZnO/Pt | RT | 4.2 | 0.05 | – | 78s | [42] |
| CNT/$SnO_2$ | 200 | 84 | 4 | – | 120s | [34] |
| Pd/CNT | 100 | 3.68 | 100 ppm | 1.8 min | 6 min | [38] |
| F-MWCNTs/ZnO/Pt | RT | 3.6 | 0.05 | – | 20s | [47] |
| Pt/GCNTs | RT | 42.6 | 4 | 2 min | – | [44] |
| Pt-Pd/MWCNTs | RT | 4 | 4 | – | – | [48] |

### 7.4.2 GRAPHENE-BASED $H_2$ SENSORS

Graphene is a monolayer of $sp^2$ hybridized carbon atoms arranged in a honeycomb lattice which was discovered by Geim and Novoselov in 2004 [49]. Graphene exhibits an excellent mechanical stability along with exclusive electrical properties. Graphene is a striking sensing material for manufacturing a sensing device payable to its intrinsic properties like higher interaction with gas molecules, large surface area and zero rest mass of its charged carriers. The large surface area per unit volume, makes them apt for the adsorption of gas molecules. The variation in resistance of the graphene sheets on exposure to analyte species lead to a variation in free-electron concentration (increase/decrease). Graphene also displays excellent carrier mobility at RT since zero rest mass of graphene carriers are near to its Dirac point [36, 49, 50].

Graphene is extensively defined as a fundamental material in hydrogen sensors due to its upgrading surface area and susceptibility of its electronic properties to the variations triggered by adsorbing atoms and molecules, including $H_2$. Still, the graphene sensitivity towards hydrogen is limited [51]. Numerous graphene hybrid structures, like graphene with transition metals, were considered in various parts of gas sensing with a motive to enhance the properties of $H_2$ sensors. Chu et al. and Shafiei et al. examined the catalytic effect of Pt NPs on the $H_2$ sensing properties of such altered layered graphene placed on a SiC. Shafiei et al. manufactured Pt/graphene sheets structures by reducing spray deposited layer of graphite oxide by means of hydrazine vapor. The sensors remained at a constant reverse bias, as the sensing characteristics was much promising for reverse bias rather than forward bias [52]. $H_2$ sensing performance of the sensors by incorporating Pt-decorated over graphene sheets and Pt-decorated over multi-walled CNTs were studied by Kaniyoor et al. Drop casting method were employed for the decoration procedure. Pt/graphene displayed a comparable response time then had double the sensitivity of Pt/CNT [53, 54]. Wong et al. developed a $H_2$ sensor based on Pt-GO, which exhibited sensitivity of 10% towards 200 ppm of $H_2$ at ambient conditions. Similarly, the Pt@rGO NCs prepared via freeze-drying method showed outstanding $H_2$ sensing [55].

Esfandir et al. prepared hybrid of the GO sheets with $TiO_2$ nanoparticles using photoreduction process and decorated the hybrid structure with catalytic nanoparticles. Maximum sensitivity of 92% with rapid response time, less than 20 s was achieved towards 500 ppm of $H_2$ were detected by the sample with low concentration of Pd (2 wt.%) decorated on the $TiO_2$/RGO sample at a comparatively low temperature (180°C). Further, Pd/$TiO_2$/RGO shown improved sensing performance related to the Pt/$TiO_2$/RGO as shown in Figure 7.10. Better rate of spill-over effect and dissociation of $H_2$ molecules on Pd are measured as probable reasons for the enhanced sensitivity in Pd/$TiO_2$/RGO [56].

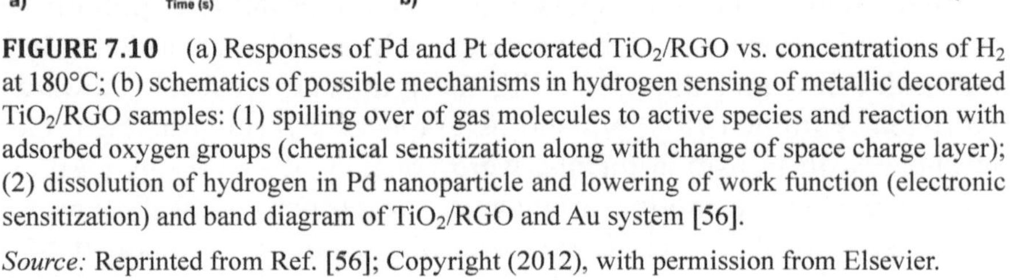


**FIGURE 7.10** (a) Responses of Pd and Pt decorated $TiO_2$/RGO vs. concentrations of $H_2$ at 180°C; (b) schematics of possible mechanisms in hydrogen sensing of metallic decorated $TiO_2$/RGO samples: (1) spilling over of gas molecules to active species and reaction with adsorbed oxygen groups (chemical sensitization along with change of space charge layer); (2) dissolution of hydrogen in Pd nanoparticle and lowering of work function (electronic sensitization) and band diagram of $TiO_2$/RGO and Au system [56].

*Source:* Reprinted from Ref. [56]; Copyright (2012), with permission from Elsevier.

Peng et al. fabricated Pt–Pd/RGO-based sensors using DEP assembly process (Figure 7.11). The Pt–Pd/RGO-based sensors showed a repeatable and stable response to $H_2$ due to carrier donation and crystal lattice expansion from metal hydride formation. They also studied in detail the gas sensing performance considering various parameters. The temperature reduces the response, but increased hydrogen concentration improves it. The flow rate of the carrier gas has negligible impact on the response. High response and slow recovery can be obtained when using nitrogen replaced by air as the carrier gas due to the involvement of oxygen. The Pt–Pd/RGO-based sensors could serve as facile hydrogen sensors with high performance, including excellent response and recovery under various environments due to the synergistic effect of Pt and Pd on rGO [57].

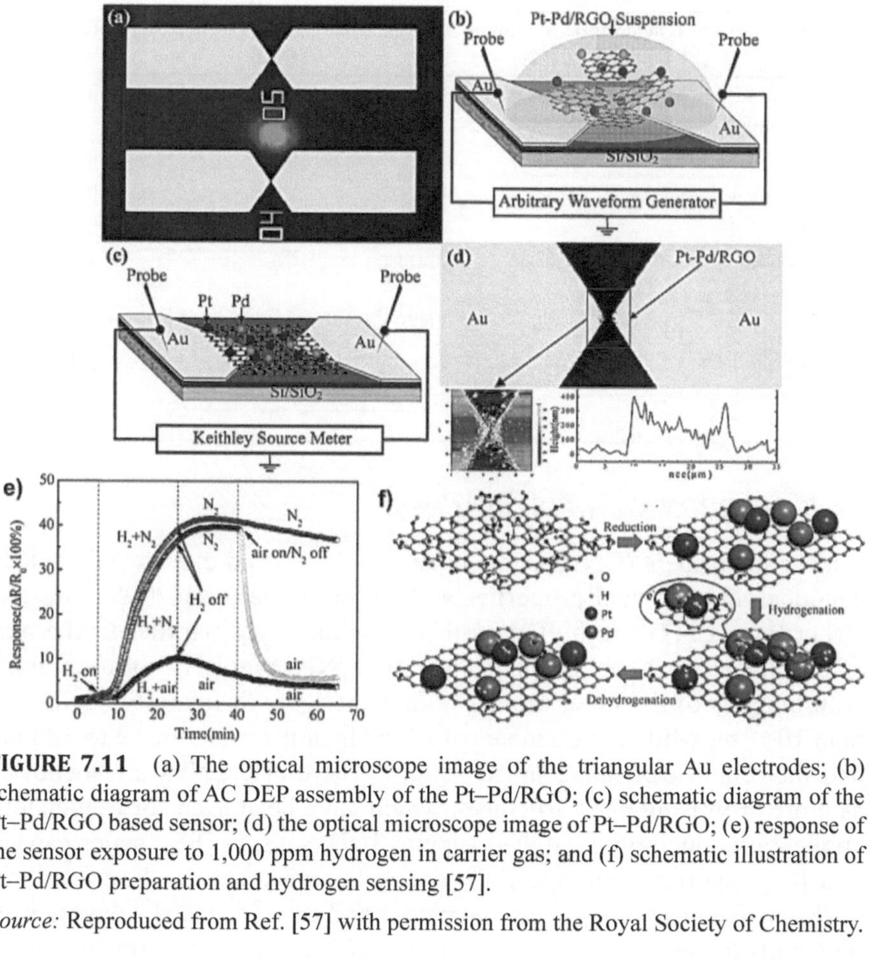


**FIGURE 7.11** (a) The optical microscope image of the triangular Au electrodes; (b) schematic diagram of AC DEP assembly of the Pt–Pd/RGO; (c) schematic diagram of the Pt–Pd/RGO based sensor; (d) the optical microscope image of Pt–Pd/RGO; (e) response of the sensor exposure to 1,000 ppm hydrogen in carrier gas; and (f) schematic illustration of Pt–Pd/RGO preparation and hydrogen sensing [57].

*Source:* Reproduced from Ref. [57] with permission from the Royal Society of Chemistry.

Phan et al. investigated bimetallic NPs, like Ni-Pd, which prevented the deep penetration of $H_2$ molecules into the Pd lattice. The Ni/Pd-Gr bimetallic composite exhibited a complete recovery and rapid response. The response was about 11 and 9% towards 1,000 ppm $H_2$ at RT for Ni/Pd ~7% and pure Pd-Gr [58]. Similarly, Sharma et al. discussed bimetal modification of graphene decorated by Pd-Ag alloy nanoparticles for $H_2$ detection which showed the maximum gas response of 16.2% towards 1,000 ppm of $H_2$. Table 7.2 shows a summary of graphene-based sensors for $H_2$ detection [59].

**TABLE 7.2** A Summary of Graphene-based $H_2$ Sensors

| Sensing Material | Temperature (°C) | Sensitivity (%) | Concentration | Response Time | Recovery Time | References |
|---|---|---|---|---|---|---|
| The Pt-NP loaded graphene aerogel | 320 | 1.6 | 1% | 0.97s | 0.72s | [60] |
| $PtPt_3Fe$/NG NPs | RT | 47 | 4 | 320s | – | [61] |
| Pt/Pd core-shell graphene | RT | 36 | 1 | 3 min | 1.2 min | [62] |
| rGO decorated Ni-doped ZnO | 150 | 63.8 | 100 ppm | 28s | – | [59] |
| Ni-Pd-Gr | RT | 11 | 1,000 ppm | 3 min | 12 min | [58] |
| Pd-Ni/carbon fibers | RT | 2 | 0.1 | 12 min | 1 min | [64] |
| Pd-graphene nanoribbon | RT | 77 | 8,000 ppm | 6s | 44s | [65] |

### *7.4.3 CARBON NANOFIBERS (CNFS)*

Carbon nanofibers (CNFs) have been widely studied due to their unique chemical and physical properties and similar structure to fullerenes and CNTs [66]. CNTs are hollow with a graphite layer parallel to the axis of the inner tube. The graphite layers of CNFs regularly form an angle with the axis of the inner tube. The diameters of CNTs are typically less than 100 nm, while the diameter of CNFs is in the range of 10 to 500 nm and the length can range up to 10 μm. Therefore, CNFs are identified as more promising nanostructures, which have been used for tremendous applications, including catalyst support [67], sensors [68], super capacitors [69], hydrogen storage [70], energy storage, etc. In the meantime, CNFs have entirely basal graphite planes and edge planes, displaying high potentials towards functionalization of CNFs in order to form functional hybrid CNF-based nanomaterials, that have been functional in the fields of energy, tissue engineering, biomedicine, nanodevices, environmental science and sensors [66, 69, 71].

CNFs can exhibit various shapes, as shown in Figure 7.12, based on the criterion of the angle of the graphene layers that compose the filament. Besides platelet (Figure 7.12(a)) and tubular or ribbon CNFs (i.e., carbon nanotubes (CNTs), Figure 7.12(b)), so-called fishbone CNFs exist in which the graphene layers are oriented under an angle between principal and perpendicular axis. The distinctive arrangements of the graphene layers depend on the geometric aspects of the metallic nanoparticle catalyst and the feedstock of the gaseous carbon (CO or hydrocarbon gas), which is introduced during the synthesis processing. It was found that the different graphene contact areas (edge planes and basal) play distinct roles within the physical and chemical behavior of CNFs [72].

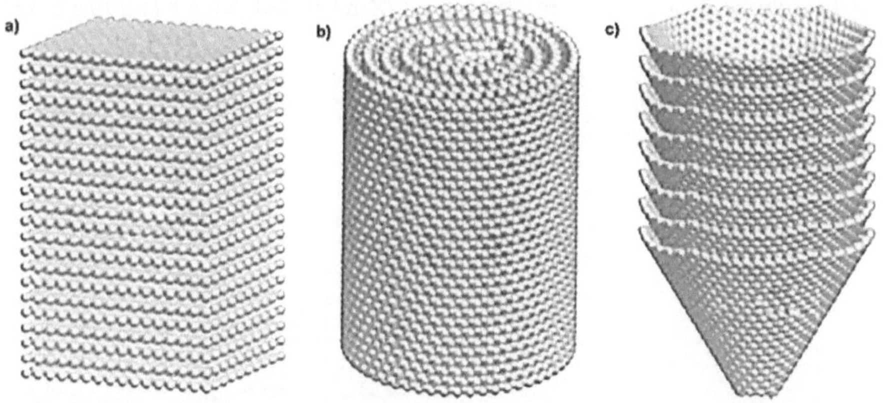


**FIGURE 7.12** Schematic illustrations of the three types of CNFs with different basal-to-edge surface area ratios: (a) platelet-type CNF; (b) tubular-type CNF; and (c) fishbone-type CNF [72].

*Source:* Reprinted from Ref. [72], Copyright (2012), with permission from Elsevier.

## 7.5 CNFS-BASED $H_2$ SENSORS

CNFs-based hydrogen sensors are hardly explored due to its sensitivity and selectivity towards $H_2$ gas. By incorporating CNFs with suitable catalytic metals, the selectivity towards $H_2$ can be resolved. Zhang et al. have prepared carbon nanofibrous mat (nano-felt) with the surface to be decorated with Pd nanoparticles by electrospinning polyacrylonitrile (PAN) nano-felt surface-functionalized with amidoxime groups, for application to detect $H_2$. The CNFs fabricated in this way had an average diameter of 300 nm, whereas the Pd nanoparticles were at the scale of tens of nanometers. Resistance of the carbon nano-felts decreased upon exposure to $H_2$ and the response was reversible. Obviously, the carbon nano-felt decorated with Pd nanoparticles had much higher sensitivity, due to the electronic interactions among $H_2$ molecules, Pd nanoparticles and the carbon nano-felt, as compared with those without Pd decoration. As the $H_2$ molecules were adsorbed onto Pd nanoparticles, the reaction between the two items would take place to form $PdH_x$. Because the $PdH_x$ had a lower work function than Pd, thus making it more effective to transfer electrons to the carbon nano-felt. As a result, a larger decrease in resistance was observed in the carbon nano-felt decorated with Pd nanoparticles [73]. Bhati et al. fabricated PAN/(PAN-b-PMMA) derived electrospun nano porous CNFs loaded on ZnO nanostructures-based hydrogen sensors using electrospinning, spin coating and RF sputtering method. CNF/ZnO-based sensor (Figure 7.13)

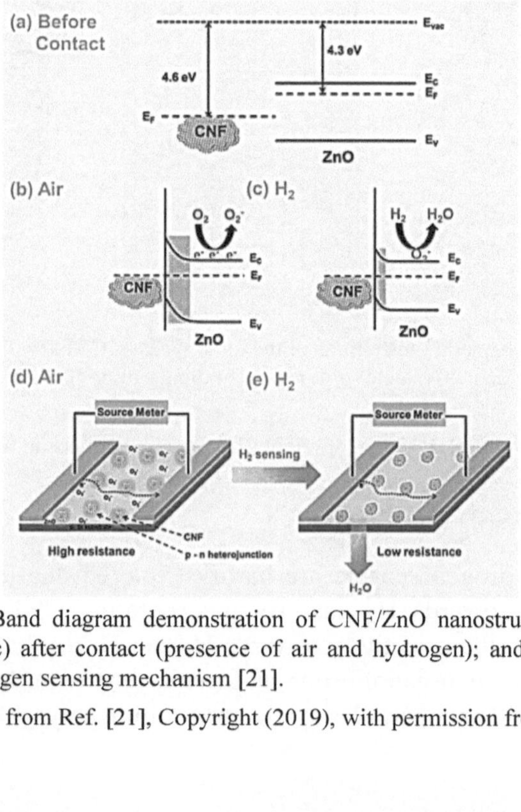


**FIGURE 7.13** Band diagram demonstration of CNF/ZnO nanostructures: (a) before contact; and (b, c) after contact (presence of air and hydrogen); and (d, e) schematic diagram for hydrogen sensing mechanism [21].

*Source:* Reprinted from Ref. [21], Copyright (2019), with permission from Elsevier.

exhibited maximum sensor response (73.54%) related to CNF (3.29%) and ZnO (44.51%) towards 100 ppm $H_2$ at 150°C. This improved response is ascribed to diffusion of $H_2$ through the nano-porous CNF, thus allowing the development of p-n heterojunction at the interface of CNF and ZnO [21].

Also, Wang et al. have discussed about CNF/$SnO_2$ composite-based $H_2$ sensors by combining electrospinning and hydrothermal method (Figure 7.14). They observed a reduction in operating temperature after making a composite with CNFs than pristine $SnO_2$. Moreover, the heterostructure of thin $SnO_2$ nanosheets vertically standing on the surface of CNFs facilitates gas sensing adequately [74].

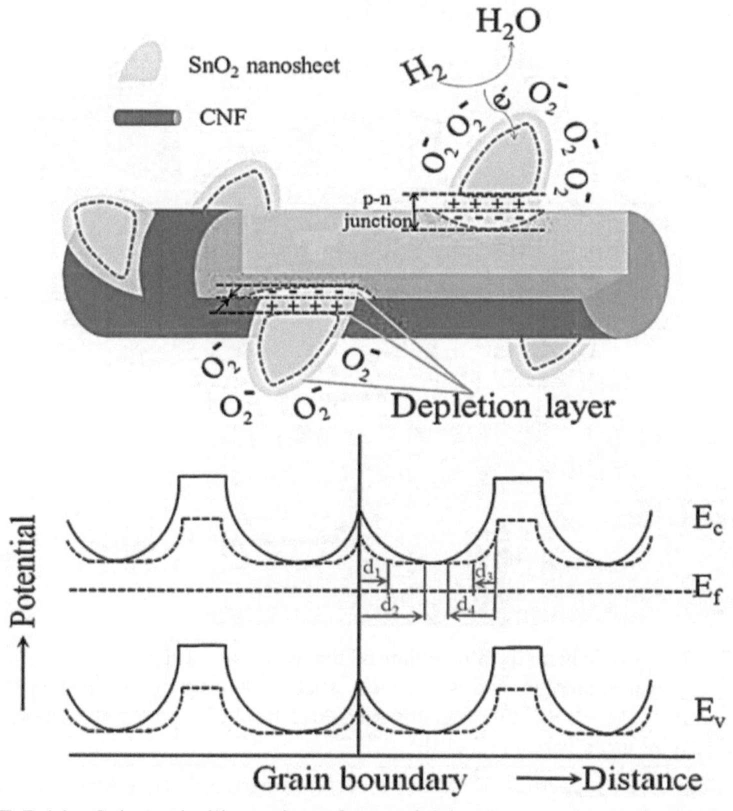


**FIGURE 7.14** Schematic illustration of potential barriers to electronic conduction at grain boundaries and at *p–n* heterojunctions for CNFs/$SnO_2$; $d_1$ and $d_3$ are depletion layer widths when exposed to hydrogen; $d_2$ and $d_4$ are depletion layer widths in air [74].

*Source:* Reproduced from Ref. [74] with permission from the Royal Society of Chemistry.

A flexible $H_2$ sensing platform was developed by Kim et al. on a single-strand nanofiber yarn which consists of high-density electrospun nanofibers, on which nanogranular Pd or Pd@Pt was coated via sputtering (Figure 7.15(a)). The method for nanofiber yarn fabrication is an advanced electrospinning technique that requires two individual nozzle systems at which polymer nanofibers are ejected to the rotating funnel collector to form core-support wire/shell-electrospun nanofibers of the yarn scaffold. Pd-deposited nanofiber yarn (yarn@Pd) and Pt-sensitized yarn@Pd (yarn@Pd@Pt) were prepared by consecutive Pd and Pt sputter deposition. The SEM images displayed that the yarn with a diameter of 600 μm consists of several hundreds of well-aligned Pd-coated nanofibers with the yarn direction (Figure 7.15(b) and (c)).

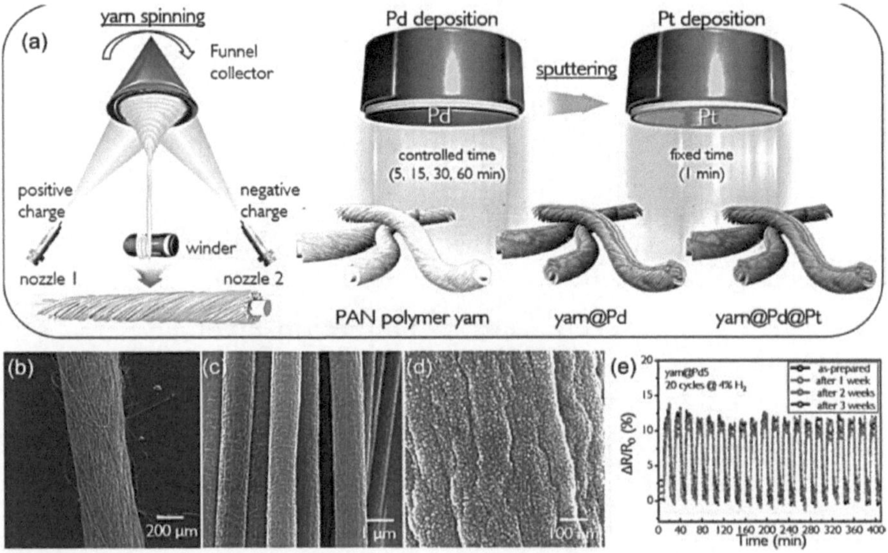


**FIGURE 7.15** (a) Schematic illustration of the synthesis of the yarn@Pd and yarn@Pd@Pt using yarn-spinning and subsequent sputter deposition of Pd and Pt (PAN: polyacrylonitrile); (b–d) SEM images of yarn@Pd15; (e) long-cycling stability tests to $H_2$ 4% for 20 cycles using yarn@Pd5 [15].

*Source:* Reprinted with permission from Ref. [15]; Copyright (2019), American Chemical Society.

The high-resolution SEM images of the yarn@Pd showed a rough surface with ultra-small Pd grains (<10 nm) (Figure 7.15(d)). Since Pd sensing layers were mechanically coupled with a flexible polymeric nanofiber, the yarn@Pd-based sensors endured severe stress during repeated volume expansion/relaxation. To demonstrate the long-term stability, the sensors were exposed to $H_2$ 4% for 20 cycles, and the cycle tests were repeated after 1, 2, and 3 weeks. As a result, the yarn@Pd-based sensors showed good reliability and long-term stability without any degradation in response (Figure 7.15(e)) [15]. Therefore, fabrication of flexible $H_2$ sensors can have significant roles in many applications where flexibility of the sensor devices is significant.

## 7.6 CONCLUSION

Pt and Pd nanoparticles were used majorly as catalytic materials towards hydrogen gas. However, a major limitation of monometallic Pd and Pt are its irreversible adsorption of hydrogen. On the other hand, the gold nanoparticle is an established active catalyst with long life. The major drawback of gold monometallic catalysts is to initiate the hydrogen adsorption process. In this chapter, carbon nanomaterial-based $H_2$ sensors were discussed in detail. The $H_2$ chemiresistors shall be classified into numerous materials: metal-based, metal oxides, and carbons nanomaterial-based. In each category, several exciting synthetic methods was advanced in order to enhance $H_2$ sensing properties. This can be either performed by combining with other materials, or with the introduction of nanocatalysts. This chapter will inspire various researchers and industrialists, and also will pay the way for carbon nanomaterials-based $H_2$ sensors, for their practical applications.

## KEYWORDS

- **chemi-resistive sensors**
- **CNF-based $H_2$ sensors**
- **CNT-based $H_2$ sensors**
- **graphene-based $H_2$ sensors**
- **hydrogen sensors**
- **metal and bimetal-based $H_2$ sensors**

## REFERENCES


1. Gu, H., Wang, Z., & Hu, Y., (2012). *Hydrogen Gas Sensors Based on Semiconductor Oxide Nanostructures, 12.*
2. Hübert, T., Boon-Brett, L., Black, G., & Banach, U., (2011). Hydrogen sensors: A review. *Sensors Actuators, B Chem., 157,* 329–352.
3. Korotcenkov, G., Han, S. D., & Stetter, J. R., (2009). Review of electrochemical hydrogen sensors. *Chemical Reviews, 109*(3), 1402–1433.
4. Hübert, T., Boon-Brett, L., Palmisano, V., & Bader, M. A., (2014). Developments in gas sensor technology for hydrogen safety. *Int. J. Hydrogen Energy, 39,* 20474–20483.
5. Buttner, W. J., Post, M. B., Burgess, R., & Rivkin, C., (2011). An overview of hydrogen safety sensors and requirements. *Int. J. Hydrogen Energy, 36,* 2462–2470.
6. Manjavacas, G., & Nieto, B., (2016). *Hydrogen Sensors and Detectors.* Elsevier Ltd.
7. Quarton, C. J., Tlili, O., Welder, L., Mansilla, C., Blanco, H., Heinrichs, H., Leaver, J., et al., (2019). The curious case of the conflicting roles of hydrogen in global energy scenarios. *Sustain. Energy Fuels, 4,* 80–95.
8. Anand, K., Singh, O., Singh, M. P., Kaur, J., & Singh, R. C., (2014). Hydrogen sensor based on graphene / ZnO nanocomposite. *Sensors Actuators B. Chem., 195,* 409–415.
9. Penner, R. M., (2017). A nose for hydrogen gas: Fast, sensitive $H_2$ sensors using electrodeposited nanomaterials. *Acc. Chem. Res., 50,* 1902–1910.
10. Ma, L., Wang, X., Zhang, Q., Tong, X., Zhang, Y., & Li, Z., (2018). Pt catalyzed formation of a Ni@Pt/reduced graphene oxide nanocomposite: Preparation and electrochemical sensing application for glucose detection. *Anal. Methods, 10,* 3845–3850.
11. Cho, S. Y., Ahn, H., Park, K., Choi, J., Kang, H., & Jung, H. T., (2018). Ultrasmall grained Pd nanopattern $H_2$ sensor. *ACS Sensors, 3,* 1876–1883.
12. Pranti, A. S., Loof, D., Kunz, S., Zielasek, V., Bäumer, M., & Lang, W., (2019). Design and fabrication challenges of a highly sensitive thermoelectric-based hydrogen gas sensor. *Micromachines.* doi: 10.3390/mi10100650.
13. Yang, M., & Dai, J., (2014). Fiber optic hydrogen sensors: A review. *Photonic Sensors, 4,* 300–324.
14. Sil, D., Hines, J., Udeoyo, U., & Borguet, E., (2015). Palladium nanoparticle-based surface acoustic wave hydrogen sensor. *ACS Appl. Mater. Interfaces, 7,* 5709–5714.
15. Kim, D. H., Kim, S. J., Shin, H., Koo, W. T., Jang, J. S., Kang, J. Y., Jeong, Y. J., & Kim, I. D., (2019). High-resolution, fast, and shape-conformable hydrogen sensor platform: Polymer nanofiber yarn coupled with nanograined Pd@Pt. *ACS Nano, 13,* 6071–6082.
16. Yang, F., Donavan, K. C., Kung, S. C., & Penner, R. M., (2012). The surface scattering-based detection of hydrogen in air using a platinum nanowire. *Nano Lett., 12,* 2924–2930.
17. Zhang, H., Li, Z., Liu, L., Xu, X., Wang, Z., Wang, W., Zheng, W., et al., (2010). Enhancement of hydrogen monitoring properties based on Pd-$SnO_2$ composite nanofibers. *Sensors Actuators, B Chem., 147,* 111–115.

18. Koo, W. T., Cho, H. J., Kim, D. H., Kim, Y. H., Shin, H., Penner, R. M., & Kim, I. D., (2020). Chemiresistive hydrogen sensors: Fundamentals, recent advances, and challenges. *ACS Nano*, doi: 10.1021/acsnano.0c05307.
19. Kim, J. J. H., Jeon, J. G., Ovalle-Robles, R., Kang, T. J., Jung, D., Han, M., Lee, G. S., et al., (2018). High-performance, flexible hydrogen sensors that use carbon nanotubes decorated with palladium nanoparticles. *Int. J. Hydrogen Energy, 43*, 6456–6461.
20. Punetha, D., Kar, M., & Pandey, S. K., (2020). A new type low-cost, flexible and wearable tertiary nanocomposite sensor for room temperature hydrogen gas sensing. *Sci. Rep., 10*, 1–11.
21. Singh, B. V., Nathani, A., Nigam, A., Sharma, C. S., & Kumar, M., (2019). PAN/(PAN-b-PMMA) derived nanoporous carbon nanofibers loaded on ZnO nanostructures for hydrogen detection. *Sensors Actuators, B Chem., 299*, 126980.
22. Zhang, D., Sun, Y., Jiang, C., & Zhang, Y., (2017). Room temperature hydrogen gas sensor based on palladium decorated tin oxide/molybdenum disulfide ternary hybrid via hydrothermal route. *Sensors Actuators, B Chem., 242*, 15–24.
23. Xiao, Z., Kong, L. B., Ruan, S., Li, X., Yu, S., Li, X., Jiang, Y., et al., (2018). Recent development in nanocarbon materials for gas sensor applications. *Sensors Actuators, B Chem., 274*, 235–267.
24. Yang, F., Taggart, D. K., & Penner, R. M., (2009). *Fast, Sensitive Hydrogen Gas Detection Using Single Palladium Nanowires That Resist Fracture, 2*, 1–6.
25. Chen, H., Li, Y., Zhang, F., Zhang, G., & Fan, X., (2011). Graphene supported Au-Pd bimetallic nanoparticles with core-shell structures and superior peroxidase-like activities. *J. Mater. Chem., 21*, 17658–17661.
26. Gupta, G., Sharma, S., & Mendes, P. M., (2016). Nafion-stabilized bimetallic Pt-Cr nanoparticles as electrocatalysts for proton exchange membrane fuel cells (PEMFCs). *RSC Adv., 6*, 82635–82643.
27. Zhang, H., & Toshima, N., (2013). Synthesis of Au/Pt bimetallic nanoparticles with a Pt-rich shell and their high catalytic activities for aerobic glucose oxidation. *J. Colloid Interface Sci., 394*, 166–176.
28. Gilroy, K. D., Ruditskiy, A., Peng, H. C., Qin, D., & Xia, Y., (2016). Bimetallic nanocrystals: Syntheses, properties, and applications. *Chem. Rev., 116*, 10414–10472.
29. Singh, H. P., Gupta, N., Sharma, S. K., & Sharma, R. K., (2013). Synthesis of bimetallic Pt-Cu nanoparticles and their application in the reduction of rhodamine B. *Colloids Surfaces A Physicochem. Eng. Asp., 416*, 43–50.
30. Ferrando, R., Jellinek, J., & Johnston, R. L., (2008). Nanoalloys: From theory to applications of alloy clusters and nanoparticles. *Chem. Rev., 108*, 845–910.
31. Li, X., Liu, Y., Hemminger, J. C., & Penner, R. M., (2015). *Catalytically Activated Palladium @ Platinum Nanowires for Accelerated Hydrogen Gas Detection.* doi: 10.1021/acsnano.5b00302.
32. Bin, J. W., Cho, S. Y., Suh, B. L., Yoo, H. W., Jeon, H. J., Kim, J., & Jung, H. T., (2019). Polyelemental nanolithography via plasma ion bombardment: From fabrication to superior $H_2$ sensing application. *Adv. Mater., 31*, 1–11.
33. Koo, W. T., Kim, Y., Kim, S., Suh, B. L., Savagatrup, S., Kim, J., Lee, S. J., et al., (2020). Hydrogen sensors from composites of ultra-small bimetallic nanoparticles and porous ion-exchange polymers. *Chem., 6*, 2746–2758.

34. Majumdar, S., Nag, P., & Devi, P. S., (2014). Enhanced performance of CNT/$SnO_2$ thick film gas sensors towards hydrogen. *Mater. Chem. Phys., 147*, 79–85.
35. Kathiravan, D., Huang, B. R., & Saravanan, A., (2017). Self-assembled hierarchical interfaces of ZnO nanotubes/graphene heterostructures for efficient room temperature hydrogen sensors. *ACS Appl. Mater. Interfaces, 9*, 12064–12072.
36. Basu, S., & Hazra, S. K., (2016). *Graphene-Based Junction Devices for Hydrogen Sensors*. Chapter 6; Progresses in chemical sensor. doi: 10.5772/62559.
37. Li, W., Zhang, L. S., Wang, Q., Yu, Y., Chen, Z., Cao, C. Y., & Song, W. G., (2012). Low-cost synthesis of graphitic carbon nanofibers as excellent room temperature sensors for explosive gases. *J. Mater. Chem., 22*, 15342–15347.
38. Kim, J. H., Jeon, J. G., Ovalle-Robles, R., & Kang, T. J., (2018). Aerogel sheet of carbon nanotubes decorated with palladium nanoparticles for hydrogen gas sensing. *Int. J. Hydrogen Energy, 43*, 6456–6461.
39. Suehiro, J., Yamane, S., & Imasaka, K., (2007). Carbon nanotube-based hydrogen gas sensor electrochemically functionalized with palladium. *Proc. IEEE Sensors,* 554–557.
40. Lin, T., & Huang, B., (2012). Sensors and actuators B: Chemical palladium nanoparticles modified carbon nanotube / nickel composite rods (Pd / CNT / Ni) for hydrogen sensing, *Sensors Actuators B. Chem., 162*, 108–113.
41. Ding, D., Chen, Z., Rajaputra, S., & Singh, V., (2007). Hydrogen sensors based on aligned carbon nanotubes in an anodic aluminum oxide template with palladium as a top electrode. *Sensors Actuators, B Chem., 124*, 12–17.
42. Dhall, S., & Jaggi, N., (2014). Accept Cr T, sensors actuators. *A. Phys.* doi: 10.1016/j.sna.2015.01.010.
43. Kumar, M. K., & Ramaprabhu, S., (2006). *Nanostructured Pt Functionalized Multiwalled Carbon Nanotube Based Hydrogen Sensor,* 11291–11298.
44. Baro, M., & Ramaprabhu, S., (2018). ScienceDirect room temperature hydrogen gas sensing properties of monodispersed platinum nanoparticles on graphene-like carbon-wrapped carbon nanotubes. *Int. J. Hydrogen Energy,* 1–9.
45. Jung, D., Han, M., & Lee, G. S., (2015). Fast-response room temperature hydrogen gas sensors using platinum-coated spin-capable carbon nanotubes. *ACS Appl. Mater. Interfaces, 7*, 3050–3057.
46. Dhall, S., Sood, K., & Jaggi, N., (2014). A hydrogen gas sensor using a Pt-sputtered MWCNTs/ZnO nanostructure. *Measurement Science and Technology, 25*(8), 085103.
47. Dhall, S., Sood, K., & Nathawat, R., (2017). Room temperature hydrogen gas sensors of functionalized carbon nanotubes-based hybrid nanostructure: Role of Pt sputtered nanoparticles. *Int. J. Hydrogen Energy,* 1–7. ScienceDirect.
48. Randeniya, L. K., Martin, P. J., & Bendavid, A., (2012). *Detection of Hydrogen from 5 ppm to 4 % Using Multi-Walled Carbon Nanotube Yarns Coated with Nanocrystalline Pd and Pd / Pt Layered Structures, 50*, 1786–1792.
49. Singh, E., Meyyappan, M., & Nalwa, H. S., (2017). Flexible graphene-based wearable gas and chemical sensors. *ACS Appl. Mater. Interfaces, 9*, 34544–34586.
50. Wang, T., Huang, D., Yang, Z., Xu, S., He, G., Li, X., Hu, N., et al., (2016). A review on graphene-based gas/vapor sensors with unique properties and potential applications. *Nano-Micro Lett., 8*, 95–119.

51. Shin, D. H., Lee, J. S., Jun, J., An, J. H., Kim, S. G., Cho, K. H., & Jang, J., (2015). Flower-like palladium nanoclusters decorated graphene electrodes for ultrasensitive and flexible hydrogen gas sensing *Sci. Rep., 5*, 1–11.
52. Shafiei, M., Spizzirri, P. G., Arsat, R., Yu, J., Plessis, J., Dubin, S., Kaner, R. B., et al., (2010). *Platinum/Graphene Nanosheet/SiC Contacts and Their Application for Hydrogen Gas Sensing,* 13796–13801.
53. Kaniyoor, A., Jafri, R. I., Arockiadoss, T., & Ramaprabhu, S., (2009). *Nanostructured Pt Decorated Graphene and Multi-Walled Carbon Nanotube Based Room Temperature Hydrogen Gas Sensor*, 382–386.
54. Nair, K. G., Ramakrishnan, V., Unnathpadi, R., Karuppanan, K. K., & Pullithadathil, B., (2020). Unraveling hydrogen adsorption kinetics of bimetallic Au-Pt nano island-functionalized carbon nanofibers for room-temperature gas sensor applications. *J. Phys. Chem. C, 124*, 7144–7155.
55. Wang, J., Rathi, S., Singh, B., Lee, I., Joh, H. I., & Kim, G. H., (2015). Alternating current dielectrophoresis optimization of Pt-decorated graphene oxide nanostructures for proficient hydrogen gas sensor. *ACS Applied Materials & Interfaces*, *7*(25), 13768–13775.
56. Esfandiar, A., Ghasemi, S., Irajizad, A., Akhavan, O., & Gholami, M. R., (2012). The decoration of $TiO_2$ / reduced graphene oxide by Pd and Pt nanoparticles for hydrogen gas sensing. *Int. J. Hydrogen Energy, 37*, 15423–15432.
57. Peng, Y., Ye, J., & Zou, K., (2016). The hydrogen sensing properties of Pt – Pd / reduced graphene oxide based sensor under different. *RSC Advances*, 24880–24888.
58. Phan, D., & Chung, G., (2014). Reliability of hydrogen sensing based on bimetallic Ni–Pd/graphene composites. *International Journal of Hydrogen Energy*, *39*(35), 20294–20304.
59. Sharma, B., & Kim, J., (2018). MEMS based highly sensitive dual FET gas sensor using graphene decorated Pd-Ag alloy nanoparticles for $H_2$ detection. *Sci. Rep.,* 1–9.
60. Harley-trochimczyk, A., Chang, J., Zhou, Q., Dong, J., Pham, T., Worsley, M. A., Maboudian, R., Zettl, A., & Mickelson, W., (2015). Sensors and actuators B: Chemical catalytic hydrogen sensing using microheated platinum nanoparticle-loaded graphene aerogel. *Sensors Actuators B. Chem., 206*, 399–406.
61. Sripada, R., Bhagavathi, P. V., Baro, M., Nagappan, N. S. P., & Sundara, R., (2015). Platinum and platinum-iron alloy nanoparticles dispersed nitrogen-doped graphene as high performance room temperature hydrogen sensor. *Int. J. Hydrogen Energy, 40*, 10346–10353.
62. Phan, D. T., Uddin, A., S. M. I., & Chung, G. S., (2015). A large detectable-range, high-response and fast-response resistivity hydrogen sensor based on Pt/Pd core-shell hybrid with graphene. *Sensors Actuators, B Chem., 220*, 962–967.
63. Bhati, V. S., Ranwa, S., Rajamani, S., Kumari, K., Raliya, R., Biswas, P., & Kumar, M., (2018). *Improved Sensitivity with Low Limit of Detection of Hydrogen Gas Sensor Based on rGO Loaded Ni Doped ZnO Nanostructures*. doi: 10.1021/acsami.7b17877.
64. Ou, Y. J., Si, W. W., Yu, G., Tang, L. L., Zhang, J., & Dong, Q. Z., (2013). *Nanostructures of Pd – Ni Alloy Deposited on Carbon Fibers for Sensing Hydrogen, 569*, 130–135.

65. Johnson, B. J. L., Behnam, A., Pearton, S. J., & Ural, A., (2010). *Hydrogen Sensing Using Pd-Functionalized Multi-Layer Graphene Nanoribbon Networks, 32611*, 4877–4880.
66. Wang, Z., Wu, S., Wang, J., Yu, A., & Wei, G., (2019). Carbon nanofiber-based functional nanomaterials for sensor applications. *Nanomaterials*. doi: 10.3390/nano9071045.
67. Karthikeyan, K. K., & Biji, P., (2016). *A Novel Biphasic Approach for Direct Fabrication of Highly Porous, Flexible Conducting Carbon Nanofiber Mats from Polyacrylonitrile (PAN)/$NaHCO_3$ Nanocomposite* (Vol. 224). Elsevier Ltd.
68. Lee, J. S., Kwon, O. S., Shin, D. H., & Jang, J., (2013). $WO_3$ nanonodule-decorated hybrid carbon nanofibers for $NO_2$ gas sensor application. *J. Mater. Chem. A, 1*, 9099–9106.
69. Zhou, X., Liu, B., Chen, Y., Guo, L., & Wei, G., (2020). Carbon nanofiber-based three-dimensional nanomaterials for energy and environmental applications. *Mater. Adv., 1*, 2163–2181.
70. Ruiz-Cornejo, J. C., Sebastián, D., & Lázaro, M. J., (2020). Synthesis and applications of carbon nanofibers: A review. *Rev. Chem. Eng., 36*, 493–511.
71. Feng, L., Xie, N., & Zhong, J., (2014). Carbon nanofibers and their composites: A review of synthesizing, properties and applications. *Materials (Basel), 7*, 3919–3945.
72. Cheng, H. Y., Zhu, Y. A., Sui, Z. J., Zhou, X. G., & Chen, D., (2012). Modeling of fishbone-type carbon nanofibers with cone-helix structures. *Carbon N. Y., 50*, 4359–4372.
73. Zhang, L., Wang, X., Zhao, Y., Zhu, Z., & Fong, H., (2012). Electrospun carbon nano-felt surface-attached with Pd nanoparticles for hydrogen sensing application. *Mater. Lett., 68*, 133–136.
74. Wang, Z., Liu, S., Jiang, T., Xu, X., & Zhang, J., (2015). *RSC Advances Carbon Nano Fibers: A Novel Hierarchical Nanostructures Based Hydrogen Sensor*, 64582–64587.

## CHAPTER 8

# The Changes in Magnetic Properties of Nanostructured Carbon Compounds Destined for Energy Conversion

V. I. KODOLOV,[1] I. N. SHABANOVA,[2] N. S. TEREBOVA,[2] and V. V. KODOLOVA-CHUKHONTSEVA[3]

[1]*M.T. Kalashnikov Izhevsk State Technical University, Izhevsk, Russia*

[2]*Udmurt Federal Research Center, Ural Division of RAS, Izhevsk, Russia*

[3]*Great Peter St. Petersburg State Polytechnic University, St. Petersburg, Russia*

**Corresponding author. E-mail: vikodolov@mail.ru*

**ABSTRACT**

The energy storage and increasing energy are possible at the unpaired electron quantity growth on the carbon cover fibers formed at the creation of metal-carbon mesoparticles. In this case, the increasing of magnetic properties takes place. This fact is explained by the negative charges quants interactions (interference) as well as by the interactions of negative and positive charges (annihilation). Special attention is devoted to the interactions of metal (Copper and Nickel) carbon mesoparticles with such oxidizers as ammonium polyphosphate (APPh) and aluminum oxide. In these reactions, it is observed the reduction of Phosphorus and Aluminum, the growth of mesocomposite metal atomic magnetic moment, and the simultaneous increasing of spin quantity on the mesocomposite carbon

Carbon Composites: Composites with Nanotubes, Nanomaterials, and Graphene Oxide.
Eduardo A. Castro, Ann Rose Abraham, & A. K. Haghi (Eds.)

shell. The introduction of the "annihilation" notion explains the increasing of magnetic characteristics. For the mesoparticle (nanostructures) reactivity estimation the proposals are done. These proposals are concluded in the using of Fractal Theory ideas (Kolmogorov–Avrami equations) together with energetic analogs of reactivity theory.

## 8.1 INTRODUCTION

The creation of magnetic properties at the quantity of unpaired electron growth into nanostructured carbon compounds is perspective. The use of f- or d-metals for increasing of this mesoparticles stability is possible. The present chapter is based on the application of chemical mesoscopic idea to the processes of magnetic mesoparticles formation. The above ideas are considered on the examples of the metal-carbon mesoscopic composites modification reactions. The hypothesis about the possibility of annihilation at the interaction of positive and negative charges quants in red-ox processes is confirmed by the examples of processes of Copper and Nickel Carbon mesocomposites modification with an application such substances as ammonium polyphosphate (APPh), silica ($SiO_2$), aluminum oxide, iron oxide, nickel oxide and copper oxide (CuO) [1–3]. When the mesoparticles modification reactions with the using APPh, $SiO_2$, metal oxides are carried out, the redox processes are realized. In these cases, the modifiers reduction reactions take place. The structures of metal-carbon mesoscopic composites with active carbon shells are defined by means of complex of methods, including X-ray photoelectron spectroscopy (XPS), transition electron microscopy with high permission, electron microdiffraction, and also EPR spectroscopy.

## 8.2 RESULTS AND DISCUSSION

Realized investigations show that the correspondent reagents reduction accompanied by the magnetic properties growth and the quantity of unpaired electrons increasing on the carbon fibers covering mesoparticles takes place. In Tables 8.1 and 8.2, the examples of metal atomic magnetic moments changes (in Born magnetons) and quantities of unpaired electrons (in spin/g) for mesoparticles modified by APPh or silica after the mechanochemical modification processes proceeding are given.

**TABLE 8.1** The Values of Copper (Nickel) Atomic Magnetic Moments in the Interaction Products for Systems: Cu-C MC–APPh (or $SiO_2$) and Ni-C MC–APPh (or $SiO_2$)

| Systems Cu-C MC–Substances | $\mu_{cu}$ | Systems Ni-C MC–Substance | $\mu_{Ni}$ |
|---|---|---|---|
| Cu-C MC–Silica | 3.0 | Ni-C MC–Silica | 4.0 |
| Cu-C MC–APPh | 2.0 | Ni-C MC–APPh | 3.0 |
| Cu-C MC–APPh, relation 1:0.5 | 4.2 | | |

**TABLE 8.2** The Unpaired Electron Values (from EPR Spectra) for Systems "Cu-C MC–Silica" and "Cu-C MC–APPh" (Relation 1:1) in Comparison with Initial Mesoparticle Cu-C MC

| Substance | Quantity of Unpaired Electrons (spin/g) |
|---|---|
| Cu-C mesocomposite | $1.2 \times 10^{17}$ |
| System "Cu-C MC–$SiO_2$" | $3.4 \times 10^{19}$ |
| System "Cu-C MC–APPh" | $2.8 \times 10^{18}$ |

The metal atomic magnetic moment growth proceeds owing to the redox processes with the above chemical compounds. In Refs. [1, 2], it's shown that the reduction reactions of Phosphorus and Silicon from correspondent substances at the interaction on the interphase boundary with mesoparticles are realized (Figures 8.1 and 8.2).

The increase of the metal atomic magnetic moment during the mixing of the Cu-C mesocomposite with chemical compounds and the change of the number of delocalized electrons on carbon structures is connected with the interaction of a nanosized particle with the environment. The assumed mechanism of the increase of the metal atomic magnetic moment is the electron exchange at the phase boundary similar to heterogeneous catalysis in multi-cells. For a large particle, the relation of the surface to mass is so small that the metal surrounding cannot influence the electron density of its d-zone; in this case, the magnetic moment of a small particle can noticeably change. It is shown in Ref. [2] that as a result of mechanochemical mixing, in the obtained systems "Cu-C–APPh" and "Cu-C–silica," phosphorus or silicon of the compound are reduced; this fact confirms the interaction presence of mesocomposite with reagents on the phase boundary.

The XPS P2p spectra for phosphorus-containing mesocomposites prepared at the ratio of presents the reagents Cu-C: APPh = 1 and Cu–C: APPh = 2 shown in Figure 8.1. The phosphorus reduction at the ratio of 2 is almost 50% more complete [1, 2] and from the comparison Figure 8.1(a) and (b).

Such a result can be explained by the rate of electron quantization at the increase in the thickness of the APPh layer.

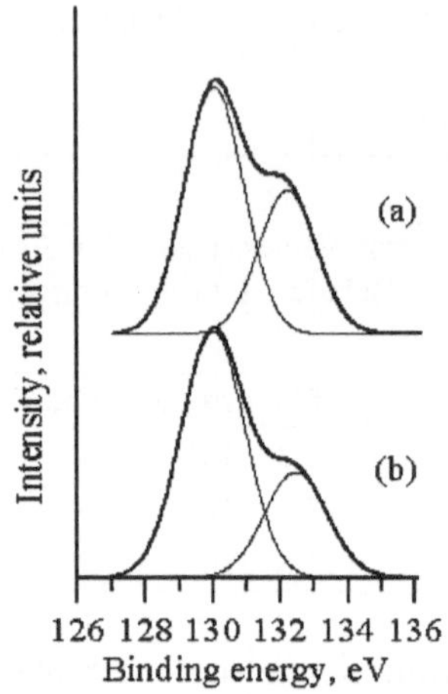


**FIGURE 8.1** The XPS P2p spectra: (a) Cu/C + APPh 1:1; (b) Cu/C + APPh 1:0.5.

In this case, the structure Phosphorus containing Copper Carbon mesocomposite is similar to the initial mesocomposite, only the increase of carbon fiber thickness is noted because of the possible disposition of phosphorus reduced between carbon atoms of carbon fibers.

In Figure 8.2, the comparison of TEM microphotographs for the initial Copper Carbon mesoscopic composite and the phosphorus-containing this mesocomposite after modification by APPh is given.

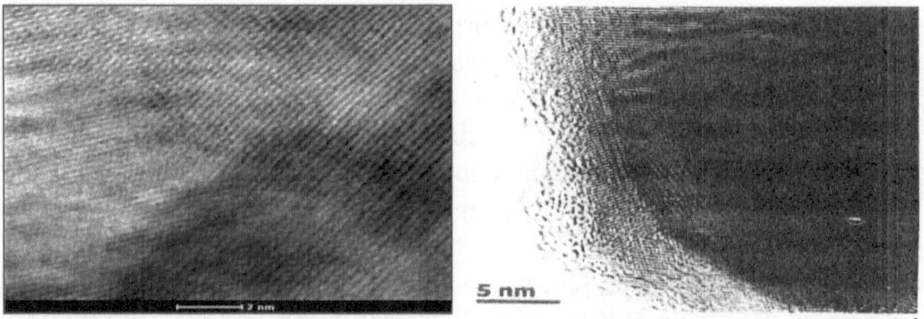


a b

**FIGURE 8.2** Comparison of TEM microphotographs for initial copper carbon mesocomposite (a); and phosphorus-containing Cu C mesocomposite (b).

The phosphorus disposition between carbon fibers can be explained by the interaction of Phosphorus electrons with the carbine double bonds electrons, which is confirmed by further studies on the reactions of meso-composite obtained with different reagents.

At the mechanochemical processes of APPh with copper carbon mesoscopic composite, the ammoniac molecules are chipped off and evaporated. Therefore, there is no nitrogen in the X-ray photoelectron spectrum of the product obtained. The presence of oxidized forms of Carbon and Phosphorus is determined by the results of XPS and electron sound micro roentgen spectroscopic analysis of element composition. The results obtained attest about the addition of acidic P-O-H groups, which are formed after the ammoniac evaporation, to the double bonds of carbon envelope with the formation C–H and P–O–C bonds. According to X-ray photoelectron spectra, the binding energy equaled to 133 eV corresponds to P-O bond, and the correspondent energy equaled to 286 eV – C–O bond. In all probability, the phosphorus is disposed between the carbon fibers of Cu–C mesoscopic carbon shell and is reduced because of the interaction with unpaired electrons, which are found on the mesocomposite carbon shell.

The reduction process may be presented by the following multi-stage scheme: the first step – the negative charges quants are directed to the positive charged nucleus of Phosphorus atom, near which there is the cloud of positive charge quants; the second step – the interaction of negative charged quants with positive charged quants, which leads to the annihilation phenomena creation; the third step – at the annihilation the inner electromagnetic field is formed which stimulates the Phosphorus reduction process; the fourth step – at the reduction process the water vapors are formed and are evaporated, that leads to C=P bond formation (132.6 eV on X-ray photoelectron spectra); the fifth step – the electromagnetic radiation formed stimulates across the coordination bonds d orbital of Copper atom that leads to the formation of unpaired electrons and the growth of Copper atomic magnetic moment; the sixth step – the fifth step explains the increasing of the spins number on the carbon cover of Copper cluster.

A similar explanation of the mechanism of the reduction-oxidation reaction is possible for the interaction of aluminum oxide with copper carbon mesoscopic composite. In this case, aluminum is reduced completely with the simultaneous growth of copper atomic magnetic moment from 1.3 to 4.8 Bohr magnetons. These changes corresponded with the increase of spin number on the Carbon shell from $10^{17}$ spin/g to $10^{19}$ spin/g [4].

Thus, it's possible to change the magnetic properties of mesoparticles and the number of unpaired electrons using the redox reactions of metal-carbon mesocomposites with different oxidizers. In other words, there is the possibility of the direct change of mesoparticles' magnetic-electric properties. This process certainly has an attitude toward energy conversion and also energy generation and storage.

## 8.3 CONCLUSION

In the present chapter, the creation of magnetic electric mesoparticles by the reduction-oxidation reactions realization is presented. The redox processes mechanism using the idea about charge quantization, interference, and annihilation is proposed.

## KEYWORDS

- **annihilation**
- **atomic magnetic moments**
- **charge quantization**
- **energy storage**
- **interference**
- **Kolmogorov–Avrami equations changed**
- **nanostructures reactivity**
- **redox reactions**

## REFERENCES

1. Kodolov, V. I., & Trineeva, V. V., (2012). Perspectives of idea development about nanosystems self organization in polymeric matrixes. In: *The Problems of Nanochemistry for the Creation of New Materials* (pp. 75–100). Torun, Poland: IEPMD.
2. Akhmetshina, L. F., Lebedeva, G. A., & Kodolov, V. I., (2012). Phosphorus-containing metal/carbon nanocomposites and their application for the modification

of intumescent fireproof coatings. *Journal of Characterization and Development of Novel Materials, 4*(4), 451–468.
3. Kodolov, V. I., Trineeva, V. V., Semakina, N. V., Yakovlev, G. I., Volkova, E. G., et al., (2008). *Patent 2337062 Russia Technique of Obtaining Carbon Nanostructures from Organic Compounds and Metal Containing Substances.*
4. Kodolov, V. I., Khokhriakov, N. V., Trineeva, V. V., & Blagodatskikh, I. I., (2008). Activity of nanostructures and its display in nanoreactors of polymeric matrixes and in active media. *Chem. Phys. & Mesoscopy, 10*(4), 448–460.

CHAPTER 9

# Factors Influencing Physical and Mechanical Properties of Polymer Composites Modified with Metal/Carbon Nanocomposites

A. M. LIPANOV,[1] V. I. KODOLOV,[2,3] M. YA. MEL'NIKOV,[4]
I. N. SHABANOVA,[2] N. S. TEREBOVA,[2] and
V. V. KODOLOVA-CHUKHONTSEVA[5]

[1]*Keldysh Institute of Applied Mathematics, Russian Academy of Sciences, Moscow, Russia*

[2]*Udmurt Federal Research Center, Ural Branch of the Russian Academy of Sciences, Izhevsk, Russia*

[3]*M.T. Kalashnikov Izhevsk State Technical University, Izhevsk, Russia*

[4]*M.V. Lomonosov Moscow State University, Moscow, Russia*

[5]*Great Peter St. Petersburg State Polytechnic University, St. Petersburg, Russia*

*Corresponding author. E-mail:*

## ABSTRACT

In the present chapter, the results of the investigation of metal/carbon nanocomposites (Met/C NC) functionalized with ammonium polyphosphate (APPh) and aerosil ($SiO_2$) are shown. The X-ray photoelectron spectroscopy (XPS) investigations have revealed the increase of the

Carbon Composites: Composites with Nanotubes, Nanomaterials, and Graphene Oxide.
Eduardo A. Castro, Ann Rose Abraham, & A. K. Haghi (Eds.)

atomic magnetic moment of the metal in the composition. A small increase in the amount of delocalized electrons on the carbon shell in a metal/carbon nanocomposite (NC) has been found by the electron paramagnetic resonance (EPR) method. Based on the above investigations, the factors influencing the physical and mechanical properties of polymer composites are determined. The change in the physical and mechanical characteristics is shown using the modification of the epoxide compound as an example.

## 9.1 INTRODUCTION

It is known that the inclusion of nanoparticles into the composition of a composite allows for a change in its physical and mechanical characteristics. It is shown [1–4] that metal/carbon nanocomposites (Met/C NC) are effective in improving the properties of the materials prepared with their use. It is important to clarify the reasons for the phenomenon and to reveal the factors influencing the material properties.

## 9.2 OBJECT OF INVESTIGATION

The combination of the method for preparing metal nanoparticles in polymers during the reduction of the metal-containing phase with the self-organization of the molecular structure of the polymer and the formation of various forms of carbon was used in fabricating Met/C NC [5]. The method does not require the presence of reducing agents; in the process of synthesis, a carbon shell is formed protecting metal nanoparticles against oxidation and aggregation.

Earlier, we have presented the investigation of metal/carbon NC with the use of various physicochemical methods [6]. It has been established that metal nanoparticles in the NC have mainly a spherical shape and are stabilized in carbon structures. The average size of the particles is 20–30 nm for copper/carbon (Cu/C) NC and 10–20 nm for nickel/carbon (Ni/C) NC. Figure 9.1 shows the micrographs of the Met/C NC.

The structure of the carbon shell was evaluated using X-ray photoelectron spectroscopy (XPS), high-resolution transmission electron microscopy (TEM) and electron diffraction (Figures 9.2 and 9.3).

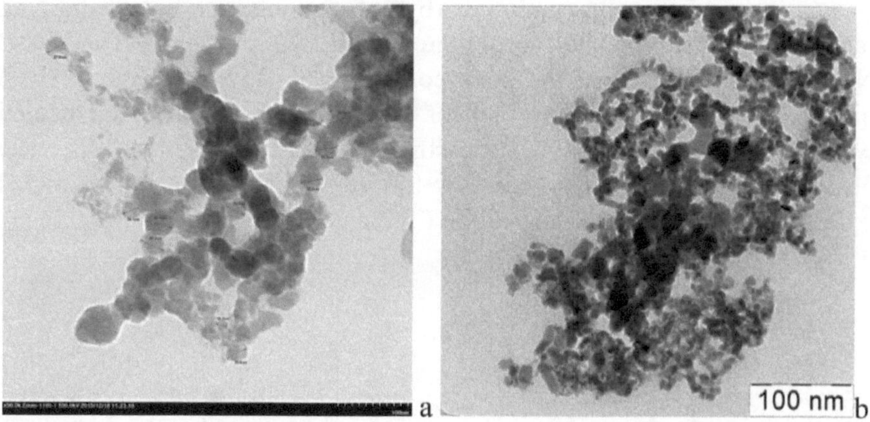


**FIGURE 9.1** Micrographs of the metal/carbon nanocomposites: (a) copper particles 10–40 nm in size in carbon structures; (b) nickel particles 5–25 nm in size in carbon structures.

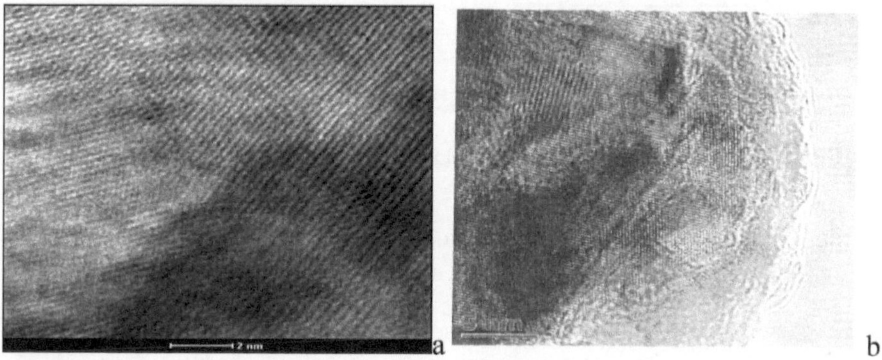


**FIGURE 9.2** TEM micrographs of the structures of the Cu/C (a); and Ni/C (b) nanocomposites.

As it follows from Figure 9.2, the structures of carbon shells on the nickel and copper clusters are close in shape – they are oriented fibers; this is confirmed by the electron diffraction data.

According to the XPS spectra (Figure 9.3), the carbon fibers can contain fragments of carbyne and polyacetylene; the peak at 285 eV is the C-H bond, and the component at 281–282 eV is ascribed to carbyne. It can be assumed that at the junction of the two fragments there are unpaired electrons. The co-ordinational interaction of metal with double bonds of corresponding fragments, mainly carbyne fragments, can be a stabilizing factor for such carbon shell. From the C1s spectra for the Cu/C and Ni/C NCs, it can be seen that the peak corresponding to C-H bonds is 30% higher for the Ni/C NC than that for the Cu/C NC. It can be connected with the decrease (shortening) of carbyne fragments in the Ni/C NC. The obtained result can be explained by the incomplete reduction of nickel oxide during the Ni/C NC formation (Table 9.1).

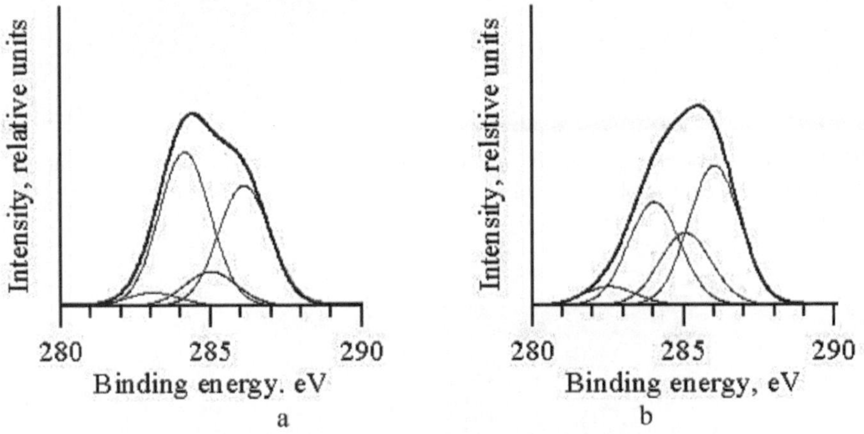


**FIGURE 9.3** C1s spectra for Cu/C (a); and Ni/C (b) nanocomposites.

Table 9.1 presents the ratio of metal-containing phases in the metal/carbon NC calculated with the use of X-ray diffraction (XRD) patterns [1].

**TABLE 9.1** The Content of Metal Containing Phases in the Metal Carbon Mesocomposites

| Phase | Cu/C Nanocomposite | Ni/C Nanocomposite |
|---|---|---|
| CuO | 1.17% | – |
| $Cu_2O$ | 5.19% | – |
| Cu | 93.64% | – |
| NiO | – | 32.15% |
| Ni | – | 67.85% |

From the above, it follows that the Ni/C NC has more junctions of the carbyne and polyacetylene fragments and a larger number of unpaired electrons, which is confirmed by the EPR spectra: for the Ni/C

NC – $2.46 \cdot 10^{23}$ spin/g, and for the Cu/C NC – $1.2 \cdot 10^{17}$. The presence of unpaired electrons on the fiber shell of the studied particles indicates the possibility of their activity in the reactions with electrophilic chemical substances.

## 9.3 EXPERIMENTAL PROCEDURE

The experimental samples were the compositions prepared by mechanochemical mixing of the Cu/C or Ni/C NC with ammonium polyphosphate (APPh) or aerosol in a triturating machine RetschRM200. The spectra were obtained on an X-ray electron magnetic spectrometer with a resolution of $10^{-4}$ and luminosity of 0.085% at the excitation by AlK$\alpha$ line of 1486.5 eV in vacuum $10^{-7}$–$10^{-8}$ Pa. Based on Van Fleck theory, a model was developed describing the relation between the multiplet splitting of the metal 3s spectra and the change of the number of unpaired d-electrons and the atomic magnetic moment [7]. The EPR spectra were registered on a radio spectrometer E-3 ("Varian") at room temperature (RT). The reference material for the g-factor determination was diphenyl picryl hydrazyl ($g\Phi = 2.0036$).

## 9.4 RESULTS AND DISCUSSION

The association of the metal and carbon phases in the metal/carbon NC leads to the change of the metal atomic magnetic moment determined with the use of XPS. The presence of $Cu^{2+}$ and $Ni^{2+}$ contributes to the change of the metal atomic magnetic moment in the metal/carbon NC. The magnetic moment value of $Cu^{2+}$ is ~1.73 $\mu_B$ and $Ni^{2+}$ ~2.3 $\mu_B$. However, the obtained values of the metal atomic magnetic moment in the metal/carbon NC cannot be explained by the presence of the metal oxide phases since there is an insufficient amount of the metal oxide.

Table 9.2 presents the calculation results for the metal atomic magnetic moment in the metal/carbon NC in comparison with that in a metal sample.

**TABLE 9.2** Atomic Magnetic Moments of the Metals in the Metal/Carbon Nanocomposites

| Metal/Carbon Nanocomposite/Metal Sample | $Cu_n/Cu_m$ | $Ni_n/Ni_m$ |
|---|---|---|
| Atomic Magnetic Moment, $\mu_B$ | 1.3/0 | 1.8/0.5 |

Using the XPS method [7], it has been established that the mechanochemical mixing of the metal/carbon NC with chemical substances results in the increase of the metal atomic magnetic moment in the metal/carbon NC.

Table 9.3 shows examples of the change of the metal atomic magnetic moment in the mixture after mechanochemical mixing.

**TABLE 9.3** The Values of the Copper and Nickel Atomic Magnetic Moments in the System "Metal/Carbon Nanocomposite–Chemical Compound" at the ratio of 1:1

| System "Cu/C–Chemical Compound" | $\mu_{Cu}$ | System "Ni/C–Chemical Compound" | $\mu_{Ni}$ |
|---|---|---|---|
| Cu/C–aerosil | 3.0 | Ni/C–aerosil | 4.0 |
| Cu/C–ammonium polyphosphate, ratio of 1:1 | 2.0 | Ni/C–ammonium polyphosphate | 3.0 |
| Cu/C–ammonium polyphosphate, ratio of 1:0.5 | 4.2 | – | |

The EPR investigations of the particles of the metal/carbon NC and the system "metal/carbon NC–chemical compound" reveal the presence of delocalized electrons on the carbon shell. The number of electrons changes in the system "metal/carbon NC–chemical compound."

The investigation results for the Cu/C NC and the systems "Cu/C–aerosil" and "Cu/C–APPh" are shown in Table 9.4.

**TABLE 9.4** The EPR Investigation Results for the Cu/C Nanocomposite and the Systems "Cu/C–Aerosil" and "Cu/C–Ammonium Polyphosphate"

| Substance | g-Factor | Number of Unpaired Electrons (spin/g) |
|---|---|---|
| Cu/C nanocomposite | 2.0036 | $1.2 \times 10^{17}$ |
| System "Cu/C–aerosil" | 2.0036 | $3.4 \times 10^{19}$ |
| System "Cu/C–ammonium polyphosphate" | 2.0036 | $2.7 \times 10^{18}$ |

The increasing of magnetic characteristics of metal-carbon mesocomposites at the mechanochemical interactions of mesocomposites with reagents (oxidizers) because of reduction-oxidation (redox) reactions with simultaneous growth of unpaired electrons. In these cases, oxidizers, for example, ammonium polyphosphate (APPh) or aerosol, are reduced.

Figure 9.4 presents the XPS P2p spectra for phosphorus containing NCs prepared at the ratio of the reagents Cu/C:APPh = 1 and Cu/C:APPh = 2.

The phosphorous reduction at the ratio of 2 is almost 50% more complete [8]; such a result can be explained by the rate of the electron quantization at the increase in the thickness of the APPh layer (Figure 9.4).

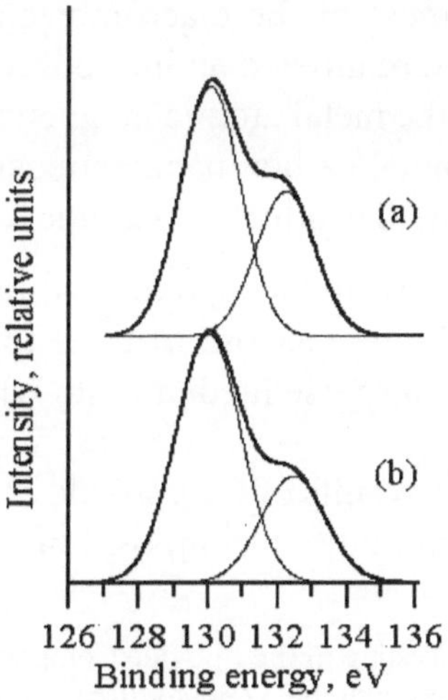


**FIGURE 9.4** The XPS P2p spectra: (a) Cu/C + APPh 1:1; (b) Cu/C + APPh 1:0.5.

When APPh is replaced by aerosil, the variation of the ratio from 1 to 2 does not lead to a considerable change in the metal atomic magnetic moment. The reduction of silicon from silicon oxide takes place by 50% independent of the used amount of the reagents (Figure 9.5); this can be explained by the decrease of the rate of electron quantization in the layer containing Si-O bonds.

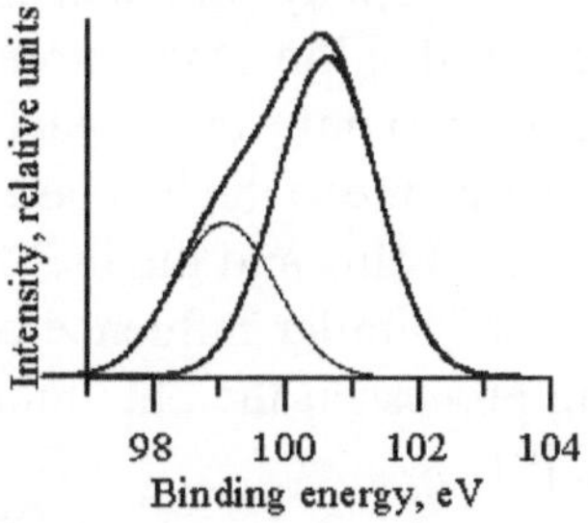


**FIGURE 9.5** The XPS Si2p spectrum for the sample Cu/C + $SiO_2$ at the ratio 1:1 (E(SiO) = 100.7 eV; E(Si) = 99 eV).

Based on the accumulated experimental data, we can conclude that the factors determining the physical and mechanical properties of polymer composite materials modified with a metal/carbon NC are as follows:

i. The presence of delocalized electrons on carbon structures in a metal/carbon nanocomposite facilitates the orientation interactions and alignment of the macromolecules or molecules of the disperse medium relative to an introduced particle;
ii. The increase of the metal atomic magnetic moment transforms the particle of the metal/carbon nanocomposite into a magnetic field source promoting the additional structurization of the medium.

Thus, the modification of the epoxide compound with the system "Cu/C–aerosil" results in the increase of the impact or adhesive strength values depending on the disperse medium into which the above system is introduced.

Table 9.5 presents the results of the epoxide compound modification in comparison with the non-modified composition.

**TABLE 9.5** The Results of Testing of the Epoxide Compound Modified with the System "Cu/C–Aerosil" in Comparison with the Non-Modified Sample

| SL. No. | Epoxide Compound EZK-11 | Impact Strength (kg·cm/cm2) | Adhesive Strength (MPa) |
|---|---|---|---|
| 1. | Reference composition | 4.03 | 7.80 |
| 2. | Modification of the hardener of amine type (C = 0.1%) | 8.33 | 9.10 |
| 3. | Modification ED-20 (C = 0.1%) | 6.90 | 15.08 |

When the modified metal/carbon NC is introduced into epoxy, the orientation of the polar segments of epoxy oligomer takes place and adhesive strength increases. At the introduction of the modified nanostructures into the hardener of the amine type, mainly, the impact strength of the obtained epoxides increases presumably due to the increase of the probability of the linear growth of the polymer chains and the increase of the average mass of the interatomic chains. The similar influence of the constant magnetic field during the hardening process at the obtaining of polymer composite material is shown in Ref. [9].

## 9.5 CONCLUSION

Thus, when the modified metal/carbon NC is introduced, the polarization and orientation processes and chemical transformations take place, providing the conditions for improving the physical and chemical properties of the material.

## KEYWORDS


- **adhesive strength**
- **electron paramagnetic resonance**
- **nanocomposites**
- **polymer**
- **transmission electron microscopy**
- **X-ray photoelectron spectroscopy**


## REFERENCES


1. Lipanov, A. M., Kodolov, V. I., Melnikov, M. Y., Pergushov, V. I., & Trineeva, V. V., (2016). Effect of small amounts of metal/carbon nanocomposites on the properties of polymer materials. *Doklady Chemistry, 466*(1), 15–17.
2. Trineeva, V. V., Pershin Yu, B., Bystrov, S. G., & Kodolov, V. I., (2015). The study of the effect of very small amounts of metal/carbon nanocomposite on the polycarbonate structure. *Chemical Physics and Mesoscopy, 17*(1), 126–131.
3. Karavayeva, N. M., Pershin Yu, V., Kodolov, V. I., Bystrov, S. G., & Deberdeev, T. R., (2018). Change of the morphology and swelling of hardened epoxy compositions at the modification with very small amounts of nickel/carbon nanocomposite // all materials. *Encyclopedia Reference Book, 11*, 8–11.
4. Kodolov, V. I., & Trineeva, V. V., (2018). Chemical mesoscopies as new trend in chemical sciences. *Polymer Sciences, 4*(2, 10), 1–7.
5. Trineeva, V. V., Kodolov, V. I., & Makhneva, T. M., (2013). The main methods for the preparation of metal/carbon nanocomposites with different morphology and composition. *Nanotekhnika, 3*(35), 3–8.
6. Trineeva, V. V., Vakhrushina, M. A., Bulatov, D. L., & Kodolov, V. I., (2012). Preparation of metal/carbon nanocomposites and investigation of their structural peculiarities. *Nanotekhnika, 4*(32), 18–20.

7. Shabanova, I. N., Terebova, N. S., Sapozhnikov, G. V., & Kodolov, V. I., (2017). X-ray photoelectron spectroscopy of the mechanism of surface functionalization of metal/carbon nanostructures with sp-elements. *Physics of the Solid State, 59*(1), 174–179.
8. Kodolov, V. I., Trineeva, V. V., Shabanova, I. N., & Terebova, N. S., (2018). On the possible mechanism of the redox synthesis of metal/carbon nanocomposites modified by p- and d-elements. *Chemical Physics and Mesoscopy, 20*(4), 606–614.
9. Cheremukhina, I. V., (2016). *Scientific and Technological Basis for Physical Modification of Reinforced Polymer Composite Materials for Engineering and Construction Application.* Dr. Diss., Yu. A. Gagarin Saratov State Technical University, Saratov.

# CHAPTER 10

# The Metal Carbon Mesocomposites as Base for the Synthesis of Magnetic Compounds by Redox Processes

V. I. KODOLOV,[1,2] V. V. KODOLOVA-CHUKHONTSEVA,[1,3]
I. N. SHABANOVA,[4] and N. S. TEREBOVA[4]

[1]*Basic Research – High Educational Center of Chemical Physics and Mesoscopics, Udmurt Federal Research Center, RAS, Izhevsk, Russia*

[2]*M.T. Kalashnikov Izhevsk State Technical University, Izhevsk, Russia*

[3]*St. Petersburg Institute of Macromolecular Compounds, Russian Academy of Sciences, St. Petersburg, Russia*

[4]*Udmurt Federal Research Center, Russian Academy of Sciences, Izhevsk, Russia*

**Corresponding author. E-mail:*

## ABSTRACT

The chapter is dedicated to the consideration of the metal-carbon mesocomposites application possibilities for the obtaining of magnetic compounds. This trend is determined by the correspondent peculiarities of content and structure of mesoscopic composites. The main peculiarities of these nano-sized particles are the following: (a) the presence of unpaired electrons on the carbon shell; (b) the presence of poly acetylene and carbine fragments in the carbon shell; (c) the growth of metal atomic magnetic moment at reduction-oxidation processes. The creation of reactive mesoscopic

Carbon Composites: Composites with Nanotubes, Nanomaterials, and Graphene Oxide.
Eduardo A. Castro, Ann Rose Abraham, & A. K. Haghi (Eds.)

materials with regulated magnetic characteristics, which can find different applications, is very topical. The present investigation has a fundamental character. It's based on the ideas concerning the change of metal-carbon mesocomposites reactivity. The use is possible as metal-carbon mesocomposites both, and they have been modified analogously.

## 10.1 INTRODUCTION

The manuscript is presented as the review of a series of papers, manuscripts, and patents on the obtaining, investigations, and applications of uncials magnetic mesoparticles, which are mesoscopic metal-carbon composites [1–16]. The correspondent structure and content of these mesocomposites are caused by the special conditions of their production. In as much of the application of mesoparticles is determined by the properties of these particles, which are defined by their structure and content. The activity of metal-carbon mesoscopic composites is caused by the structure and composition of corresponding mesocomposites, which contain the delocalized electrons and double bonds on the surface of the carbon shell that leads to the growth of metal atomic magnetic moments and the spin quantities on carbon shell. Therefore, the initial mesoscopic composites early participate in reactions, especially radical processes, and reduction-oxidation processes. This activity may be used in modification processes mesocomposites accompanied by the magnetic characteristics changes in modified mesoscopic composites. Such development of possibilities processes opens a new era for further investigations and expansion of metal-carbon mesoscopic composite application fields.

## 10.2 THE SHORT SURVEY OF INVESTIGATIONS ON THE MAGNETIC CHARACTERISTICS CHANGES FOR METAL CARBON MESOSCOPIC COMPOSITES

The investigation of Copper- and Nickel-Carbon mesocomposites (Cu-C and Ni-C MC) interactions with such reagents as ammonium polyphosphate (APPh), silica ($SiO_2$), aluminum oxide, iron oxide, nickel oxide, and copper oxide (CuO) are carried out. In correspondent reactions, the element reduction for reagents and Nickel or Copper atomic magnetic moments growth in mesoparticles take place. In Tables 10.1 and 10.2, the examples of metal atomic magnetic moment changes (in Boron magnetons) and quantities of unpaired electrons (in spin/g) for mesoparticles modified by APPh or silica are given.

**TABLE 10.1** The Values of Copper (Nickel) Atomic Magnetic Moments in the Interaction Products for Systems: Cu-C MC–APPh (or $SiO_2$) and Ni-C MC–APPh (or $SiO_2$)

| Systems Cu-C MC–Substances | $\mu_{cu}$ | Systems Ni-C MC–Substance | $\mu_{Ni}$ |
|---|---|---|---|
| Cu-C MC–Silica | 3.0 | Ni-C MC–Silica | 4.0 |
| Cu-C MC–APPh | 2.0 | Ni-C MC–APPh | 3.0 |
| Cu-C MC–APPh, relation 1:0.5 | 4.2 | | |

**TABLE 10.2** The Unpaired Electron Values (from EPR Spectra) for Systems "Cu-C MC–Silica" and "Cu-C MC–APPh" (Relation 1:1) in Comparison with Initial Mesoparticle Cu-C MC

| Substance | Quantity of Unpaired Electrons (spin/g) |
|---|---|
| Cu-C mesocomposite | $1.2 \times 10^{17}$ |
| System "Cu-C MC–$SiO_2$" | $3.4 \times 10^{19}$ |
| System "Cu-C MC–APPh" | $2.8 \times 10^{18}$ |

The experimental results, including results of XPS spectroscopy (Figures 10.1 and 10.2), confirm the explanation of mesocomposites magnetic characteristics growth with the use of interference and annihilation notions for quants radiated by metal-carbon mesocomposites.

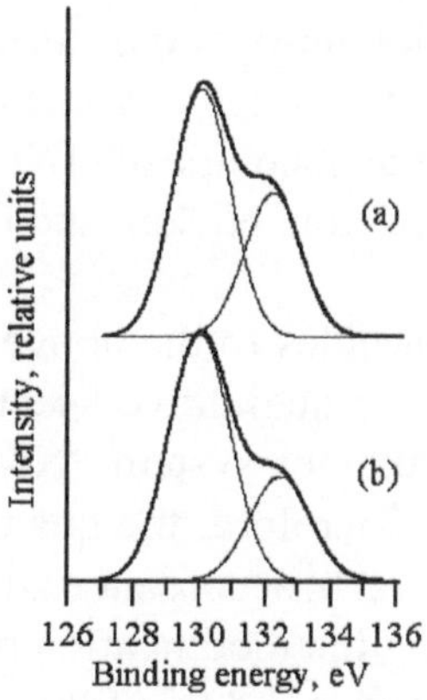


**FIGURE 10.1** The XPS P2p spectra: (a) Cu/C + APPh 1:1; (b) Cu/C + APPh 1:0.5.

In the case, when APPh is replaced by aerosol (or silicon), the reduction of silicon from silicon oxide takes place by 50% independent of the used amount of the reagents (Figure 10.2); this can be explained by the decrease of the rate of the electron quantization in the layer containing Si-O bonds.

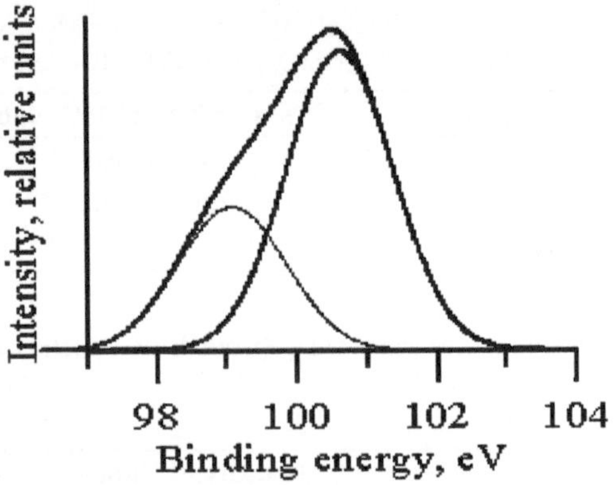


**FIGURE 10.2** The XPS Si2p spectrum for the sample Cu/C + $SiO_2$ at the ratio 1:1 (E(SiO) = 100.7 eV; E(Si) = 99 eV).

The early created method [3, 4] on the base of the X-ray photoelectron spectra is used for the Copper Carbon mesocomposite modification mechanism investigation. The inner levels spectra for C1s, O1s, Si2p, P2p, S2p Cu3s are studied to investigate the mechanism of chemical bond formation between metal or carbon atoms and silicon or phosphorus atoms. It's noted the presence of changes in the Cu3s spectra in mesocomposites modified in comparison with the initial mesocomposite.

Below, the results of investigations by means of X-ray electron spectroscopy for Nickel Carbon nanocomposite (NC) interactions with oxides of such metals as Aluminum, Iron, Nickel, and Copper are given in Tables 10.3 and 10.4.

The relations of mesoparticles to the above oxides are changed from 1:1 to 1:0.2 depending on the qualitative spectra obtained; for example, the relations of 1:1 and 1:0.5 for system "Ni-C MC–$Al_2O_3$" leads to a full mask of mesoparticle. Therefore, the quantity of aluminum oxide is decreased to the relation 1: 0.2. In accordance with Al3s spectra, Aluminum is completely reduced during the modification process, and Nickel atomic magnetic moment is increased to 4.8 $\mu_B$ (Table 10.3).

**TABLE 10.3** Parameters of Multiple Splintering Me3s Spectra in System Ni–C + $Al_2O_3$

| Sample | $I_2/I_1$ (Ni) | $\Delta_{Ni}$, eV | $\mu_{Ni}$, $\mu_Б$ |
|---|---|---|---|
| Ni/C | 0.3 | 3.0 | 1.8 |
| Ni/C + $Al_2O_3$ = 1:0.2 | 0.9 | 1.0 | 4.8 |

*Note:* $I_2/I_1$: the relation of multiple splintering lines maximums intensities;

Δ: Energetic distance between the multiple splintering maximums in Me3s spectra.

In this case, the reduction process is related not only to aluminum oxide but also to nickel oxide from a metal cluster of mesoparticle (Ni-C MC). Therefore, the reduction processes are stipulated by the electron transport from the carbon shell of mesoparticle in direction to $Al^{3+}$ and $Ni^{2+}$ of atoms in correspondent oxides.

The next researches with using of X-ray photoelectron spectroscopy (XPS) are carried out for the following systems: Ni-C MC–$Fe_2O_3$ and Ni-C MC–CuO (Table 10.4).

**TABLE 10.4** Parameters of Multiple Splintering Me3s Spectra in Systems "Ni-C MC–$Fe_2O_3$" and "Ni-C MC–CuO"

| Sample | $I_2/I_1$ (Ni) | $\Delta_{Ni}$, eV | $\mu_{Ni}$ $\mu_Б$ | $(I_2/I_1)_{Fe}$ | $\Delta_{Fe}$, eV | $\mu_{Fe}$,$\mu_Б$ |
|---|---|---|---|---|---|---|
| Ni-C MC + ($Fe_2O_3$) | 0.5 | 2.4 | 2.5 | 0.6 | 2.0 | 3.0 |
| Ni-C MC + CuO | 0.5 | 3.0 | 2.5 | 0.4 | 3.0 | 2.0 |

*Note:* $I_2/I_1$: the relation of multiple splintering lines maximums intensities;

Δ: Energetic distance between the multiple splintering maximums in Me3s spectra.

It's necessary to note that the iron oxide is sufficiently easily transformed in $Fe_3O_4$ (magnetite). At the heating, the further reduction of iron with its implantation within the carbon shell is observed. When CuO is added to Ni-C mesocomposite in the relation 1:1 at the grinding, the relative intensity of multiple maximum for Ni3s and Cu3s spectra are correlated with the atomic magnetic moments values (Table 10.4).

Thus, when the mesoparticles modification reactions with using APPh, $SiO_2$, and metal oxides are carried out, the redox processes are realized. In these cases, the modifiers reduction reactions take place.

## 10.3 CONCLUSION

According to the above short survey of investigations with the application of XPS and EPR spectra, the reduction-oxidation processes with the participation of metal-carbon mesocomposites are accompanied by the growth metal atomic magnetic moments that are explained by the electron shift on high energetic levels. Electron shift takes place at the action electromagnetic radiation formed because of the annihilation phenomenon. The energy which is evolved at this can be used in chemical reactions, including connection to mesoparticle. On the base of studies results, the hypothesis of the possibility of two phenomena (interference and annihilation) in redox processes is proposed. The changes for metal atomic magnetic moments in dependence on the number of electrons that participated in the redox process are observed.

## KEYWORDS

- **ammonium polyphosphate**
- **atomic magnetic moments**
- **copper oxide**
- **mesocomposites**
- **mesoparticles**
- **photoelectron spectroscopy**

## REFERENCES

1. Mustakimov, R. V., Kodolov, V. I., Shabanova, I. N., & Terebova, N. S., (2017). Modification of copper carbon nanocomposites with the use of ammonium polyphosphate for the application as modifiers of epoxy resins. *Chemical Physics & Mesoscopics, 19*(1), 50–57.
2. Kodolov, V. I., Trineeva, V. V., Kopylova, A. A., et al., (2017). Mechanochemical modification of metal-carbon nanocomposites. *Chemical Physics & Mesoscopics, 19*(4), 569–580.
3. Shabanova, I. N., Kodolov, V. I., Terebova, N. S., & Trineeva, V. V., (2012). *X-Ray Electro Spectroscopy in Investigation of Metal/Carbon Nanosystems and Nanostructured Materials* (p. 252). Izhevsk–Moscow: Publ. "Udmurt University."
4. Shabanova, I. N., Terebova, N. S., Kodolov, V. I., et al., (2013). The investigation of metal or carbon nanocomposites electron structure by X-ray photoelectron spectroscopy. In: *Nanostructure, Nanosystems and Nanostructured Materials:*

*Theory, Production and Development* (pp. 177–230). Toronto, Canada – New Jersey, USA: Apple Academic Press – 559p.

5. Kodolov, V. I., & Trineeva, V. V., (2013). Fundamental definitions for domain of nanostructures and metal/carbon nanocomposites. In: *Nanostructure, Nanosystems and Nanostructured Materials: Theory, Production and Development* (pp. 1–42, 559). Toronto, Canada – New Jersey, USA: Apple Academic Press.
6. Kodolov, V. I., Akhmetshina, L. F., Chashkin, M. A., et al., (2013). *The Functionalization of Metal or Carbon Nanocomposites or the Introduction of Functional Groups in Metal/Carbon Nanocomposites* (pp. 147–175). Ibid.
7. Kodolov, V. I., & Trineeva, V. V., (2012). Perspectives of idea development about nanosystems self organization in polymeric matrixes. In: *The Problems of Nanochemistry for the Creation of New Materials* (pp. 75–100). Torun, Poland: IEPMD.
8. Akhmetshina, L. F., Lebedeva, G. A., & Kodolov, V. I., (2012). Phosphorus-containing metal/carbon nanocomposites and their application for the modification of intumescent fireproof coatings. *Journal of Characterization and Development of Novel Materials, 4*(4), 451–468.
9. Kodolov, V. I., Trineeva, V. V., Semakina, N. V., Yakovlev, G. I., Volkova, E. G., et al., (2008). *Patent 2337062 Russia Technique of Obtaining Carbon Nanostructures from Organic Compounds and Metal Containing Substances.*
10. Kodolov, V. I., Khokhriakov, N. V., Trineeva, V. V., & Blagodatskikh, I. I., (2008). Activity of nanostructures and its display in nanoreactors of polymeric matrixes and in active media. *Chem. Phys. & Mesoscopy, 10*(4), 448–460.
11. Nikolaeva, O. A., Kodolov, V. I., Zakharova, G. S., et al., (2004). *Method of Obtaining Carbon-Metal-Containing Nanostructures.* Patent of the RF 2225835, registered on 20.03.2004.
12. Kodolov, V. I., Trineeva, V. V., Pershin Yu, V., et al., (2020). *Method of Metal Carbon Nanocomposites Obtaining from Metal Oxides and Polyvinyl Alcohol.* Pat. RU 2018122 001.
13. Kodolov, V. I., Trineeva, V. V., Terebova, N. S., et al., (2018). The change of electron structure and magnetic characteristics of modified copper carbon nanocomposites. *Chemical Physics & Mesoscopics, 20*(1), 72–79.
14. Kodolov, V. I., et al., (2019). Modified magnetic metal-carbon mesoscopic composites with bio active substances. *Chemical Physics & Mesoscopics, 21*(3), 446–454.
15. Mustakimov, R. V., (2019). *The Method of Modification of Metal Carbon Nanostructures by Ammonium Poly Phosphate.* Pat. N 2694092.
16. Kodolov, V. I., & Kodolova-Chukhontseva, V. V., (2019). *Fundamentals of Chemical Mesoscopics* (p. 218). Monograph. - Izhevsk: Publisher – M.T. Kalashnikov Izhevsk State Technical University.

# CHAPTER 11

# How Mesoscopic Physics Explains the Redox Synthesis of Metal/Carbon Nanocomposites Within Polymeric Matrices Nanoreactors

V. I. KODOLOV[1,2*] and V. V. KODOLOVA-CHUKHONTSEVA[3]

[1]*BRHEC of Chemical Physics and Mesoscopics, Udmurt Federal Research Center, RAS, Izhevsk, Russia*

[2]*M.T. Kalashnikov Izhevsk State Technical University, Izhevsk, Russia*

[3]*Great Peter St. Petersburg State Polytechnic University, St. Petersburg, Russia*

**Corresponding author. E-mail: vikodolov@mail.ru*

## ABSTRACT

The use of mesoscopic physics' main principles for the explanation of metal/carbon nanocomposites (Met/C NC) formation within polymeric matrices nanoreactors is discussed. It's shown that the nanocomposites (NCs) redox synthesis is started owing to the transport of electrons from the nanoreactors walls with correspondent potentials across mesoscopic metal particles to opposite walls with other potentials. It's assumed that these electrons' transport direction promotes the metal reduction reaction and, simultaneously, the oxidation of hydrocarbon groups of polymeric matrices nanoreactor walls. The conditions of Met/C NC redox synthesis by the interaction of the metal-containing phase in polymer solutions or

Carbon Composites: Composites with Nanotubes, Nanomaterials, and Graphene Oxide.
Eduardo A. Castro, Ann Rose Abraham, & A. K. Haghi (Eds.)

by the intercalation of the metal-containing phase to the polymeric phase are proposed.

## 11.1 INTRODUCTION

With the development of nanostructure synthesis and also nanostructured systems, the necessity to change the theoretical and experimental investigations of processes with nanoparticle participation arises. The increased attention to the fundamental principles of fractal theory [1], mesoscopic physics [2, 3], and nanochemistry [4] in consideration of processes with the participation of nanosystems explains by a more detailed study of these processes' peculiarities. Between these scientific trends, the likeness is more than distinctions, and the differences lead to the full understanding of phenomena in nanosize fields.

Self-organization in a system and self-similarity (image) is the ground notions in mesoscopic physics as well as in fractal theory and nanotechnology.

Therefore, it is interesting in this chapter to consider the application of these disciplines for the synthesis of nanostructures, including metal/carbon nanocomposites (Met/C NC), and for the modification of polymeric compositions by nanostructures (Met/C NC) obtained.

Mesoscopic physics deals with mesoscopic particles, which have a linear size from 1,000 to 0.1 nm. This interval is proposed: because the length of the coherence wave equals 1,000 nm, and the electron wavelength in different materials changes from 100 to 0.1 nm. For these particles, the following particularities are known as interference, spectrum quantization, and charge quantization [2].

The conditions when these appearances take place correspond to particles with a size equal to less than 20 nm (approximately) and to a temperature near to 0.1 K [3]. At these conditions, the particles can only vibrate, and also electro transport across mesoscopic particles is possible. The application principles of mesoscopic physics are possible, if the mesoscopic particle (nanoparticle) will be established within a small volume space, for instance, nanoreactor, in which this nanoparticle is coordinated on active groups of nanoreactor walls in correspondence with realized conditions.

Thus, the nanoreactor can be presented as the nanosized hollow or the limited free volume in polymeric matrix. Chemical particles are directly self-organized in this hollow for the creation of transitional (mesoscopic) state before the nanoproduct formation. Also, the nanoreactor definition can be proposed as the specific porous nanostructure, in which the distance between walls changes from 1 nm to 100 nm. These sizes are proposed because the electron transport across mesoscopic particles from one to other walls (branches) takes place when it corresponds to the electron wavelengths in matrices.

## 11.2 THE FORMATION OF MESOREACTORS, FILLED BY METAL CONTAINING PHASE

In the presented chapter, the following metal salts as metal chlorides in which metals are Fe, Co, Ni, and Cu with similar geometrical forms of clusters (fractals). As functional polymers are used polymers with side functional groups –OH, –O(O)CCH$_3$, –Cl (polyvinyl alcohol: PVA, polyvinyl acetate: PVAc, polyvinyl chloride: PVC). The consideration of processes in detail are discussed on the example of the "PVA–Cu or Ni Chloride" water solutions interactions.

It is proposed that mesoscopic metal particles can be formed from metal ions within PVA nanoreactor (spiral macromolecule with HO functional groups) for the case of dilute solutions PVA and metal chlorides. In this case, the formation of complex salts and acids takes place.

Owing to that the HO functional groups are displaced within spiral macromolecules; the stability of spiral form and summer negative charge are guaranteed. The metal ions are involved inside spiral macromolecules of PVA, and complex anions form external shell with result acidic medium (pH = 1).

## 11.3 FUNDAMENTAL NOTIONS AND THEORETICAL BASIS FOR THE MESOSCOPIC PHYSICS APPLICATION TO THE NANOSTRUCTURES FORMATION WITHIN NANOREACTORS

For the beginning, let us give some definitions:

- *What is mesoscopic physics?*

  Mesoscopic Physics deals with mesoscopic particles which have linear size from 1,000 to 0.1 nm.

- *Why is indicated from 1,000 nm, and no 100 nm on analogy with the definition of nanotechnology?*

  In this case, the answer follows: because the length of the coherence wave equals to 1,000 nm, and the electron wavelength in different materials changes from 100 to 0.1 nm. For these particles, the following particularities are known as the interference, spectrum quantization and charge quantization [1].

The mesoscopic physics principles are used when the mesoparticles are found within a small volume, for instance, mesoreactor, in which corresponding mesoparticle is interacted with mesoreactor walls with the activated complex formation.

- *What is nanoreactor and what nanoreactors can be used?*

  The nanoreactor can be presented as the nanosized hollow or the limited free volume in polymeric matrix; chemical particles are directly self-organized in this hollow for the creation of transitional (mesoscopic) state before the nanoproduct formation.

  Also, the nanoreactor definition can be proposed as the specific porous nanostructure, in which the distance between walls changes from 1 nm to 100 nm.

- *Why these sizes are proposed?*

  According to mesoscopic physics basis, the electron transport across mesoscopic particles from one to other walls (branches) takes place when it corresponds to the electron waves lengths in matrices.

There are some examples which confirm these sizes. The first example is the UV polymerization of butadiene and its analogous in the thiocarbamide (thiourea) crystal hollows (nanoreactors, pores) at–85°C. The distance between walls in these hollows equals to approximately 1 nm (Figure 11.1) [6].

**FIGURE 11.1** The scheme of polybutadiene synthesis in nanoreactor (hollow) of thiourea crystals.

The second example is the interface polycondensation between the dichloranhydride of phthalic acid in benzene solution and the sodium ethylene glycolate in water solution [7].

The size of the layer, in which the interaction of active molecules proceeds, corresponds to the sum of active molecule radii.

In these cases, the transport of electrons (Case 1) or electrons and ions (Case 2) is observed. Both examples are classic cases of nanostructures synthesis in nanoreactors (in pores or interface layers).

Also, the double electric layer may be imputed to the nanoreactors for redox reactions at the electrolysis [8]. On a level with indicated nanoreactors definitions there are some examples of nanoreactors definitions which are given in chapter [8].

Now let us consider the nanoreactors as intermolecular layers or space between submolecular structures within polymeric matrices. The minimal distance between walls in these nanoreactors shall be corresponded to the sum of Van der Waals radii for atoms (groups) in walls.

Among these nanoreactors for us it is interesting the nanoreactors of polymeric matrices in which it is possible the formation of Met/C NC. This is a new scientific trend, which is actively developed.

Based on our observations, the formation of Met/C NC within nanoreactors of polymeric matrices could be related to the photography processes. First of all, while metal containing phase is unbalanced with the polymeric phase, then the nanoreactor within polymeric matrix is expected to be filled up by metal clusters of metal-containing phase. Meanwhile, it should be noted that the process is accompanied with cluster coordination on active centers of nanoreactor walls. This is in fact, the photography process. Afterward, during the process of xerogels formation, the development will circulate. And then during the heating process of obtained product, the fixation is expected.

Thus, at the metal/carbon nanocomposites process, the following procedures are expected: the coordination reactions, the redox reactions and certainly the self-organization processes.

At this stage, we focus on the nanostructure's formation. Therefore, the active chemical metal, which contains particle, is taken root in the free nanosized hollow of polymeric matrices. It coordinates with functional groups of nanoreactor walls and loses the ability to diffusion and rotation. Now, this muddled particle can be characterized by vibration motion, and to be characterized by electron motion, including the transport electrons across it. Nevertheless, while the electro-transport across our particle (cluster) will be carried out, the redox reactions will be stimulated and the metal reduction will happen. This process is obviously direct and will lead to self-organization which may be observed by the "Atomic Force Microscopy" or "AFM" method. For the further stimulation of the (metal/carbon) nano-composites formation process, the regular regime of unfinished product heating is required.

Accordingly, the redox synthesis of (metal/carbon) nano-composites can be explained by basic principles of "mesoscopic physics."

A detailed study on the redox synthesis of (Metal/Carbon) nanocomposites in nanoreactors of polymeric matrices is presented in three main stages [9, 10]:

1. Modeling of nanoreactors filled by metal-containing phase and quantum chemical modeling of processes within nanoreactors.
2. Experimental models on nano-rectors filled by metal-containing phase with using two methods:
   i. Mixture of salt solution with the solution of functional polymer (as an example; polyvinyl alcohol solution); and
   ii. The common degeneration of polymeric phase with metal-containing phase (metal oxides) in active medium (such as water).
3. The properly redox synthesis of (Metal/Carbon) nanocomposites in nanoreactors of polymeric matrices (obtained xerogels) at narrow temperature intervals.

As a matter of fact, the first and second stages are related to preparatory stages. Meanwhile, in the second stage, the functional groups in nanoreactor walls participate in coordination reactions with metal ions (please refer to method 2.1 above) or with the clusters of metal-containing phase (please refer to method 2.2 above 2.2).

The computational experiment was carried out with software products Gamess and HyperChem, with visualization. The definite result of the computational experiment is obtained stage by stage. Any transformations at the initial stages are taken into account at further stages. The prognostic possibilities of the computational experiment consist in defining the probability of the formation of nanostructures of definite shapes. The optimal dimensions and shape of the internal cavity of nanoreactors, the optimal correlation between metal-containing and polymeric components for obtaining the necessary nanoproducts are found with the help of quantum-chemical modeling [9].

If method 2.1 is used, for instance, the interaction of PVA and metal chloride solutions where metals are iron, cobalt, nickel, and copper, then the colored xerogels is formed as follows: iron – brown-red, nickel – pale-green, cobalt – blue, and copper – yellow-green. Consequently, PVA interacts with metal ions with the formation of complex compounds.

For the corresponding correlations "polymer–metal containing phase," the dimensions, shape, and energetic characteristics of nanoreactors are found with the help of coordination, the structure and relief of xerogel surface change. The comparison of phase contrast pictures on the corresponding films indicates a greater concentration of the extended polar structures in the films containing copper, in comparison with the films containing nickel and cobalt (Figure 11.2).

Benzene

C(O)Ar- film withdrawal

↓Cl______ Cl ___________ interface board → -[-O($CH_2$)$_2$O(O)CArC(O)-]$_n$-

Na polyethylene terephtalate

O $CH_2CH_2$ -

Water

**FIGURE 11.2** Pictures of phase contrast of PVA surfaces containing copper (a); nickel (b), and cobalt (c).

The processing of the pictures of phase contrast to reveal the regions of energy interaction of cantilever with the surface in comparison with the background produces practically similar results with optical transmission microscopy [11, 12].

Corresponding to data of AFM, the sizes of nanoreactors obtained from solutions of metal chlorides and the mixture of PVA with polyethylene polyamine (PEPA) [11] are determined (Table 11.1).

**TABLE 11.1** Sizes of Nanoreactors Found with the Help of Atomic Force Microscopy [11]

| Composition | Sizes of AFM Formations | | | | |
|---|---|---|---|---|---|
| | Length | Width | Height | Area | Density |
| PVA:PEPA: $CoCl_2$ = 2:1:1 | 400–800 | 150–400 | 30–40 | 60–350 | 5.5 |
| PVA:PEPA: $NiCl_2$ = 2:1:1 | 80–100 | 80–100 | 25–35 | 6–12 | 120 |
| PVA:PEPA: $CuCl_2$ = 2:1:1 | 80–100 | 80–100 | 20–30 | 6–20 | 20 |
| PVA:PEPA: $CoCl_2$ = 2:2:1 | 600–900 | 300–600 | 100–120 | 180–500 | 3.0 |
| PVA:PEPA: $NiCl_2$ = 2:2:1 | 40–80 | 40–60 | 10–30 | 2–4 | 350 |

The results of AFM investigations of xerogels films obtained from metal oxides and PVA [13, 14] are distinguished (Figure 11.3) in comparison with previous data, that testify to the difference in reactivity of metal chlorides and metal oxides.

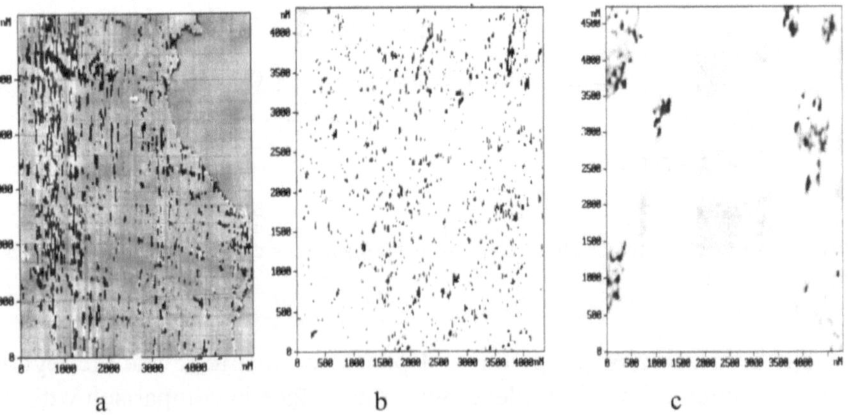

**FIGURE 11.3** Phase contrast pictures of xerogels films PVA–Ni (a); and PVA–Cu (b).

According to AFM results investigation, the addition of Ni/C NC in PVA leads to more strong coordination in comparison with the analogous addition Cu/C NC. The mechanism of formation of nanoreactors filled with metals was found with the help of IR spectroscopy.

Depending on metal coordinating ability and conditions for nanostructure obtaining (in liquid or solid medium with minimal content of liquid), we obtain "embryos" of future nanostructures of different shapes, dimensions, and composition. It is advisable to model coordination processes and further redox processes with the help of quantum chemistry apparatus.

The availability of d-metal in polymeric matrix results, in accordance with modeling results, in its regular distribution in the matrix and self-organization of the matrix.

At the same time, the metal orientation proceeds in interface regions and nanopores of polymeric phase, which stipulates further direction of the process to the formation of Met/C NC. In other words, the birth and growth of nanosized structures occur during the process in the same way as known from the macromolecule physics and fractal theory [16].

The mechanism of formation of nanoreactors filled with metals was found with the help of IR spectroscopy. Thus, at the second stage the coordination of metal-containing phase and corresponding orientation in nanoreactor take place.

To investigate the processes at the second stage of obtaining Met/C NC X-ray photoelectron spectroscopy (XPS), transmission electron microscopy (TEM) and IR spectroscopy are applied. The sample for IR spectroscopy was prepared when mixing metal/carbon NC powder with 1 drop of Vaseline oil in agate mortar to obtain a homogeneous paste with further investigation of the paste obtained on the appropriate instrument. As the Vaseline oil was applied when the spectra were taken, we can expect strong bands in the range 2,750–2,950 $cm^{-1}$.

At the third stage, it is required to give the corresponding energy impulse to transfer the "transition state" formed into carbon/metal NC of definite size and shape. To define the temperature ranges in which the structuring takes place, DTA-TG investigation is applied.

To define the temperature ranges in which the structuring takes place, DTA-TG investigation is applied. It is known that small changes of weight loss (TG curve) at invariable exothermal effect (DTA curve) testify to the self-organization (structural formation) in the system. Below typical curves of DTA-TG investigation lead.

According to data of DTA-TG investigation optimal temperature field for film nanostructure obtaining is 230–270°C, and for spatial nanostructure obtaining – 325–410°C.

It is found that in the temperature range under 200°C nanofilms, from carbon fibers associated with metal phase as well, are formed on metal or metal oxide clusters. When the temperature elevates up to 400°C, 3D nanostructures are formed with different shapes depending on the coordinating ability of the metal.

Two types of NCs (in Vaseline oil) were investigated by IR spectroscopy. In case of Cu/C NC a weak absorption is found at 720 $cm^{-1}$. In case of Ni/C NC, except for this absorption, the absorption at 620 $cm^{-1}$ is also observed.

In IR spectrum of copper/carbon NC two bands with a high relative area are found:

- at 1,323 $cm^{-1}$ (relative area – 9.28);
- at 1,406 $cm^{-1}$ (relative area – 25.18).

These bands can be referred to as skeleton oscillations of polyarylene rings. In IR spectrum of nickel/carbon NC the band mostly appears at 1,406 $cm^{-1}$ (relative area – 14.47).

According to the investigations with TEM, the formation of carbon nanofilm structures consisting of carbon threads is characteristic for copper/carbon NC. In contrast, carbon fiber structures, including nanotubes, are formed in nickel/carbon NC.

Assuming that the NCs obtained can be considered as oscillators transferring their oscillations onto the medium molecules, we can determine to what extent the IR spectrum of liquid medium will change. It is proposed to consider the obtaining of Met/C NC in nanoreactors of polymeric matrixes as self-organization process similar to the formation of ordered phases.

The perspectives of this investigation are looked through in an opportunity of thin regulation of processes and the entering of corrective amendments during processes.

This process can be considered as the recrystallization and can be described by the fractal theory with the application of Avrami-Kolmogorov equations. However, these equations were adopted to the conditions of Met/C NC redox synthesis. At the beginning let us discuss the possibilities of Avrami equations for the determination of the heating regime conditions at the redox synthesis of Met/C NC.

The difference of potentials between the interacting particles and object walls stimulating these interactions is the driving force of self-organization processes (formation of nanoparticles with definite shapes). The potential jump at the boundary "nanoreactor wall-reacting particles" is defined by the wall surface charge and reacting layer size. If we consider the redox process as the main process preceding the nanostructure formation, the work for charge transport corresponds to the energy of the nanoparticle formation process in the reacting layer. The equation of energy conservation for mesoreactor during the mesoparticle formation:

$$nF\Delta\phi = RT\ln\left(\frac{N_p}{N_r}\right) \tag{1}$$

where; $n$ is the number reflecting the charge of chemical particles moving inside "the mesoreactor;" $F$ is the Faraday number; $\Delta\varphi$ is the difference of potentials between the mesoreactor walls and flow of chemical particles; $R$ is the gas constant; $T$ is the process temperature; $N_{\mathrm{p}}$ is the mol share of nanoparticles obtained; $N_{\mathrm{r}}$ is the mol share of initial reagents from which the nanoparticles are obtained.

Using the above equation, we can determine the values of equilibrium constants when reaching the certain output of nanoparticles, sizes of nanoparticles and shapes of nanoparticles formed with the appropriate equation modification. The internal cavity sizes or nanoreactor reaction zone and its geometry significantly influence the sizes and shapes of nanostructures.

The sequence of the processes is conditioned by the composition and parameters (energy and geometry) of nanoreactors. To accomplish such processes, it is advisable to preliminarily select the polymeric matrix containing the nanoreactors in the form of nanopores or crazes as process appropriate. Such selection can be realized with the help of computer chemistry. Further, the computational experiment is carried out with the reagents placed in the nanoreactor with the corresponding geometry and energy parameters. Examples of such computations were given in Ref. [3]. The experimental confirmation of polymer matrix and nanoreactor selection to obtain carbon/metal-containing nanostructures was given in Refs. [4, 5]. Kolmogorov–Avrami equations are widely used for such processes which usually reflect the share of a new phase produced.

These equations are successfully applied [6] to forecast per molecular structures and prognosticate the conditions on the level of parameters resulting in the obtaining of nanostructures of definite size and shape. The equation was also used to forecast the formation of fibers [7]. The application of Avrami equations in the processes of nanostructure formation:

**a. Embryo formation and crystal growth in polymers [6]:**

$$1 - v = \exp\left[-k\tau^n\right] \quad (2)$$

where; υ is the crystallinity degree; τ is the duration; k is the value corresponding to the process specific rate; n is the number of the degrees of freedom changing from 1 to 6.

**b. Graphitization process with the formation of carbon nanostructures [8, 9]:**

$$v = 1 - \exp\left[-B\tau^n\right] \quad (3)$$

where; υ is the volume share that was changed; τ is the duration; B is the index connected with the process rate; n is the value determining the process directedness.

**c. Process of fiber formation [7]:**

$$\omega = 1 - \exp\left[-z\tau^n\right] \quad (4)$$

where; ω is the share of the fiber formed; τ is the process duration; z is the statistic sum connected with the process rate constant; n is the number of the degrees of freedom (for the fiber n equals 1). Instead of k, B, or z, based on the previous considerations, the activity of nanoreactors can be used (a).

At the same time, the metal orientation proceeds in interface regions and nanopores of polymeric phase which conditions further direction of the process to the formation of metal/carbon NC. In other words, the birth and growth of nanosize structures occur during the process in the same way as known from the macromolecule physics [6], in which Avrami equations are successfully used. The application of Kolmogorov–Avrami equations to the processes of nanostructure formation was previously discussed in the papers dedicated to the formation of ordered shapes of macromolecules [6], formation of carbon nanostructures by electric arc method [8], obtaining of fiber materials [7].

Then the share of the product being formed (W) in nanoreactor will be expressed by the following equation:

$$W = 1 - \exp\left(-a\tau^{n}\right) = 1 - \exp\left[-\left(\frac{\varepsilon_S}{\varepsilon_V}\right)\tau^{n}\right] = 1 - \exp\left\{-\left[\left(\frac{\varepsilon_S^0 d}{\varepsilon_V}\right)\frac{S}{V}\right]\tau^{n}\right\} \quad (5)$$

where; $a$ is the nanoreactor activity, $a = \varepsilon_S/\varepsilon_V$; $\varepsilon_S$ is the surface energy reflecting the energy of interaction of reagents with nanoreactor walls, $\varepsilon_S = \varepsilon^0{}_S dS$; $\varepsilon_V$ is the nanoreactor volume energy, $\varepsilon_V = \varepsilon^0{}_V V$; $\varepsilon^0{}_S d$ is the multiplication of surface layer energy by its thickness; $\varepsilon^0{}_V$ is the energy of nanoreactor volume unit; S is the surface of nanoreactor walls; V is the nanoreactor volume.

Then the equation for defining the share (W) of nanostructures formed can be written down as follows by analogy with the aforesaid equations:

$$1 - W = \exp\left[-\beta a\tau^{n}\right] \quad (6)$$

where; β is the coefficient taking into account the changes in the activity in the process of nanoproduct formation.

Substituting the value of $a$, we get the dependence of the share of nanostructures formed upon the ratio of the energy of nanoreactor internal surface to its volume energy:

$$W = 1 - \exp\left[-\beta\left(\frac{\varepsilon_S}{\varepsilon_V}\right)\tau^{n}\right] = 1 - \exp\left[-\beta\left(\frac{\varepsilon_S^0 S}{\varepsilon_V^0 V}\right)\tau^{n}\right] \quad (7)$$

If the nanoreactor internal walls become the shells of nanostructures during the process, the nanostructures so obtained are the nanoreactor mirror reflections.

The formation processes of metal-containing nanostructures in carbon or carbon-polymeric shells in nanoreactors can be related to one type of reaction series using the terminology of the theory of linear dependencies of free energies (LFE) [10]. Then it is useful to introduce definite critical values for the volume, surface energy of nanoreactor internal walls, as well as the temperature critical value. When the ration lg $k/k_c$ is proportional – $\Delta\Delta F/RT$, the ratio $W/W_c$ can be transformed into the following expression:

$$\frac{W}{W_c} = b\exp\left\{-\left(\frac{k}{k_c}\right)\left(\frac{\tau}{t_c}\right)^n\right\} = b\exp\left\{-\left(\frac{\tau}{t_c}\right)^n \exp\left(-\frac{\Delta\Delta F}{RT}\right)\right\}$$
$$= b\exp\left\{\left(\frac{\tau}{\tau_c}\right)^n \left[\exp k_T k_{VS}\left(\frac{\varepsilon_V}{\varepsilon_{Vk}} - \frac{\varepsilon_S}{\varepsilon_{Sk}}\right)\frac{Q}{T}\right]\right\} \tag{8}$$

where; b is the proportionality coefficient considering the temperature factor; $k_{VS}$ is the coefficient considering correlations $\varepsilon_V/\varepsilon_{Vk}$ and $\varepsilon_S/\varepsilon_{Sk}$; $\varepsilon_V$ and $\varepsilon_{Vk}$ is the volume energies of nanoreactor and "equilibrium" nanoreactor calculated via the ratios of their volumes; $\varepsilon_S$ and $\varepsilon_{Sk}$ is the surface energy and its equilibrium value; T and θ is the temperature of the process and temperature of the equilibrium process; τ is the time required to develop the process of nanostructure formation; n is the index of the process directedness to the formation of nanostructures of definite shapes. The values of volume and surface energies are given after the transformation of ΔΔF in accordance with Ref. [11], in which the physical sense of Taft constants is substantiated using the indicated energies.

When the metal ion moves inside the nanoreactor with redox interaction of ion (mol) with nanoreactor walls, the balance setting in the pair "metal containing-polymeric phase" can apparently be described with the following equation:

$$zF\Delta\varphi = RT\ln K = RT\ln\left(\frac{N_p}{N_r}\right) = RT\ln(1-W) \tag{9}$$

where; z is the number of electrons participating in the process; Δφ is the difference of potentials at the boundary "nanoreactor wall-reactive mixture"; F is the Faraday number; R is the universal gas constant; T is the process temperature; K is the process balance constant; $N_p$ is the number of moles of the product produced in nanoreactor; $N_r$ is the number of moles of reagents or atoms (ions) participating in the process which filled the nanoreactor; W is the share of nanoproduct obtained in nanoreactor.

In turn, the share of the transformed components participating in phase interaction can be expressed with the equation which can be considered as a modified Avrami equation:

$$W = 1 - \exp\left[-\tau^n \exp\left(\frac{zF\Delta\varphi}{RT}\right)\right] \tag{10}$$

where; $\tau$ is the duration of the process in nanoreactor; n is the number of degrees of freedom changing from 1 to 6. When "n" equals 1, one-dimensional nanostructures are obtained (linear nanostructures, nanofibers). If "n" equals 2 or changes from 1 to 2, flat nanostructures are formed (nanofilms, circles, petals, wide nanobands). If "n" changes from 2 to 3 and more, spatial nanostructures are formed as "n" also indicates the number of degrees of freedom. The selection of the corresponding equation recording form depends on the nanoreactor (nanostructure) shape and sizes and defines the nanostructure growth in the nanoreactor.

In case of nanostructure interaction with the medium molecules, the medium self-organization is effective with the proximity or concordance of the oscillations of separate fragments of nanostructures and chemical bonds of the medium molecules. At the same time, it is possible to use quantum-chemical computational experiment with the known software products.

According to chapter [12], the process vibration nature is discovered. This fact corroborates by IR spectroscopic investigations of nanostructures aqueous soles. Self-organization processes in media and compositions can be compared with the processes of crystalline phase origin and growth. At the same time, the growth can be one-, two-, and three-dimensional. For such processes, Avrami equations are widely used which usually reflect the share of the new phase appearing. In this case, the degree of nanostructure influence on active media and compositions is defined by the number of nanostructures, their activity in this composition and interaction duration. The temperature growth during the formation of new phases in self-organizing medium prevents the process development.

Different shapes of nanostructures appear during the formation of lamellar or linear substances. For instance, graphite, clay, mica, asbestos, many silicates have lamellar structure and, consequently, they contain nanostructures with the shape of rotation bodies. It can be explained by the fact that bands of lamellar structure are rolled into clews and spirals with further sewing between them and the formation of spheres, cylinders, ellipsoids, cones, etc. Nanostructures formed can be classified based on complexity. The formation of film structures resembling petals is possible for carbon-containing structures that are formed mainly from hexagons with a certain number of pentagons and an insignificant number of heptagons. Under certain energy actions, the distortion of coplanar (flat) aromatic rings is known, when $\pi$-electrons are shifted and charges are separated on the ring, the ring polarity goes up. For polymeric chains and bands formed from macromolecules the possibilities of formation of supermolecular structures, corresponding to the possibility of taking a definite shape, can be predicted with the help of Kolmogorov–Avrami's equation [6] for one-, two-, and three-dimensional crystallization of macromolecules. Here, energy exchange during the formation of ordered nanostructures and without energy exchange with the surroundings, i.e., under a thermal and thermal embryo-formation, is considered.

As mentioned before, at two-dimensional growth circles that form petals and further more complex nanostructures resembling flowers can be formed [13]. When being rolled, the circles formed can produce distorted semi-spheres or can serve as a basis for the transformation into "a beady."

During the redox process connected with the coordination process, the character of chemical bonds changes. Therefore, correlations of wave numbers of the changing chemical bonds can be applied as the characteristic of the nanostructure formation process in nanoreactor:

$$W = 1 - \exp\left\{-\tau^{n} \cdot \frac{\nu_{is}}{\nu_{fs}}\right\} \tag{11}$$

where; $\nu_{in}$ corresponds to wave numbers of the initial state of chemical bonds, and $\nu_f$ is the wave numbers of chemical bonds changing during the process. By the analogy with the above calculations, the parameters *a* in the Eqn. (5) should be considered as a value that reflects the transition from the initial to the final state of the system and represents the ratios of activities of system states.

The experimental modeling of obtaining nanofilms after the alignment of copper compounds with PVA at 200°C revealed that optimal duration when the share of nanofilms approaches 100% equals 2.5 hours. This corresponds to the calculated value based on the aforesaid Avrami equation. The calculations are made supposing the formation of copper nanocrystals on the nanofilms. It is pointed out that copper ions are predominantly reduced to metal. Therefore, it was accepted for the calculations that *n* equals 2 (two-dimensional growth), potential of redox process during the ion reduction to metal ($\Delta\phi$) equals 0.34 V, temperature ($T$) equals 473

K, Faraday number ($F$) corresponds to 26.81 (A·hour/mol), gas constant R equals 2.31 (W·hour/mol·degree). The analysis of the dimensionality shows the zero dimension of the ratio $-\frac{zF\Delta\varphi}{RT}$. The calculations are made when changing the process duration with a half-hour increment. Results of calculations on the Eqn. (10) are brought in the table.

If nanofilms are scrolled together with copper nanowires, β is taken as equaled to 3, the temperature increases up to 400°C, the optimal time when the transformation degree reaches 99.97%, corresponds to the duration of 2 hours, thus also coinciding with the experiment. According to the calculation results, if following the definite conditions of the system exposure, the duration of the exposure has the greatest influence on the value of nanostructure share.

The selection of the corresponding equation form depends upon the shape and sizes of the nanoreactor (nanostructure) and defines the nanostructure growth in nanoreactor or the influence distribution of the nanostructure on the structurally changing medium. At the same time $\nu_{is}$ corresponds to the frequency of skeleton oscillations of the C–C bond at 1,100 cm$^{-1}$, $\nu_{fs}$ is the symmetrical skeleton oscillations of the C=C bond at 1,050 cm$^{-1}$. In this case, the equation looks as Eqn. (11).

For the example discussed the content of nanofilms in % will be changing together with the changes in the duration as follows in table.

Under the aforesaid conditions, the linear sizes of copper (from ion radius to atom radius) and carbon-carbon bond (from C–C to C=C) are changing during the process. Apparently, the structure of copper ion and electron interacts with electrons of the corresponding bonds forming the layer with linear sizes $r_i + l_{C-C}$ in the initial condition and the layer with the size $r_a + l_{C=C}$ in the final condition. Then the equation for the content of nanofilms can be written down as follows:

$$W = 1 - \exp\left\{-\tau^n \cdot \frac{r_a + l_{C=C}}{r_i + l_{C-C}}\right\} \tag{12}$$

At the same time, $r_i$ for $Cu^{2+}$ equals 0.082 nm, $r_a$ for four-coordinated copper atom corresponds to 0.113 nm, bond energy C–C equals 0.154 nm, and C=C bond – 0.142 nm. Representing the ratio of activities as the ratio of corresponding linear sizes and taking the value $n$ as equal to 2, at the same time changing τ in the same intervals as before, we get the following change in the transformation degree based on the process duration (Table 11.2).

**TABLE 11.2** The Changes of Nanofilms Content Calculated Dependent on the Duration of Redox Synthesis

| **Duration (Hours)** | | 0.5 | 1.0 | 1.5 | 2.0 | 2.5 |
|---|---|---|---|---|---|---|
| **Content of Nanofilms (%)** | Eqn. (10) | 22.5 | 63.8 | 89.4 | 98.3 | 99.8 |
| | Eqn. (12) | 23.0 | 64.9 | 90.5 | 98.5 | 99.9 |
| | Eqn. (13) | 23.7 | 66.0 | 91.2 | 98.7 | 99.9 |

Thus, with the help of Kolmogorov–Avrami equations or their modified analogs we can determine the optimal duration of the process to obtain the required result. It opens up the possibility of defining other parameters of the process and characteristics of nanostructures obtained (by shape and sizes).

Modified Kolmogorov Avrami equations were tested to prognosticate the duration of the processes of obtaining metal/carbon nanofilms in the system "Cu–PVA" at 200°C [2]. The calculated time (2.5 hours) correspond to the experimental duration of obtaining carbon nanofilms on copper clusters.

The methods of optical spectroscopy and XPS allow to determine the energy of the interaction of the chemical particles in the nanoreactors with the active centers of the nanoreactor walls, which stimulate reduction-oxidation processes.

Depending on the nature of the metal salt and the electrochemical potential of the metal, different metal reduction nanoproducts in the carbon shells differing in shape are formed. Based on this result, we may speak about a new scientific branch – nanometallurgy. The stages of nanostructure synthesis may be represented in Figure 11.4.

Otherwise, as follows from this scheme, the First step (stage) of metal/carbon NC formation is the growth of carbon fibers associated with metal clusters. The Second stage is the formation of Carbon film which covers metal clusters. And then this film transforms in the 3D nanostructures with metal-containing clusters inside.

The possible ways for obtaining metallic nanostructures in carbon shells have been determined. The investigation results allow us to speak about the possibility of the isolation of metallic and metal-containing nanoparticles in the carbon shells differing in shape and structure. However, there are

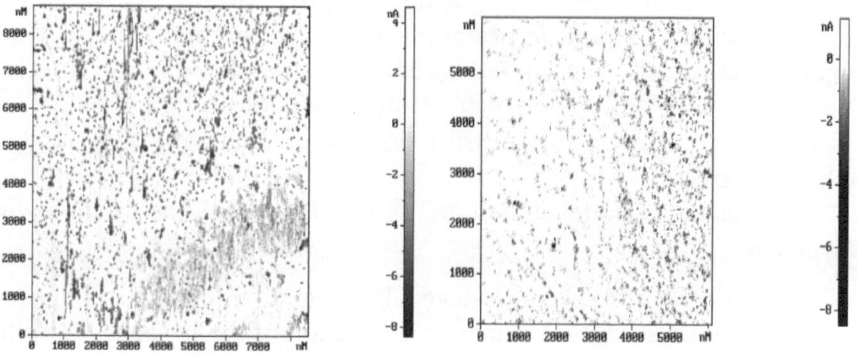

*a* *b*

**FIGURE 11.4** Scheme of copper/carbon nanostructures obtained from copper ions and polyvinyl alcohol.

still problems related to the calculation and experiment because using the existing investigation methods. It is difficult unambiguously to estimate the geometry and energy parameters of nanoreactors under the condition of 'erosion' of their walls during the formation of metallic nanostructures in them.

It is interesting that the scheme (Figure 11.4) and the succession of stages on fractal dimension correspond to experimental morphology of NCs obtained after corresponding heating stages (Figure 11.5).

It is noted that the different shapes of metal-containing nanostructures (mesoparticles) are formed in the dependence of the temperature growth (200°C → 400°C) and the metal activity (Figure 11.5).

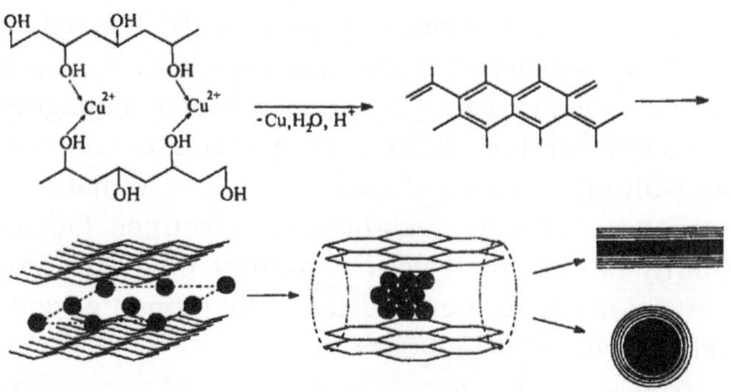


**FIGURE 11.5** Microphotographs obtained with the help of transmission electron microscopy. Cu/C nanocomposite.

Two types of NCs rather widely applied during the modification of various polymeric materials were investigated. These were: copper/carbon NC and nickel/carbon NC specified below. In turn, the nanopowders obtained were tested with the help of high-resolution TEM, electron microdiffraction, laser analyzer, XPS, and IR spectroscopy.

The method of metal/carbon NC synthesis applied has the following advantages:

- Originality of stage-by-stage obtaining of Metal/ Carbon nanocomposites with intermediary evaluation of the influence of initial mixture composition on their properties.
- Wide application of independent modern experimental and theoretical analysis methods to control the technological process.
- Technology developed allows synthesizing a wide range of metal/ carbon nanocomposites depending on the process conditions.
- Process does not require the use of inert or reduction atmospheres and specially prepared catalysts.
- Method of obtaining metal/carbon nanocomposites allows applying secondary raw materials.

So, the nanoreactor walls, containing the functional groups, have different potentials. When the mesoscopic particles (in our case, ions, or metal oxide clusters) are coordinated with functional groups of nanoreactor walls, their potentials change too. Now, according to mesoscopic physics principles, the conditions are created for electrons transport across mesoscopic particles (clusters) in which there are reducers (positively charged metal or metal ions). The redox processes are begun and the d-metal electron structure changes with the growth of unpaired electron number. d-Metal becomes paramagnetic. At the same time, the hydrocarbon shells, corresponding to nanoreactor walls, are oxidized with the aromatic cycles or hexagon formation in which $\pi$ electrons are interacted with metal unpaired electrons. Otherwise, there is the possibility of unpaired electrons formation on the carbon shell of metal/carbon NC obtained. Let us discuss the properties and especially energetic characteristics of Met/C NC.

## 11.4 METAL/CARBON NANOCOMPOSITES (MET/C NC) PROPERTIES, ESPECIALLY ENERGETIC CHARACTERISTICS

Met/C NC will be more active than carbon or silicon nanostructures because their masses are bigger at identical sizes and shapes. Therefore, the vibration energy transmitted to the medium is also high.

Metal/carbon nanocomposite (Me/C) represents metal nanoparticles stabilized in carbon nanofilm structures [12–14]. In turn, nanofilm structures are formed with carbon amorphous nanofibers associated with metal-containing phase. As a result of stabilization and association of metal nanoparticles with carbon phase, the metal chemically active particles are stable in the air and during heating as the strong complex of metal nanoparticles with carbon material matrix is formed. The test results of NCs obtained are given in Table 11.3.

**TABLE 11.3** Characteristics of Metal/Carbon Nanocomposites (Met/C NC)

| Type of Met/C NC | Cu/C | Ni/C | Co/C | Fe/C |
|---|---|---|---|---|
| Composition, metal/carbon [%] | 50/50 | 60/40 | 65/35 | 70/30 |
| Density [g/cm³] | 1.71 | 2.17 | 1.61 | 2.1 |
| Average dimension [nm] | 20(25) | 11 | 15 | 17 |
| Specific surface [m²/g] | 160 (average) | 251 | 209 | 168 |
| Metal nanoparticle shape | Close to spherical, there are dodecahedrons | There are spheres and rods | Nanocrystals | Close to spherical |
| Caron phase shape (shell) | Nanofibers associated with metal phase forming nanocoatings | Nanofilms scrolled in nanotubes | Nanofilms associated with nanocrystals of metal-containing phase | Nanofilms forming nanobeads with metal-containing phase |
| Atomic magnetic moment (metal) [8] [μB] | 0.0 | 0.6 | ~1.0 | 2.1 |
| Atomic magnetic moment (nanocomposite) [μB] | 0.6 | 1.6 | 1.7 | 2.3 |

For the corresponding correlations "polymer–metal containing phase" the dimensions, shape, and energy characteristics of nanoreactors are found with the help of AFM [9, 15]. Depending on a metal participating in coordination, the structure and relief of xerogel surface change.

To investigate the processes at the second stage of obtaining Met/C NC XPS, TEM, and IR spectroscopy are applied.

In turn, the nanopowders obtained were tested with the help of high-resolution TEM, electron microdiffraction, laser analyzer, XPS, and IR spectroscopy.

The distinctive feature of the considered technique for producing Met/C NC is a wide application of independent, modern, experimental, and theoretical analysis methods to substantiate the proposed technique and investigation of the composites obtained (quantum-chemical calculations, methods of TEM and electron diffraction, method of XPS, X-ray phase analysis, etc.). The technique developed allows synthesizing a wide range of Met/C NC by composition, size, and morphology depending on the process conditions. In its application it is possible to use secondary metallurgical and polymer raw materials. Thus, we can adjust the NC structure to extend the function of its application without pre-functionalization. Controlling the sizes and shapes of nanostructures by changing the metal-containing phase, we can, to some extent, apply completely new, practicable properties to the materials which sufficiently differ from conventional materials.

The essence of the method [11] consists in coordination interaction of functional groups of polymer and compounds of 3D-metals as a result of grinding of metal-containing and polymer phases. Further, the composition obtained undergoes thermolysis following the temperature mode set with the help of thermogravimetric and differential thermal analyzes. At the same time, we observe the polymer carbonization, partial or complete reduction of metal compounds and structuring of carbon material in the form of nanostructures with different shapes and sizes.

The metal-carbon mesocomposites are metal clusters stabilized by an envelope consisting of carbon nanofibers. These structures are stable in the air and during the heating, but show the activity at the interaction with oxidizers and radical particles.

Figure 11.6 demonstrates the microphotographs of TEM specific for different types of Met/C NC.

If the Met/C NC are considered as supermolecules, their surface energies analogously energy of usual molecules consists portions of energy which correspond to progressive, rotation, and vibration motions and also electronic motion:

$$\varepsilon_S^{NC} = \varepsilon_{prog} + \varepsilon_{rot} + \varepsilon_{vib} + \varepsilon_{elm} \tag{13}$$

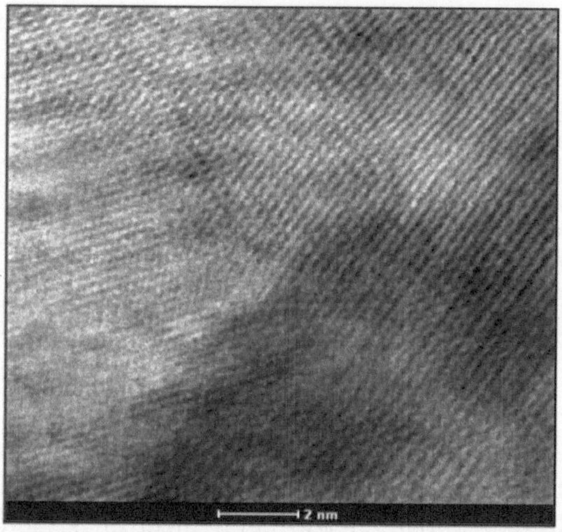


**FIGURE 11.6** Microphotographs of metal/carbon nanocomposites: (A) Cu/C; (B) Ni/C; (C) Co/C; and (D) Fe/C.

where; $\varepsilon_s^{NC}$ is the surface energy of NC; $\varepsilon_{prog}$ is the NC surface energy portion which corresponds to the progressive motion; $\varepsilon_{rot}$ is the NC surface energy portion which corresponds to the rotation motion; $\varepsilon_{vib}$ is the NC surface energy portion which corresponds to the vibration motion; $\varepsilon_{elm}$ is the NC surface energy portion which corresponds to electron motion for the interactions of NCs with surroundings molecules.

When the NC progressive motion portion increases the nanoparticles diffusion processes importance grows in media that leads to their coagulation with the decreasing of the surface energy aggregate obtained. The coagulation of nanoparticles increases at the increasing of their quantities in a medium which has little viscosity.

The decreasing of medium viscosity is arrived at the addition of nanoparticles with the great portion of surface energy which corresponds to the rotation motion. If the relatively small interaction of nanoparticles with medium molecules takes place, the NC rotation increasing leads to the local formation of free volumes as well as viscosity decreasing and coagulation of nanoparticles.

The decreasing of active nanoparticles quantities in media with sufficiently biggest viscosity and polarity, the energetic portions of progressive motion and rotation can be considerably decreased, and in the same time, the surface energy portion of vibration can be essentially increased.

From the Met/C NC Raman and IR spectra analysis it follows that the skeleton vibration of them on the vibrations frequencies corresponds to ultrasonic vibrations. The NC vibrations energy values are determined by the corresponding nanoparticles sizes and masses. The mass of nanoparticle (metal/carbon NC) depends on their content and types of metal-containing phase clusters.

Usually Met/C NC have the great dipole moment. Therefore, it is possible the proposition that NC is a vibrator which radiates electromagnetic waves. The NC vibration emission in the medium is determined by their dielectric characteristics and the corresponding functional groups presence in the medium.

At the Met/C NC obtaining the interaction of polymeric matrix with metal-containing phase leads to the formation of metal clusters covered by carbon shells accompanied by metal electron structure changes. In some cases, the medium characteristics influence on NCs throw into the increasing of NC surface energy portion, which concerns with changes of their electron structure and equally with electron structure of medium.

In these cases, the growth of metal atomic magnetic moment is observed that corresponds to the unpaired electron number increasing. The considerable changes of metal atomic magnetic moments in NCs proceed when the phosphorus atoms include to carbon shells of NCs (Table 11.4).

**TABLE 11.4** The Comparison of Atomic Magnetic Moments of Some 3D Metals in Metal/Carbon Nanocomposites and Their Functionalized Analogs [6, 13]

| Metal in Nanocomposite | Cu | Ni | Co | Fe |
|---|---|---|---|---|
| Metal atomic magnetic moment for nanocomposite | 0.6 | 1.6 | 1.7 | 2.3 |
| Metal atomic magnetic moment for phosphorus-containing nanocomposite | 2.0 | 3.0 | 2.5 | 2.5 |

The appearance or increase of paramagnetic properties is linked with the possibility of chemical bonds formation, and also the growth of Met/C NC reactivity into polar media.

Vibration portion of surface energy of Met/C NC can be presented as:

$$\varepsilon_{vib} = \frac{mu_{vib}^2}{2} \tag{14}$$

where; $m$ is the mass of metal/carbon NC, which includes summary mass of carbon or carbon, polymeric shell and cluster of metal-containing phase; $u_{vib}$ is the vibration velocity correspondent to the product of vibration amplitude on vibration frequency.

Usually, the fine dispersed suspensions or sols of nanostructures are used for the modification of polymeric compositions (matrixes) to distribute nanoparticles into polymeric matrix. The liquids, which are used for polymeric materials production, are represented as dispersion medium. For the obtaining of fine dispersed suspensions small and super small quantities of Met/C NC are introduced into dispersed media at the mechanical and further the ultrasonic mixing.

The NC vibration amplitude is the product of wave number on velocity of light. Therefore, the Eqn. (1) can be written:

$$\varepsilon_{vib} = \frac{(m_{ci} + m_{sh})dvc}{2} \tag{15}$$

where; $m_{cl}$ is the mass of metal-containing phase; $m_{sh}$ is the mass of carbon or carbon polymeric shell; d is the average linear size of metal/carbon NC nanoparticles; $v$ is the wave number of NC skeleton vibration; c is the velocity of light.

Under the influence of electromagnetic radiation arising at the ultrasonic vibration of NC the self-organization of polymeric compositions macromolecules is possible. According to the Eqn. (2), the NC vibration energy is determined by the average mass and size of NC (Table 11.5).

**TABLE 11.5** Energetic Characteristics of Metal/Carbon Nanocomposites [11–13]

| **Metal in Metal/Carbon Nanocomposites** | **Cu** | **Ni** | **Co** | **Fe** |
|---|---|---|---|---|
| Summary mass of metal/carbon nanocomposite (Au) | 36.75 | 40.01 | 42.50 | 42.69 |
| Specific surface of metal/carbon nanocomposite, S, $m^2/g$ | 160 | 251 | 209 | 163 |
| Skeleton vibration frequency of metal/ carbon nanocomposite, $v_{vib}$, $s^{-1}$ | $4\times10^{11}$ | $4.2\times10^{11}$ | $4.1\times10^{11}$ | $4\times10^{11}$ |
| Average vibration velocity of metal/ carbon nanocomposites, $u_{vib}$, cm/s | $8.5\times10^{5}$ | $4.6\times10^{5}$ | $6.1\times10^{5}$ | $6.9\times10^{5}$ |
| Average vibration energy of metal/carbon nanocomposites, $\varepsilon_{vib}$, erg | $1.6\times10^{13}$ | $4.3\times10^{12}$ | $7.6\times10^{12}$ | $9.9\times10^{12}$ |

The vibration motion portion energy essentially determines by the vibration amplitude because the NC skeleton vibration frequencies practically equal (the proximity of metal nature).

If the vibration amplitudes and the NC sizes are correlated, the NCs sizes show the great influence on the vibration energy and also on the vibration velocity. Therefore, according to the increasing of vibration energy as well as the increasing of Met/C NC linear sizes, the correspondent nanocomposites (M/C NC) are displaced in the following order:

$$\varepsilon_{vib}\left(\frac{Cu}{C}nc\right) > \varepsilon_{vib}\left(\frac{Fe}{C}nc\right) > \varepsilon_{vib}\left(\frac{Co}{C}nc\right) > \varepsilon_{vib}\left(\frac{Ni}{C}nc\right)$$

This disposition of NCs on the vibration energy values is possible, when the vibration motion portion is greatly bigger on the comparison with other portions of surface energy. However, the common surface energy of NCs obtained differs from previous row and corresponds to values of specific surface:

$$S_{sp}\left(\frac{Ni}{C}nc\right) > S_{sp}\left(\frac{Co}{C}nc\right) > S_{sp}\left(\frac{Fe}{C}nc\right) > S_{sp}\left(\frac{Cu}{C}nc\right)$$

Certainly, in dependence on methods and conditions of the Met/C NC synthesis the changes of forms and contents for NCs are possible. At the same time, the fundamentals of the metal/carbon NC activity open the new perspectives for their interaction prognosis in the different media.

The NCs described above were investigated with the help of IR spectroscopy by the technique indicated above. In this chapter, the IR spectra of Cu/C and Ni/C NCs are discussed (Figures 11.7 and 11.8), which find a wider application as the material modifiers.

On IR spectra of two NCs the common regions of IR radiation absorption are registered. Further, the bands appearing in the spectra and having the largest relative area were evaluated. We can see the difference in the intensity and number of absorption bands in the range 1,300–1,460 $cm^{-1}$, which confirms the different structures of composites.

In the range 600–800 $cm^{-1}$ the bands with a very weak intensity are seen, which can be referred to the oscillations of double bonds (π-electrons) coordinated with metals.

**FIGURE 11.7** IR spectra of copper/carbon nanocomposite powder.

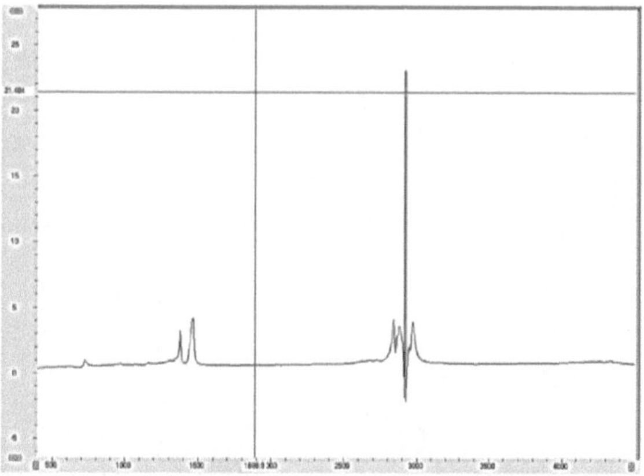

**FIGURE 11.8** IR spectra of nickel/carbon nanocomposite powder.

In case of Cu/C NC a weak absorption is found at 720 $cm^{-1}$, and in case of Ni/C NC the absorption at 620 $cm^{-1}$ is also observed.

In the accordance with investigations results realized by IR spectroscopy and transmission electron microscopy, the Copper and Nickel Carbon mesocomposites contain Carbon threads with unpaired electrons. There are several absorption bands in the range 2,800–3,050 $cm^{-1}$, which are attributed to valence oscillations of C-H bonds in aromatic and aliphatic compounds. These absorption bonds are connected with the presence of Vaseline oil in the sample. It is difficult to find the presence of metal in the composite as the metal is stabilized in carbon nanostructure. At the same time, it should be pointed out that apparently NCs influence the structure of Vaseline oil in different ways. The intensities and number of bands for Cu/C and Ni/C NCs are different:

- for copper/carbon nanocomposite in the indicated range – 5 bands, and total intensity corresponds by the relative area – 64.63;
- for nickel/carbon nanocomposite in the same range – 4 bands with total intensity (relative area) – 85.6.

The distribution of nanoparticles in water, alcohol, and water-alcohol suspensions prepared based on the above technique are determined with the help of laser analyzer. In Figures 11.9 and 11.10, you can see distributions of copper/carbon NC in the media with different polarity and dielectric penetration.

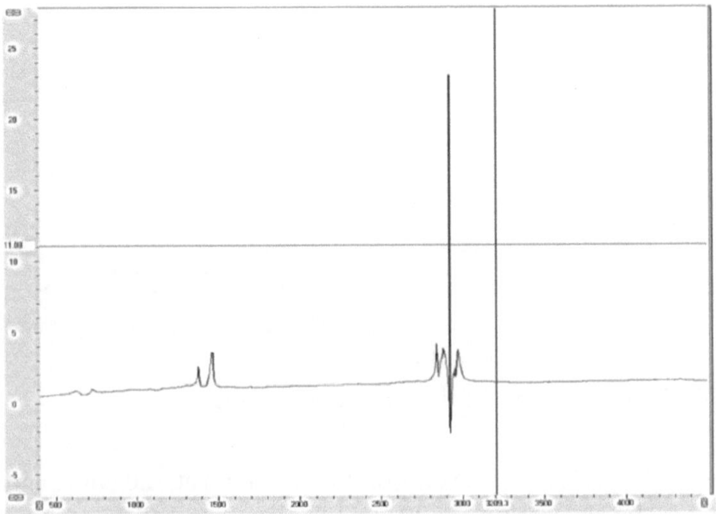

**FIGURE 11.9** Distribution of copper/carbon nanocomposites in alcohol.

When comparing the figures, we can see that ultrasound dispergation of one and the same NC in media different by polarity results in the changes of distribution of its particles. In water solution, the average size of Cu/C NC equals 20 nm, and in alcohol medium – greater by 5 nm.

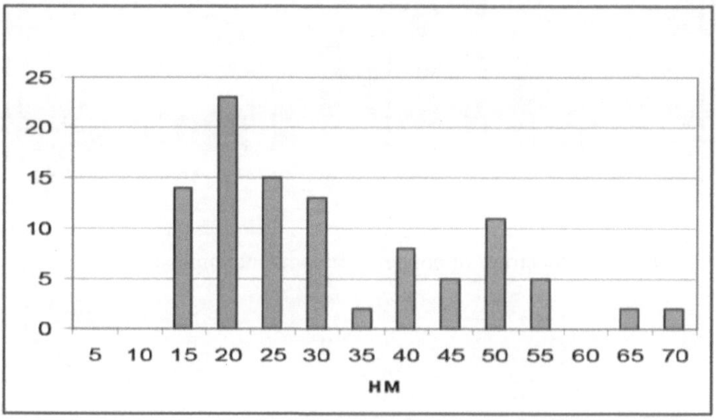


**FIGURE 11.10** Distribution of copper/carbon nanocomposites in water.

Assuming that the NCs obtained can be considered as oscillators transferring their oscillations onto the medium molecules, we can determine to what extent the IR spectrum of liquid medium will change, e.g., PEPA applied as a hardener in some polymeric compositions, when we introduce small and super small quantities of NC into it.

The introduction of a modifier based on Met/C NC into the composition results in medium structuring, a decrease in the number of defects, thus improving the material's physical and mechanical characteristics [5, 6]. The availability of metal compounds in NCs can provide the final material with additional characteristics, such as magnetic susceptibility and electric conductivity. Data of EPR spectra investigations for Copper/Carbon and Nickel/carbon NCs are given below.

EPR spectrum of Copper/Carbon NC is presented as a singlet spectrum in which the distance between points of incline maximum $\Delta H_{pp}$ = 6.8 Hz, g-factor equals to g-factor of diphenyl picryl hydrazyl (g = 2.0036), and the unpaired electrons number corresponds to value $1.2\times10^{17}$ spin/g (Figure 11.11). In the comparison with this spectrum, the EPR spectrum for Nickel/Carbon NC has $\Delta H_{pp}$ corresponding to 2,400 Hz, g – 2.46 and the unpaired electrons number – $10^{22}$ spin/g (Table 11.6).

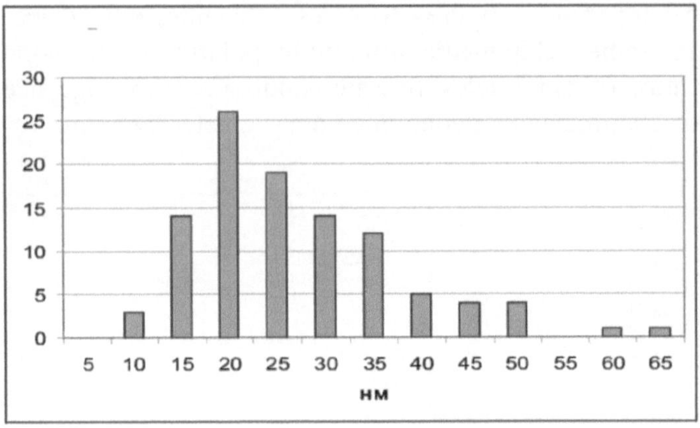


**FIGURE 11.11** EPR spectrum of copper/carbon nanocomposite.

**TABLE 11.6** Data of EPR Spectra for Copper/Carbon and Nickel/Carbon Nanocomposites

| Type of Metal/Carbon Nanocomposite | g-Factor | Number of Unpaired Electrons (spin/g) |
|---|---|---|
| Copper/carbon nanocomposite | 2.0036 | $1.2\times10^{17}$ spin/g |
| Nickel/carbon nanocomposite | 2.46 | $10^{22}$ spin/g |

It is possible, the spectra difference explains the Carbon shape difference for these NCs. Thus, the Met/C NC are stable radicals with the migration of unpaired electrons from metal to carbon shell and back. These nanoparticles can stimulate the transport of electrons within media.

## 11.5 CONCLUSION

In this chapter, the possibilities of developing new ideas on base of Mesoscopic Physics principles for self-organization processes during Redox synthesis within nanoreactors of polymeric matrices are discussed on the examples of Met/C NC. It is proposed to consider the obtaining of Met/C NC in nanoreactors of matrices as the self-organization process similar to the formation ordered phases analogous to crystallization. The perspectives of investigations are looked through in an opportunity of thin regulation of processes and the entering of corrective amendments during processes stages.

According to the analysis of Met/C NC characteristics, which are determined by their sizes and content, their activities are stipulated the correspondent dipole moments and vibration energies.

The introduction of phosphorus in NCs by method similar to redox synthesis in nanoreactors leads to the growth of dipole moments and metal magnetic moments for corresponded NCs. It's shown that the NCs vibration energies depend on their average masses.

It's noted that the specific surface of Met/C NC particles changes in dependence on the nature of NC in other order than the correspondent order of the vibration energies. Therefore, the energetic characteristics of NCs are more important for activity determination in comparison with their size characteristics.

## KEYWORDS


- **electrons transport**
- **mesoscopic physics**
- **metal-containing phase**
- **metal salts solutions**
- **metal/carbon nanocomposite**
- **nanoreactor walls**
- **polymer solutions**
- **polymeric phase**
- **redox synthesis**


## REFERENCES


1. Imri, I., (2009). *Introduction in Mesoscopic Physics* (p. 304). M.: Physmatlit.
2. Moskalets, M. V., (2010). *Fundamentals of Mesoscopic Physics*. Khar'kov: NTU KhPI.
3. Morozov, A. D., (2002). *Introduction in Fractal Theory* (p. 160). M.-Izhevsk: ICT.

4. Kolmogorov, A. N., & Fomin, S. V., (2009). *Introductory Real Analysis* (p. 403). USA, Portland: Prentice Hall.
5. Wunderlikh, B., (1979). *Physics of Macromolecules* (Vol. 2, p. 422). M.: Mir.
6. Brown, J. F. Jr., & White, D. M., (1960). *J. Am. Chem. Soc., 82*, 5671.
7. Morgan, P. W., (1970). Condensation polymers by interfacial and solution methods (translated) L. *Chemistry*, 448.
8. Buchachenko, A. L., (2003). Nanochemistry – direct way to high technologies – *Uspechi Chimii, 72*(5), 419–437.
9. Kodolov, V. I., & Khokhriakov, N. V., (2009). *Chemical Physics of Formation and Transformation Processes of Nanostructures and Nanosystems* (Vol. 1, 2, pp. 361, 415). Izhevsk: Publ. IzhSACA.
10. Shabanova, I. N., Kodolov, V. I., Terebova, N. S., & Trineeva, V. V., (2012). *X-Ray Photoelectron Spectroscopy Investigations of Metal/Carbon Nanosystems and Nanostructured Materials* (p. 252). M.: Izhevsk: Publ. "Udmurt university."
11. Kodolov, V. I., Khokhriakov, N. V., Trineeva, V. V., & Blagodatskikh, I. I., (2008). Nanostructure activity and its display in nanoreactors of polymeric matrices and in active media. *Chemical Physics and Mesoscopy, 10*(4), 448–460.
12. Kodolov, V. I., (2009). The addition to previous paper. *Chemical Physics and Mesoscopy, 11*(1), 134–136.
13. Trineeva, V. V., Vakhrushina, M. A., Bulatov, D. I., & Kodolov, V. I., (2012). The obtaining of metal/carbon nanocomposites and investigation of their structure phenomena. *Nanotechnics, 4*, 50–55.
14. Trineeva, V. V., Lyakkhovich, A. M., & Kodolov, V. I., (2009). Forecasting of the formation processes of carbon metal-containing nanostructures using the method of atomic force microscopy. *Nanotechnics, 4*(20), 87–90.
15. Kodolov, V. I., Blagodatskikh, I. I., Lyakhovich, A. M., et al., (2007). Investigation of the formation processes of metal-containing carbon nanostructures in nanoreactors of polyvinyl alcohol at early stages. *Chemical Physics and Mesoscopy, 9*(4), 422–429.
16. Kodolov, V. I., Khokhriakov, N. V., & Kuznetsov, A. P., (2006). To the issue of the mechanism of the influence of nanostructures on structurally changing media at the formation of "intellectual" composites. *Nanotechnics, 3*(7), 27–35.
17. Kodolov, V. I., Khokhriakov, N. V., Trineeva, V. V., & Blagodatskikh, I. I., (2010). Problems of nanostructure activity estimation, nanostructures directed production and application. *Nanomaterials Yearbook – 2009: From Nanostructures, Nanomaterials and Nanotechnologies to Nanoindustry* (pp. 1–18). N.Y.: Nova Science Publishers, Inc.
18. Fedorov, V. B., Khakimova, D. K., Shipkov, N. N., & Avdeenko, M. A., (1974). To thermodynamics of carbon materials. *Doklady AS USSR, 219*(3), 596–599.
19. Fedorov, V. B., Khakimova, D. K., Shorshorov, M. H., et al., (1975). To kinetics of graphitization *Doklady AS USSR, 222*(2), 399–402.
20. Khokhriakov, N. V., & Kodolov, V. I., (2005). Quantum-chemical modeling of nanostructure formation. *Nanotechnics, 2*, 108–112.
21. Lipanov, A. M., Kodolov, V. I., Khokhriakov, N. V., et al., (2005). Challenges in creating nanoreactors for the synthesis of metal nanoparticles in carbon shells. *Alternative Energy and Ecology, 2*(22), 58–63.

22. Kodolov, V. I., Didik, A. A., Yu Volkov, A., & Volkova, E. G., (2004). Low-temperature synthesis of copper nanoparticles in carbon shells. HEIs' news. *Chemistry and Chemical Engineering, 47*(1), 27–30.
23. Serkov, A. T., (1975). *Theory of Chemical Fiber Formation* (p. 548). M.: Himiya.
24. Palm, V. A., (1967). *Basics of Quantitative Theory of Organic Reactions* (p. 356). L: Himiya.
25. Kodolov, V. I., (1965). On modeling possibility in organic chemistry. *Organic Reactivity* (Vol. 2, No.4. pp. 11–18). Tartu: TSU.
26. Chashkin, M. A., Kodolov, V. I., Zakharov, A. I., et al., (2011). Metal/carbon nanocomposites for epoxy compositions: Quantum-chemical investigations and experimental modeling. *Polymer Research Journal, 5*(1), 5–19.
27. Kodolov, V. I., & Trineeva, V. V., (2013). Fundamental definitions for domain of nanostructures and metal/carbon nanocomposites. In: *Nanostructure, Nanosystems and Nanostructured Materials: Theory, Production and Development* (pp. 1–42). Toronto-New Jersey: Apple Academic Press.
28. Kodolov, V. I., Trineeva, V. V., Blagodatskikh, I. I., Yu Vasil'chenko, M., Vakhrushina, M. A., & Yu Bondar, A., (2013). The nanostructures obtaining and the synthesis of metal/carbon nanocomposites in nanoreactors. In: *Nanostructure, Nanosystems and Nanostructured Materials: Theory, Production and Development* (pp. 101–145). Toronto-New Jersey: Apple Academic Press.
29. Shabanova, I. N., & Terebova, N. S., (2012). Dependence of the value of the atomic magnetic moment of d metals on the chemical structure of nanoforms. In: *The Problems of Nanochemistry for the Creation of New Materials* (pp. 123–131). Torun, Poland: IEPMD.

CHAPTER 12

# New Trends in Chemical Mesoscopics

V. V. KODOLOVA-CHUKHONTSEVA[1,2*] and R. V. MUSTAKIMOV[1]

[1]*Basic Research – High Educational Center of Chemical Physics and Mesoscopics, UFRC, RAS, Izhevsk, Russia*

[2]*Peter Great St. Petersburg Polytechnic University, St. Petersburg, Russia, E-mail: vera_kodolova@mail.ru*

**Corresponding author. E-mail: rostmust@mail.ru*

## ABSTRACT

In the present mini-review, the definition of the new scientific trend, "chemical mesoscopics," which is based on the ideas of mesoscopic physics (mesoscopics) and also the latest mesoscopic chemistry, is given. This new scientific trend includes quantum notions on mesoparticle reactivity in chemical reactions. In these cases, the negative charges quants interactions (interference) are considered as well as the interactions of negative and positive charges (annihilation). The schemes of the chemical bond formation at the reactions flowing into mesoscopic reactors are presented. The (Copper, Nickel–Carbon) mesocomposites interaction with such oxidizers as ammonium polyphosphate (APPh) or aluminum oxide leads to the metal atomic magnetic moments and the spin quantity increasing. There are proposals for the reactivity estimation at the modification of polymeric materials by the correspondent mesocomposites.

Carbon Composites: Composites with Nanotubes, Nanomaterials, and Graphene Oxide.
Eduardo A. Castro, Ann Rose Abraham, & A. K. Haghi (Eds.)

## 12.1 INTRODUCTION

Chemical mesoscopic, a new trend in chemical sciences, appeared from such scientific trends as synergetics (self-organization), fractal theory (self-similarity), theories of chemical kinetics, and catalysis [1–3].

All of these trends are used for the description of mesoscopic particles (or nanostructures) behavior in the different media and at the various conditions changes. Therefore, mesoscopic physics and later appeared mesoscopic chemistry can be presented as the basis of chemical mesoscopics. The above new trend is very near to chemical physics on the considered objects and also on the phenomena and particularities of the various reactions and processes at the changes of conditions of their realization.

However, the basic aim of this advanced trend has appeared in the investigation of nanostructures (or mesoparticles) reactivity in the various media and at different changed conditions.

## 12.2 CHEMICAL MESOSCOPICS IDEAS IN THE CHEMICAL REACTIONS OF MATERIAL OBTAINING

According to known notions, the reactions can be divided into reactions realized without the element's oxidation states changes and reduction-oxidation reactions in which the change of oxidation states for elements participated in the process take place.

The reactions which flow without of oxidation states changes for elements participating in the processes are concerned with the chemical bond's formation or its destruction.

The chemical bonds formation owing to the electron wave's interference phenomenon was proposed by K. Ruedenberg in his book "Physical Nature of Chemical Bond" [4].

According to Ruedenberg, the bond energy ($E_b$) can be expressed by three energetic components (Eqn. (1)):

$$E_b = E^0 + E'' + E'' \quad (1)$$

where; $E^0$ is the Coulomb component; E' is the interference energy component; E" is the energy component which is concerned with the energy of interaction between the different bonds and which is near to zero because of various directions of interactions. The last energy component cannot be taken into account on the above reason.

In the reactions which flow without the changes of elements oxidation states the hybridization phenomena take place at the negative charges' quants radiation (or electromagnetic radiation). The correspondent radiation can be single measured, two or three measured that can be compared with the hybridization types as sp, $sp^2$, and $sp^3$ and, consequently, with the molecule's forms. These phenomena influence on polarization of the correspondent substances and media. The hybridization phenomena and the changes of electron structure are investigated by the main method of chemical mesoscopics – X-ray photoelectron spectroscopy (XPS). This method is also used for the determination of substances nature as well as metal atomic magnetic moments.

In the practice, it's possible that the chemical reactions flow in the mesoscopic (nanosized) reactors or multi-plates which are described as the space between the same banks (maybe parts of reagent or catalyst molecules). Mechanism-based on the Chemical Mesoscopics notions for electron transport across mesoparticle (for example, metal cluster) from one to another bank (mesoscopic reactor walls) is proposed (Figure 12.1).

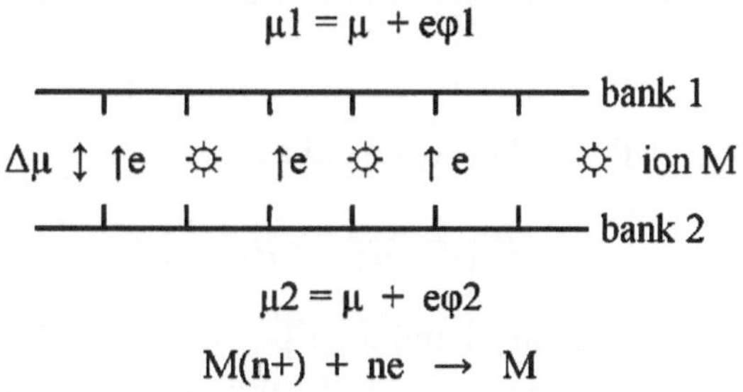


**FIGURE 12.1** Scheme of electron transport and the redox reactions within mesoscopic reactors [5].

At the reduction-oxidation reaction, the annihilation phenomenon is appeared. The process of annihilation creation is occurred by the following actions: the negative charge quants are directed to positive charged atom, near the nucleus where the positive charge quants are located, and interact with them. As a result, the annihilation with the electromagnetic direct field formation takes place. This field stimulates the electron shift with the activation of negative charge quants and then the interference of these quants (electron waves) with the new chemical bonds' formation. Thus, the reduction-oxidation processes can be explained by two phenomena: annihilation and interference following one after another.

The introduction of the correspondent mesoparticles in media or compositions leads to their polarization of them. Therefore, the modification conditions for the different materials can differ from one another.

According to the scheme of possible polarization, the increasing of medium (material) density owing to the regular orientation of material fragments with the creation of supermolecular and crystalline structures takes place (Figure 12.2).

MC δe δe δe δe

☼ → → → →

——♀——♀——♀——♀——♂——

**FIGURE 12.2** Scheme of polarization at charge quantization with expansion of quant influence on materials polar groups. [Designations: MC (☼): mesocomposite; δe: negative charges quant (electron); →: the polarization direction; ♀——♀: macromolecule fragment with functional groups].

Polarization growth can be expressed as:

$$P_{com} = \Sigma p_{fg} + p_{NC} \tag{2}$$

where; $P_{com}$ is the common (summary) polarization; $\Sigma p_{fg}$ is the sum of functional groups polarizations; $p_{NC}$ is the polarization (or dipole moment) of mesocomposite.

The polarization extent depends on the quant's electromagnetic radiation phase velocity. It's necessary to note that this velocity will be decreased in the media with high dielectric constant according to the following formula:

$$v = c / \sqrt{\varepsilon}, \tag{3}$$

where; v is the phase velocity of electromagnetic radiation; c is the light velocity; ε is the dielectric constant.

When the dielectric constant is increased, the decrease of mesocomposite influence on the media arises and the self-organization process is finished.

Depending on the development of self-organization process (single measured: 1D, double measured: 2D, third measured: 3D) the supermolecular structures (mesoparticles) of correspondent forms and sizes are organized. The surface energy of embryos increased influences on the mesoparticles formation.

This energy can be express as the sum of energetic components for the realization of different movements:

$$E_S = E_{(tr)} + E_{(r)} + E_{(osc)} + E_{(em)}, \tag{4}$$

where; $E_S$ is the surface energy of macromolecule (mesoparticle); $E_{(tr)}$ is the part of translational motion energy; $E_{(r)}$ is the part of rotary motion energy; $E_{(osc)}$ is the part of oscillatory motion energy; $E_{(em)}$ is the part of electron motion in surface layer.

In accordance with the formation of mesoparticles which have the identical orientation with each other, the parts of translational motion energy and of rotary motion energy will be near to zero [5, 6]. Therefore, the main contribution to the mesoparticle surface energy will be bring the oscillatory motion and transport of electrons in the surface layer of macromolecules (mesoparticles). Then the change of character of quants radiation wave propagation from 2D (in the surface plane) to 3D (in the space field at surface) takes place.

## 12.3 THE CONFIRMATIONS OF ABOVE NOTIONS ON THE EXAMPLES OF THE METAL CARBON MESOSCOPIC COMPOSITES MODIFICATION

The above ideas are considered on the examples of the metal-carbon mesoscopic composites modification reactions and also different polymeric materials. The hypothesis about possibility of annihilation at the interaction of positive and negative charges quants in redox processes is confirmed by the examples of processes of Copper and Nickel Carbon mesocomposites modification with application such substances as polyethylene polyamine (PEPA), ammonium iodide, ammonium polyphosphate (APPh), silica ($SiO_2$), aluminum oxide, iron oxide, nickel oxide and copper oxide (CuO) [7, 8]. In this case, when PEPA and ammonium iodide are applied, the connection reactions take place. At the interactions of PEPA with mesoparticles the C=N bond formation is explained by the interference of negative charges quants. At the modification of metal-carbon mesocomposites by oxidizers, the redox processes take place which are confirmed by the complex of methods including X-ray photoelectron spectroscopy, transition electron microscopy, electron microdiffraction and electron paramagnetic spectroscopy.

In Tables 12.1 and 12.2, the examples of metal atomic magnetic moments changes (in Boron magnetons) and quantities of unpaired electrons (in spin/g) for mesoparticles modified by APPh or silica after the mechanochemical modification processes proceeding are given.

**TABLE 12.1** The Values of Copper (Nickel) Atomic Magnetic Moments in the Interaction Products for Systems: Cu-C MC–APPh (or $SiO_2$) and Ni-C MC–APPh (or $SiO_2$)

| Systems Cu–C MC–Substances | $\mu_{cu}$ | Systems Ni–C MC–Substance | $\mu_{Ni}$ |
|---|---|---|---|
| Cu–C MC–Silica | 3.0 | Ni–C MC–Silica | 4.0 |
| Cu–C MC–APPh | 2.0 | Ni–C MC–APPh | 3.0 |
| Cu–C MC–APPh, Relation 1:0.5 | 4.2 | – | |

**TABLE 12.2** The Unpaired Electron Values (from EPR Spectra) for Systems "Cu–C MC–Silica" and "Cu–C MC–APPh" (Relation 1:1) in Comparison with Initial Mesoparticle Cu–C MC

| Substance | Quantity of Unpaired Electrons (spin/g) |
|---|---|
| Cu-C mesocomposite | $1.2 \times 10^{17}$ |
| System "Cu-C MC–$SiO_2$" | $3.4 \times 10^{19}$ |
| System "Cu-C MC–APPh" | $2.8 \times 10^{18}$ |

Owing to the use of methods investigation complex the explanation of the magnetic properties growth for the modified metal-carbon mesocomposites is presented. It's established that the phosphorus atom is displaced between the carbon fibers (threads) layers at the mechanochemical modification of Cu-C mesocomposite with the application of ammonium polyphosphate.

The XPS P2p spectra for phosphorus-containing mesocomposites prepared at the ratio of presents the reagents Cu-C: APPh = 1 and Cu-C: APPh = 2 shown in Figure 12.3. The phosphorus reduction at the ratio of 2 is almost 50% more complete [7, 8] and from the comparison from Figure 12.3(a) and (b).

Such result can be explained by the rate of the electron quantization because of the increasing the APPh layer thickness.

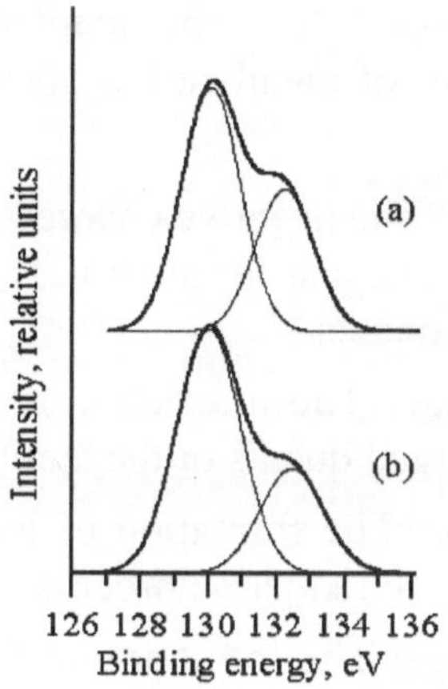


**FIGURE 12.3** The XPS P2p spectra: (a) Cu/C + APPh 1:1; (b) Cu/C + APPh 1:0.5 .

In this case, the structure phosphorus-containing copper carbon mesocomposite is similar to the initial mesocomposite, only the increasing of carbon fibers thickness is noted because of the possible disposition of phosphorus reduced between carbon atoms of carbon fibers.

In Figure 12.4, the comparison of TEM microphotographs for the initial copper carbon mesoscopic composite and the phosphorus-containing this mesocomposite after modification by APPh.

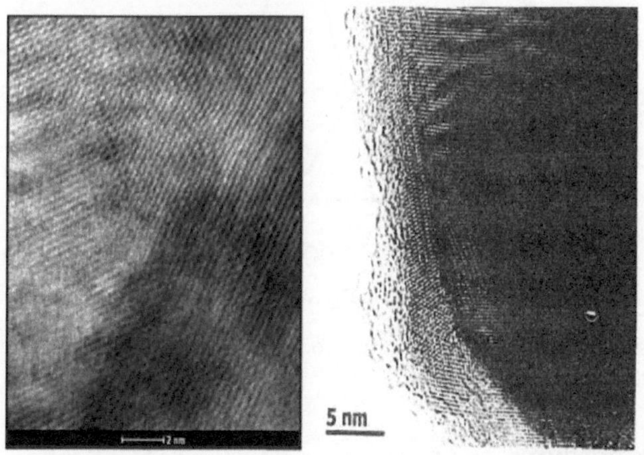


**FIGURE 12.4** Comparison of TEM microphotographs for initial copper carbon mesocomposite (a); and phosphorus-containing Cu C mesocomposite (b).

From the analysis of presented experimental results may be proposed the multistage mechanism of metal-carbon mesocomposite modification process:

1. **The First Stage:** The negative charged quants are directed to the positive charged Phosphorus atom near which there is a cloud of positive charged quants.
2. **The Second Stage:** The interaction of negative charged quants with positive charged quants or the annihilation phenomena.
3. **The Third Stage:** The formation of inner electromagnetic field which stimulates the reduction process.
4. **The Fourth Stage:** The formation of C=P bond (132.6 eV) and unpaired electrons on carbon fibers with the increasing spins number on carbon cover of metal cluster.

In the case, when APPh is replaced by aerosol (or silicon), the reduction of silicon from silicon oxide takes place by 50% independent of the used amount of the reagents (Figure 12.5); this can be explained by the decrease of the rate of the electron quantization in the layer containing Si-O bonds.

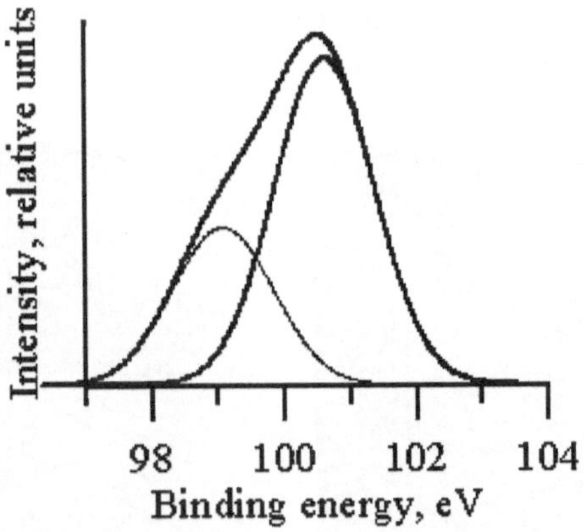


**FIGURE 12.5** The XPS Si2p spectrum for the sample Cu/C + $SiO_2$ at the ratio 1:1 (E(SiO) = 100.7 eV; E(Si) = 99 eV) .

At the XPS, TEM, and EPR investigations the mechanism of chemical bonds formation between metal, carbon, phosphorus, and silicon atoms investigates.

## 12.4 THE INFLUENCE OF METAL CARBON MESOCOMPOSITES ON THE POLYMERIC

The media (or polymeric compositions) properties changes under the mesoparticles influence can be achieved at equal distribution of these particles in composition volume and at its coagulation absence. Last is possible at the following conditions:

- Certain polarity and dielectric constant of medium;
- Minute concentration of mesoparticles;
- Ultrasound action on the correspondent suspension for the proportional distribution of mesoparticles.

The assignment of active nanostructures (mesoparticles) during the composition's modification is concluded in the activation of matrices self-organization in needful direction. For the realization of this goal, the determination of organized phase part is necessary.

In some papers [11–19], the positive results on materials properties improvement are presented when the minute quantities of metal-carbon mesocomposites are introduced in these materials. In Ref. [16], the hypothesis about nanostructures influence transmission on macromolecules of polymeric matrices is proposed. This hypothesis is complied with mesoscopic physics principles which consider quantum effects at the certain conditions of mesoparticle existence.

The composition polarization is possible because of there is the charge quantization with the wave expansion on polar (functional) groups of media (for example, polymer macromolecule). The quantum charge wave expansion leads to the functional groups' polarization (dipole moments) change as well as the extinction increasing. Last bring growth of peaks intensities in IR spectra. The individual peaks growth effects in IR spectra are observed at the introduction of mesocomposites minute quantities (Table 12.3). It's necessary to note, that the peaks intensity growth in IR spectra is observed when the quantity of introduced mesocomposite is decreased. This fact is complied with fundamental principles of chemical mesoscopic. In illustrated case the instance of fine dispersed suspension Cu-C mesocomposite (hardener for epoxy resin). According to data of Table 12.3, the decreasing of mesocomposite quantity to 0.001% leads to the growth of some peak's intensity in IR spectra.

At the second day of that suspension existence the floccules are formed and peaks intensity sharply drops. However, the suspension activity can be increased with the use of ultrasound treatment. The treatment optimal duration determined as 7 minutes. In this case, the IR spectra peak intensity in is increased in 2–4 times (Table 12.4).

**TABLE 12.3** The Peaks Intensity Change in Dependence of Cu–C Mesocomposite Concentration

| N | ν ($cm^{-1}$) | $I_1/I_0$ | $I_{0,01}/I_0$ | $I_{0,001}$ | Chemical Bonds/Groups |
|---|---|---|---|---|---|
| 1 | 1,050 | 1,235 | 1,411 | 1,686 | C–O–C st |
| 2 | 1,450 | 1,179 | 1,590 | 1,744 | C–H |
| 3 | 1,776 | 1,458 | 1,347 | 1,691 | C=O st as |
| 4 | 1,844 | 1,463 | 1,412 | 1,678 | C=O st sy |
| 5 | 2,860–3,090 | 1,182 | 1,545 | 1,750 | C–H |

**TABLE 12.4** The Peaks Intensity Changes in IR Spectrum of Cu–C Mesocomposite Depending on the Duration of Ultrasound Treatment

| N | ν ($cm^{-1}$) | $I_7/I_0$ | $I_{10}/I_0$ | Chemical Bonds/Groups |
|---|---|---|---|---|
| 1 | 1,776 | 3.7932 | 0.7574 | C=O st as |
| 2 | 1,844 | 2.5065 | 0.9115 | C=O st sy |
| 3 | 3,039 | 2.3849 | 0.9589 | C–H |

The charge (electron) quantization should lead to the macromolecule electron structure change and, as a corollary, to change the sub-molecular structures of polymeric substances.

In the monograph [6], it is established that the quants electron wave, which initiates the self-organization process in polymeric composition is expensed from carbon fibers of cover associated with metal cluster in the metal-carbon mesoscopic composite. The last leads to the correspondent orientation of sub-molecular structures in nanostructured composite surface layers.

The self-organization mechanism for polymeric compositions modified by the Metal Carbon mesocomposite minute quantities is concluded in the condition's creation for the composition polarization, which brings the great change of electron and sub-molecular structures of materials.

In this case, it's possible the following explanation: with the mesoparticle quantity decreasing, the creation of diffused radiation of electron quants directed to polymeric polar functional groups is possible that leads to $sp^3$ hybridization and 3D form of polymer surface layer. In other words, it's possible there is a necessity to take into account the forms of quants flows directed to active groups of polymers modified. Certainly, these changes influence on the modified materials properties.

## 12.5 THE PROPOSAL FOR THE ESTIMATION OF NANOSTRUCTURES REACTIVITY ESTIMATION

According to the fractal theory [3], any system can be presented as an aggregate of elements similar to the whole system. These elements have their own energetic and geometric (volume) parameters owing to which they are found within the system. The change of these parameters because of the action of external factors leads to disturbance of system balance. In this case, the system is destructed or transformed. The estimation of these changes is possible with the use of the relative parameters in which the energetic and volume values are compared with the definite standard values for the correspondent elements (fragments) in the definite reaction series. This approach to reactivity consideration is near to Taft and Pal'm theoretical works [22]. For the relative energetic parameters, the following formula $(\varepsilon-\varepsilon_0)/\varepsilon_0$ is proposed [23], where $\varepsilon_0$ corresponds to the surface energy for standard chemical fragment. In turn, analogous relative parameters are proposed [24] for the volume characteristics $(V-V_0)/V_0$.

The development of Chemical Mesoscopics in this direction is connected with the research of the size and energetic characteristics of chemical particles. The size of mesoparticles is denoted as approximately 10 nm, and the motion freedom of nanostructures (mesoparticles) is limited by the vibration with high frequency and electron transport across them. The peculiarities of mesoparticles consist in the radiation of energy quants of negative or positive charges. This radiation is the main reason of the stimulation of chemical processes. At the imposition of the negative charge quants the interference takes place, and the chemical bonds are formed. In turn, the imposition of the negative and positive quants together the phenomenon of annihilation is created. In this case, the direct electromagnetic field is appeared that leads to the stimulation of negative charge quants moving and the growth of chemical bonds formation. The phenomena of interference and annihilation are reasons of start for self-organization process with reservation confirmation order, which determine the finished product structure. For the process explanation the equation of Kolmogorov–Avrami [5] can be used:

$$W = 1 - exp\left(-k\tau^{n}\right), \tag{5}$$

where; W is the part of obtained product (for instance, polymer); k is the process rate constant; τ is the duration of process; n is the fractal dimension (for one measured process n = 1).

For the comparative estimation of reagents (or nanostructures) in one reaction series, it's possible the application of the theory of free energy linear dependences. In this case, the reactions are considered with using one of reagent as the standard compound for which W is fixed $W_0$. The estimation of reactivity can be proposed on the difference $W–W_0$, where W is calculated on formula 2, and $W_0$ is defined on analogous formula with changes $k_0\tau_0^n$. It is noted, the fractal dimension n does not change because the comparison is carried out for one type of reaction. The following equation for difference $W – W_0$ can be written:

$$W - W_0 = exp\left(-k_0\tau_0\right) - exp\left(-k\tau^{n}\right) \tag{6}$$

and after Eqn. (2) transformation:

$$W_0 = k / k_0\left(\tau / \tau_0\right)^{n} - 1 \tag{7}$$

If lg $k/k_0$ is defined as:

$$lg\ k / k_0 = -2{,}3\ RT\left\{\left[\left(\varepsilon_0 - \varepsilon\right) / \varepsilon_0\right]a + \left[\left(V_0 - V\right) / V_0\right]b\right\}, \tag{8}$$

and then this expression after transformation stands in the Eqn. (3), then the Eqn. (5) is received:

$$lg\left(W - W_0\right) = \left(\tau / \tau_0\right)^{n} exp\left\{\left[\left(\varepsilon_0 - \varepsilon\right) / \varepsilon_0\right]a + \left[\left(V_0 - V\right) / V_0\right]b\right\}, \tag{9}$$

where; values a and b are parameters, which correct the influence polar and steric (spaced) effects on reactivity in polymerization, the relation $(\varepsilon_0 - \varepsilon)/\varepsilon_0$ [23, 24] correspondents to Taft constant σ (polarity constant), and the relation $(V_0 - V)/V_0$ – Taft constant $E_s$ (steric or spaced constant) [24].

The application of the above notions for the Chemical Mesoscopics development is very perspective because it stimulates the mathematic apparatus creation for the chemical processes flowing direction prediction.

## 12.6 CONCLUSION

On the base of presented materials it is possible to make the following conclusions:

1. In a mini-review, the peculiarities of the new scientific trend "chemical mesoscopics" at the mesoparticles reactivity description in their modification processes as well as in the polymeric compounds' modification with the mesoparticles using are considered. The schemes of reactions mechanisms into mesoscopic reactors and also the mechanisms of media (compositions) polarization under the flow action of negative charges quants directed from the nanostructures or mesoparticles are discussed.
2. For the explanation of the chemical bonds' formation in the reactions without the oxidation state change the notion "interference" is introduced, and for the explanation of reduction-oxidation (Red-Ox) reactions mechanisms the notion "annihilation" is introduced. The application of these notions is considered on the examples of the metal-carbon mesocomposites modification by the different substances, including oxidizers. With the "annihilation" notion application the oxidizer reduction as well as the mesocomposite metal atomic magnetic moment growth and also the spin quantity increasing on mesocomposite carbon shell are explained.
3. On the base of experimental results, the scheme of mechanism of Copper Carbon mesocomposite modification by APPh is given. In this process, the phosphorus is reduced and is disposed between the carbon fibers of mesoparticle carbon shell. The process is accompanied by the growth of copper atomic magnetic moment from 1.3 to 4.2 Bohr magnetons (at the reagents relation equaled to 2). Simultaneously it is noted the spin increasing in 10 times on the mesoscopic carbon shell. Analogous results are observed at the interaction this mesocomposite with Aluminum Oxide.
4. The experimental results concerning to the epoxy resin suspension modification by the minute quantities of Copper Carbon mesocomposites and their modified compounds with IR spectroscopic investigations using are discussed. The minute quantities of nanostructures (mesoparticles) application at the polymeric substances' modification is substantiated.

5. The nanostructures (mesoparticles) reactivity estimation is proposed with using Kolmogorov–Avrami equations and the energetic characteristics (polarity and volume or spatial constants) for the fragments of mesoscopic compounds.

The development new scientific trend, "chemical mesoscopics," supposes the theory development for mechanisms of mesoscopic systems formation as well as mechanisms of self-organization in media and compositions. The peculiarities of structures and energetic characteristics of mesocomposites obtained cause their possibilities for applications in different fields.

The examples of applications in radical, redox, and addition processes as catalysts, reagents, and also inhibitors as well as additives and modifiers improving properties of materials (inorganic and organic polymeric materials), adhesives and glues, fireproof systems, corrosion inhibitors, medicine magnetic transport remedies, stimulators of plant growth are presented in the review [5]. The metal-carbon mesoscopic composites owing to their magnetic characteristics, can be used in electromagnetic radiation focal systems.

This unique scientific trend discovers a new era in the development of new theories in the natural sciences and in practice, for instance, novel nanostructures application widening.

## KEYWORDS

- **annihilation**
- **atomic magnetic moments**
- **charge quantization**
- **chemical reactions**
- **interference**
- **Kolmogorov–Avrami equations changed**
- **nanostructures reactivity**
- **redox reactions**

## REFERENCES

1. Kodolov, V. I., & Trineeva, V. V., (2017). New scientific trend – Chemical Mesoscopics – *Chemical Physics & Mesoscopics, 19*(3), 454–465.
2. Kodolov, V. I., Kodolova-Chukhontseva, V. V., Terebova, N. S., & Shabanova, I. N., (2019). The explanation of magnetic metal-carbon mesocomposites synthesis peculiarities by means of mesoscopic notions. *Academ. J. Polym. Sci., 3*(2), 555613.
3. Moskaletz, M. V., (2010). *Fundamentals of Mesoscopic Physics* (Vol. 2, p. 180). Khar'kov: NTU KhPI.
4. Ruedenberg, K., (1964). *Physical Nature of Chemical Bond* (p. 162). M.: Publ "Mir."
5. Kodolov, V. I., Kodolova-Chukhotseva, V. V., Shabanova, I. N., et al., (2020). Review: Possible fields of metal-carbon mesocomposites application. In: *Innovation and Challenges in Modern Physical Chemistry: Research and Practice* (pp. 57–122, 195). N. Y.: Nova Sciences Publishers, Inc.
6. Kodolov, V. I., Trineeva, V. V., Pershin Yu, V., et al., (2020). *Method of Metal Carbon Nanocomposites Obtaining from Metal Oxides and Polyvinyl Alcohol*. Pat. RU 2018122 001.
7. Kodolov, V. I., & Kodolova–Chukhontseva, V. V., (2019). *Fundamentals of Chemical Mesoscopics* (p. 218). Monograph. - Izhevsk: Publisher – M. T. Kalashnikov Izhevsk State Technical University.
8. Mustakimov, R. V., Kodolov, V. I., Shabanova, I. N., & Terebova, N. S., (2017). Modification of copper carbon nanocomposites with the use of ammonium polyphosphate for the application as modifiers of epoxy resins. *Chemical Physics & Mesoscopics, 19*(1), 50–57.
9. Kodolov, V. I., Trineeva, V. V., Terebova, N. S., et al., (2018). The change of electron structure and magnetic characteristics of modified copper carbon nanocomposites. *Chemical Physics & Mesoscopics, 20*(1), 72–79.
10. Shabanova, I. N., Kodolov, V. I., Terebova, N. S., & Trineeva, V. V., (2012). *X-Ray Electro Spectroscopy in Investigation of Metal/Carbon Nanosystems and Nanostructured Materials* (p. 252). Izhevsk–Moscow: Publ. "Udmurt University."
11. Shabanova, I. N., Terebova, N. S., Kodolov, V. I., et al., (2013). The investigation of metal or carbon nanocomposites electron structure by X ray photoelectron spectroscopy. In: *Nanostructure, Nanosystems and Nanostructured Materials: Theory, Production and Development* (pp. 177–230). Toronto, Canada – New Jersey, USA: Apple Academic Press.
12. Kodolov, V. I., Lipanov, A. M., Trineeva, V. V., et al., (2013). *The Changes of Properties of Materials Modified by Metal/Carbon Nanocomposites* (pp. 327–373). Ibid.
13. Kodolov, V. I., & Trineeva, V. V., (2013). Theory of modification of polymeric materials by super small quantities of metal/carbon nanocomposites. *Chemical Physics & Mesoscopy, 15*(3). 351–363.
14. Kodolov, V. I., & Trineeva, V. V., (2012). Perspectives of idea development about nanosystems self organization in polymeric matrixes. In: *The Problems of Nanochemistry for the Creation of New Materials* (pp. 75–100). Torun, Poland: IEPMD.

15. Akhmetshina, L. F., Lebedeva, G. A., & Kodolov, V. I., (2012). Phosphorus containing metal/carbon nanocomposites and their application for the modification of intumescent fireproof coatings. *Journal of Characterization and Development of Novel Materials, 4*(4). 451–468.
16. Kodolov, V. I., Kovyazina O. A., Trineeva, V. V., Vasilchenko Yu, M., Vakhrushina, M. A., & Chmutin, I. A., (2010). On the production of metal/carbon nanocomposites, water and organic suspensions on their basis. *VII International Scientific-Technical Conference Nanotechnologies to the Production – 2010: Proceedings* (pp. 52, 53). Fryazino.
17. Chashkin, M. A., Kodolov, V. I., & Trineeva, V. V., (2011). Metal/carbon nanocomposites– epoxy compositions quantum chemical investigation and experimental modeling. *Polymers Research Journal, 5*(1), 5–19.
18. Kodolov, V. I., Trineeva, V. V., Semakina, N. V., Yakovlev, G. I., Volkova, E. G., et al., (2008). *Patent 2337062 Russia Technique of Obtaining Carbon Nanostructures from Organic Compounds and Metal Containing Substances.*
19. Kodolov, V. I., Trineeva, V. V., Kovyazina, O. A., & Vasilchenko Yu, M., (2012). Production and application of metal/carbon nanocomposites. In: *The Problems of Nanochemistry for the Creation of New Materials* (pp. 23–36). Torun, Poland: IEPMD.
20. Akhmetshina, L. F., Kodolov, V. I., Tereshkin, I. P., & Korotin, A. I., (2010). The influence of carbon metal containing nanostructures on strength properties of concrete composites. *Internet Journal "Nanotechnologies in Construction," 6*, 35–46.
21. Pershin Yu, V., & Kodolov, V. I., (2012). Polycarbonate modified with Cu-C nanocomposite. In: *The Problems of Nanochemistry for the Creation of New Materials"* (pp. 173–179, 250). Poland – Torun: Publ. IEPMD.
22. Pal'm, V. A., (1974). *The Introduction in Theoretical Organic Chemistry* (p. 446). M.: Publ. "High School."
23. Kodolov, V. I., (1965). Possibilities of modeling in organic chemistry. *Organic Reactivity, 2*(4, 6), 11–17.
24. Kodolov, V. I., & Spasskiy, S. S., (1976). Parameters in Alfrey-price and Taft equations. *Vysokomol. Soed., 18*(9), 1986–1992.

# CHAPTER 13

# Carbon Nanotube as a Promising Nanomaterial for Water Treatment

NEENAMOL JOHN, BONY K. JOHN, JINCY MATHEW, and BEENA MATHEW*

*School of Chemical Sciences, Mahatma Gandhi University, Priyadarsini Hills P.O., Kottayam – 686560, Kerala, India*

**Corresponding author. E-mail: beenamathew@mgu.ac.in*

## ABSTRACT

Water is an exquisite source that is essential to every living thing. Augmented amount of water usage would make the increased rate of water pollution. Wastewater treatment is one of the challenging missions of the modern world. Carbon nanotube-based nanostructures can provide effective and simple decontamination of water. Carbon nanotubes (CNTs) are allotropes of the carbon family and are known for its characteristic properties. CNTs are graphitic sheets that have been rolled into a cylindrical shape. CNTs are considered as unique materials with promising future applications due to their structure, morphology, and excellent chemical, mechanical, and electrical properties. Adsorption and photocatalysis are two advanced technologies applicable for water treatment. The open structure of CNTs offers higher adsorption of pollutants from wastewater, and the high surface area of CNTs provides effective improvement of activity of conventional photocatalysts used for water purification. This chapter

Carbon Composites: Composites with Nanotubes, Nanomaterials, and Graphene Oxide.
Eduardo A. Castro, Ann Rose Abraham, & A. K. Haghi (Eds.)

mainly discusses the role of CNTs and CNT-incorporated nanocomposite (NC) in the adsorption and photocatalysis process for water treatment.

## 13.1 INTRODUCTION

One of the most abundant natural resources on the planet is water. Only a fraction of that resource is suitable for human use. The major environmental issue confronting humanity is the rising contamination of freshwater systems with thousands of industrial and natural chemical substances. Heavy metals, medicines, dyes, pesticides, insecticides, phenols, pesticides, and detergents are among the many toxins entering water systems as a result of industrialization and human activities [1]. So, researchers tried to develop new technologies for wastewater treatment, such as ozonation, membrane separation, coagulation, electrodialysis, reverse osmosis, chemical reaction, etc. But these methods have different drawbacks like low efficiency, high energy, and a large amount of sludge formation [2]. Therefore, these methods are considered to be ineffective for the complete purification of wastewater. Nowadays, adsorption and photocatalysis are two recognized processes for wastewater treatment due to their easy recovery, simplicity, reusability, and high efficiency [3, 4].

Nanotechnology is the study of nanoscale materials and their applications. Nanomaterials are the tiniest structures with a diameter of less than 100 nanometers and are classified into nanotubes, films, quantum dots, nanowires, colloids, and particles. All of these nanomaterials are frequently employed for different applications like antibacterial activity, drug delivery, wastewater treatment, biosensors bioimaging, food packaging, drug delivery, therapeutics, biosensors, and so forth due to their excellent chemical, physical, and electrical properties. In wastewater treatment processes, cost-effective, eco-friendly, and efficient nanomaterials have been fabricated with unique features for the effective removal of pollutants from water [5, 6]. Among all the nanomaterials, CNTs are most effective for wastewater treatment because of their size, structure, and physicochemical properties.

CNTs are one of the allotropic forms of carbon, composed of graphite sheets that are wrapped up into a tube-like structure. On the basis of the number of graphitic sheets, CNTs are classified into two: single-walled carbon nanotubes (SWCNTs) and multi-walled carbon nanotubes

(MWCNTs). SWCNTs have a cylindrical shape and are made up of a single shell of graphene, whereas MWCNTs are made up of many layers of graphene sheets. Both SWCNTs and MWCNTs have been employed for water treatment process due to their excellent physical, chemical properties and large surface area [7, 8]. CNTs can be employed in adsorption and photocatalytic processes, which are highly advanced wastewater treatment approaches.

Adsorption has gained importance as a separation, detoxification, and purification procedure on an industrial scale in recent decades. Adsorption with solid materials (adsorbents) is a simple, convenient, and effective technology for pollutant removal from wastewater. The physical adhesion or attachment of molecules (or ions) to the surface of another substance is known as adsorption. The current focus is to develop innovative adsorbent materials with high adsorption capacities and removal efficiencies. To remove contaminants from wastewater, a variety of materials like zeolite, AC, and clay have been utilized as adsorbents. The removal efficiency of adsorbents is limited by their active sites, surface area, adsorption kinetic, and non-selectivity. Smart adsorbents with good adsorption capability are required to remove contaminants to the ppb level. Nano-adsorbents can solve the drawbacks of traditional adsorbents, and carbon-based nanomaterials are highly efficient adsorbents. CNTs are extremely effective adsorbents for water treatment [9–14].

Photocatalysis is a chemical process based on nanotechnology, and this process is carried out under the presence of light. This is one of the advanced methods for wastewater treatment without the need for immense facilities. The advantage of photocatalysis is no secondary waste generation, and carbon dioxide, water, and some mineral acids are the product. Photocatalysis based on semiconductors has been extensively explored for environmental applications. Conventionally used photocatalysts have various drawbacks like large bandgap, low absorption capacity, poor stability, etc. So, semiconducting materials are modified in various ways to overcome these constraints. Coupling with carbon-based materials like graphene oxide (GO), graphitic carbon nitride, CNTs are generally used cocatalysts for improving the activity of semiconducting materials. CNTs are especially appealing due to their large specific surface area, excellent electrical conductivity, and high mechanical robustness. CNT can serve as an electron reservoir for facilitating charge transit and that will decrease the charge carrier recombination. The catalytic activity will be improved by reducing electron-hole pair recombination. From several studies revealed that CNTs are one of the best modification materials for improving photocatalytic activity. According to various investigations, CNTs have been found to be one of the best modification materials for enhancing photocatalytic activity [15–18]. In this chapter, we discuss the adsorption and photocatalytic activity of CNTs for wastewater treatment.

## 13.2 CNT-BASED ADSORPTION FOR WASTEWATER TREATMENT

The increased amount of anthropogenic waste entering the water systems through human activity is a serious concern. The conditions become even more severe when the refractory contaminants are non-biodegradable, difficult to separate from water sources, and generate public health anxieties even at low concentrations [19, 20]. Adsorption of organic pollutants is an efficient strategy for eliminating toxins from wastewater. It is a surface process in which an adsorbate gets attached to an adsorbent surface by either chemical or physical forces. The development of outstanding adsorbents is the key to the success of this technique. Compared to other conventional water treatment techniques, adsorption is efficient, sustainable, cost-effective, simple, and involves a smooth operation [21]. Activated carbon (AC) was one of the prominent adsorbents used earlier. Even though its wide surface area and porous structural features are incredibly advantageous for adsorption, its trouble in regeneration and non-cost effectiveness limits the practical application of AC for wastewater treatment [22]. The presence of non-deep surface pores makes adsorption and desorption processes quick in CNTs [23].

CNTs comprise a striking class of materials exhibiting unique properties and numerous applications. These non-metallic materials show outstanding electronic, electrical, mechanical, and structural properties [24]. CNTs are formed by rolling graphene sheets in a cylindrical shape. The cylindrical CNTs are hollow and possess greater surface area and high hydrophobicity. Therefore, they can be used in wastewater treatment [25]. CNTs can be used to remove organic pollutants in wastewater by adsorption. CNTs, upon appropriate modification, show remarkable enhancement in the adsorptive elimination of organic toxins in the water. An essential characteristic of CNT-based composites is that they are easy to recover and regenerate after wastewater treatment [26].

The functionalization of CNTs can be done in various ways, such as oxidation, by the incorporation of magnetic particles, integration with other materials and grafting with polymers, etc. The adsorption of organic contaminants by CNT-based composites depends on factors such as pH of the medium, temperature, contact time between the adsorbate and the adsorbent, concentration of the pollutant solution, structure of the pollutant molecule, etc. The adsorption of pollutants by CNT-based composites occurs mainly through non-covalent interactions like hydrogen bonding, hydrophobic effects, $\pi$-$\pi$ stacking, and electrostatic interactions [27–29].

CNTs are reported to be efficient for the removal of pesticides [30, 31], polycyclic aromatic hydrocarbons (PAHs) [32–36], substituted benzenes [37], and phenolic compounds [29, 38]. Even though CNTs exhibit excellent adsorption of pollutants, certain limitations prevent the usage of pristine CNTs in wastewater treatment. The limitations of CNTs are associated with their difficulty in separation and dispersion. The particle size of CNTs is ultrafine, and they have the tendency for agglomeration, which decreases the surface area of CNTs. These limitations are overcome, and CNTs are applied in practical applications by introducing modifications [39].

### *13.2.1 ADSORPTION OF POLLUTANTS IN WATER BY UNMODIFIED CNTS*

Studies on the adsorptive elimination of organic dyes onto CNTs confirmed that the adsorptive process depends on the physical properties of CNTs [40–42]. The role of experimental environments and properties of adsorbents on the adsorption process can be clearly understood from the works of Yao et al. [40]; and Shahryari et al. [41]. The methylene blue dye removal onto CNTs showed an adsorption capacity of 41.63 mg/g at a temperature of 333 K. When MWCNTs were used for the same adsorption experiment, 132.6 mg/g adsorption capacity has displayed for methylene blue removal at a relatively less temperature of 310 K. The reason for the considerable increase in the adsorptive removal of the same dye can be concluded to be because of the higher surface area of MWCNTs.

MWCNTs with a greater surface area and enhanced adsorptive sites were extensively exploited to study the adsorptive confiscation of organic dyes. MWCNTs also have carboxyl functional groups and negative charges on their surface [43]. MWCNTs prepared through diverse methods were used for the removal of various dyes. Zare et al. [44] investigated the adsorption of Congo red dye by MWCNTs under different experimental conditions. Around 92% removal efficiency was achieved in 60 min, and the Congo red adsorption capacity of MWCNTs exhibited a rise with the decreasing temperature. Yao et al. [45] followed the chemical vapor deposition (CVD) technique to fabricate MWCNTs, and its adsorption capacity of methyl orange (MO) dye was studied. The adsorption capacity was related to the pH of the solution and the initial dye concentration. The adsorption isotherm was in agreement with a Langmuir isotherm. The kinetic studies revealed the reaction to follow pseudo-second-order kinetics. Another work investigated the adsorptive confiscation of different dyes like blue 116, red 159, and yellow 81 using MWCNTs [46]. Alkaim et al. [47] studied the adsorptive confiscation of maxilon blue by hydrothermally fabricated MWCNTs. The experimental data were well-fitted into a Freundlich adsorption isotherm. Electrostatic interactions were the major interactive force between the dye molecule and MWCNTs. Compared to MWCNTs, SWCNTs, and double-walled carbon nanotubes (DWCNTs) possess greater surface area [48] and pore volume [49]. A study by Ji et al. [30] revealed the enhanced efficiency of SWCNTs in removing pollutants by adsorption.

A comparative study on the adsorption efficiency of AC and MWCNTs for M-2BE dye removal confirmed the greater pore volume and higher diffusability of dye molecules in MWCNTs [51]. That is, MWCNTs were proved to be more efficient for the adsorptive removal of dyes [52, 53]. Aromatic pollutants are more easily adsorbed onto CNTs because of the $\pi$-$\pi$ stacking interaction between the graphitized CNT layer and the aromatic benzene ring. The sorption sites in CNTs are easily accessible, and the pores are larger. This helps in efficient adsorption of big organic pollutants like pharmaceuticals and antibiotics onto CNTs [54]. Wang and coworkers [43, 55] investigated the adsorption efficiency of MWCNTs in removing dyes from a binary dye mixture in a series of studies. The adsorption of antibiotics by MWCNTs was studied. Antibiotics like ciprofloxacin (CIP) [56], tetracycline (TC) [57], ceftazidime [58], amoxicillin [59], metronidazole [60], sulfamethoxazole [61], etc., were showed good adsorptive removal onto MWCNTs.

### *13.2.2 WASTEWATER TREATMENT BY ADSORPTION USING MODIFIED CNT SYSTEMS*

Various methods introduced modifications in CNTs, and the effect of the modification on pollutant adsorption was studied. The oxygen content in MWCNTs affect their adsorption properties [62]. The oxidation of MWCNTs introduces carboxylic acid functionalization [63]. A study by Sobhanardakani et al. [64] confirmed that the specific surface area, average pore volume, and pore diameter are greater for $HNO_3$-treated MWCNTs than pristine MWCNTs. The method of oxidation of the CNTs influences the thermodynamic behavior of adsorption. Oxidized and functionalized CNTs were also reported to be effective for the removal of heavy metals like arsenic [65], lead [66, 67], mercury [68, 69], etc., having high potential toxicity from water by adsorption.

Magnetized CNTs are another vital class of modified CNTs. The negative charge, pore diameter, and surface area were greater for magnetically modified MWCNTs than unmodified ones [70], and therefore they were found efficient for cationic dye removal [71]. Qu et al. [72] studied the adsorptive removal of dyes using magnetically modified acid-treated MWCNTs using $Fe_2O_3$ nanoparticles. Magnetite-MWCNT composite fabricated by one-pot solvothermal strategy was used for removing methylene blue dye by adsorption [73]. The dye adsorption onto the composite was due to electrostatic and $\pi$-$\pi$ interactions. A scheme for the mechanism of the same is given in Figure 13.1. Gabal et al. [74] prepared zinc-substituted cobalt ferrite/MWCNT hybrid by gelatin process. The protonation of the hybrid surface at low pH offered the maximum adsorption of anionic dyes. The increase in temperature, dye concentration, and composite dosage resulted in a rise in the adsorption efficiency of methylene blue using $CoFe_2O_4$/MWCNT composite [75]. Magnetic MWCNTs fabricated by Fazelirad et al. [76] gave good results for the adsorption of amoxicillin antibiotics from water.

The functionalization of MWCNTs can also be done using polymers. The removal of MO by adsorption using poly-2-hydroxyethyl methacrylate and chitosan-functionalized MWCNT was reported by Mahmoodian et al. [77]. In the composite fiber of calcium alginate and MWCNT, the adsorption capacity and rate of dyes showed an increase with increasing the MWCNT content [78]. The novel composite obtained by integrating CNTs into diatomite cavities gave better adsorption of anionic dyes than cationic ones [79]. The successful adsorption of direct red 23 and direct blue 86 was attained by poly(propylene imine) dendrimer-modified MWCNTs [80]. The significant removal efficiency was reported for various dyes in aqueous media by CNTs functionalized with glycine-β-cyclodextrin [81]. A recent study by Mallakpour and Rashidimoghadam [82] followed an ultrasonic method to fabricate MWCNT composite with glutaraldehyde-cross-linked poly(vinyl alcohol) and vitamin C. The composite was found effective even after five regeneration times using ethanol. Magnetically polymerized CNTs were also reported for their better adsorption properties [83]. A composite of MWCNT with iron oxide and soluble starch of better biocompatibility and hydrophilicity for MO and methylene blue adsorptive removal was reported by Chang et al. [84] Chitosan/$\gamma$-$Fe_2O_3$/MWCNT was another composite that showed excellent efficiency in MO confiscation [85].

Glutaraldehyde-crosslinked chitosan/$SiO_2$/MWCNT is a magnetic nanocomposite (NC) useful in the adsorptive elimination of direct blue 71 and reactive blue 19 dyes [86]. A magnetically distinguishable hybrid of oxidized-MWCNT modified with carrageenan exhibited better adsorption capacity for methylene blue, and the adsorption ability was greater than that of unmodified oxidized-MWCNT/$Fe_2O_3$ composite [87]. A ternary composite of guar gum grafted MWCNT/$Fe_3O_4$ was reported by Yan et al. [88]. It was utilized for the confiscation of methylene blue and neutral red dyes. The better adsorption capacity resulted from the excellent biocompatibility and hydrophilicity of the hybrid. Carboxylic acid-functionalized MWCNTs/gelatin/$Fe_3O_4$ magnetic NC were reported to display excellent performance in the adsorptive elimination of direct red 80 and methylene blue dyes from water. The removal efficiency for direct red 80 was 93.1%, and methylene blue was 76.1% [89]. Regenerative magnetic alginate/MWCNT was recently synthesized by Boukhalfa et al. [90] and applied to remove methylene blue dye, and better adsorption efficiency was achieved. Xu et al. [39] reported a novel shell-core structured CNT-based composite adsorbent which showed 90% adsorptive removal of 2-naphthol. The adsorption of the composite towards 2-naphthol has enhanced with the increase in CNT loading in the composite. A recent study developed novel magnetic CNTs for the adsorptive removal of microplastics from aqueous media. Here, the complete elimination of the contaminant was achieved in 300 min [91]. Adsorptive removal of anionic azo dyes was done using magnetic polymer-MWCNT composite. The experimental results were best fitted with Langmuir adsorption isotherm [92]. The adsorptive removal and separation efficiency of dyes showed a substantial hike while using a composite of MWCNT incorporated with poly(1-glycidyl-3-methylimidazolium chloride) and ferro-ferric oxide moieties because of the hydroxyl and amino group presence and the substantial surface area of the hybrid [93].

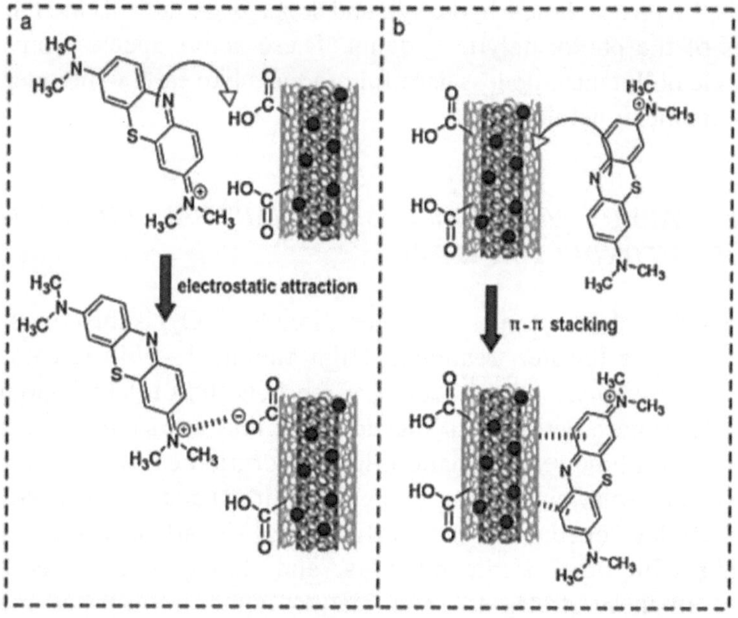


**FIGURE 13.1** Electrostatic interaction and π-π stacking between negatively charged MWCNTs and cationic methylene blue dye.

*Source:* Reproduced with permission [73]; Copyright (2011), Elsevier.

## 13.3 PHOTOCATALYTIC ACTIVITY OF CNTS FOR WASTE WATER TREATMENT

Photocatalysis is an effective wastewater treatment methodology due to its proficient degradation of diverse pollutants, and this is an eco-friendly and competent process to resolve water pollution problems [94]. A carbon nanotube is widely used as a scaffold for the assembly of nanoparticles due to the retardation capacity to enhance the photocatalytic activity of conventional photocatalysts like metal oxides, silver-based materials by the reduction of electron-hole pair recombination [95]. Carbon nanomaterials like CNTs interact with wastewater pollutants and change their toxicity, bioavailability, and mobility [96]. In the photocatalytic process, light is irradiated on the surface of photocatalysts, then electrons in the valence band of the semiconducting catalyst are transferred to the conduction band. The holes present in the conduction band will convert the $OH^-$ to hydroxyl radical, and electrons are capable of converting $O_2$ to superoxide radicals. Electrons, holes, hydroxyl, and superoxide radicals are the active species of the photocatalytic systems. These active species convert the very toxic pollutants into less harmful products like carbon dioxide, water, and inorganic mineral acids.

### 13.3.1 CARBON NANOTUBE-BASED BINARY NANOCOMPOSITES (CNT/SEMICONDUCTOR TYPE)

In the field of photocatalysis, titanium dioxide ($TiO_2$) is the conventional catalyst for wastewater treatment. High chemical stability, easy availability, nontoxicity, and low cost are the benefits of $TiO_2$. Along with these advantages, $TiO_2$ has some demerits that are small surface area, rapid electron-hole recombination, low absorption capacity and ineffective usage of solar energy. To tackle these drawbacks, several strategies have been developed, such as modification with carbon-based materials, coupling with other semiconductors, and doping with other metals. Among all these strategies, modification with carbon-based materials is very effective due to the distinctive condensation and adsorption characteristics, which accelerates the removal of organic pollutants from wastewater. One-dimensional hollow tubular structure, high thermal conductivity, mechanical strength, good thermal stability, unique electronic properties, and large surface area are the attractive properties of SWCNTs and MWCNTs for the modification of conventional semiconductors like $TiO_2$, ZnO, etc.

Payan et al. [97] described the fabrication of titanate nanotube/SWCNT NCs via the hydrothermal method and tested for the elimination of 4-chlorophenol under UV and solar irradiation. Murgolo and coworkers [98] reported an SWCNT/$TiO_2$ NC for the mineralization of sulfamethoxazole, diclofenac (DCF), iopamidol, diatrizoic acid, iopromide, and triclosan in ultrapure water. Chlorophenol can be considered as very dangerous species because most of them are hardly biodegradable and toxic and are difficult to eliminate from the environment. Mohammadi and Sabbaghi [99] studied the degradation of 2,4-dichlorophenol (2,4-DCP) using MWCNT/$TiO_2$ NC under UV-visible irradiation. This NC was fabricated by the modified sol-gel method. The result showed that 93% and 87% 2,4-DCP were removed under UV and solar irradiation, respectively within 120 min and at pH = 11. Sampaio et al. [100] explored the $TiO_2$/multi-walled carbon nanotube ($TiO_2$/CNT) synthesized by hydration–dehydration (HD) and sol-gel (SG) methods. NC produced by SG process exhibited higher efficiency for the degradation of methylene blue than those synthesized by HD method, and this CNT-loaded $TiO_2$ is very effective for the removal of phenolic compounds like 4-methoxyphenol (MP) and 4-aminophenol (AP).

TC is the most widely used antibiotic for medical applications in animals and humans. Higher concentrations of TC (100–500 mg/L) was detected in effluents of pharmaceutical manufacturing wastewaters and hospitals. So, researchers developed various techniques like electro-coagulation, adsorption, photo-peroxi-coagulation, and photocatalysis for the removal of TC from wastewater. Ahmadi et al. [101] developed MWCNT/$TiO_2$ nanocatalyst for the photocatalytic degradation of TC. After the degradation process, the chemical oxygen demand (COD) concentration of pharmaceutical wastewater reduced from 2,267 mg/L to 342 mg/L within 240 min. Antibiotic sulfamethoxazole and antiepileptic carbamazepine (CBZ) are persistent pharmaceuticals in wastewater. The average removal efficiency in water treatment plants is low. Hence, Awfa et al. [102] synthesized magnetic carbon nanotube-$TiO_2$ (MCNT-$TiO_2$) composites for eliminating sulfamethoxazole and CBZ under solar irradiation. Due to the extended visible light absorption and high surface area, MCNT-$TiO_2$ NC showed higher photocatalytic activity compared to pure $TiO_2$. Réti et al. [103] reported $TiO_2$/MWCNT composites synthesized by combined sol-gel/hydrothermal method for the decomposition of salicylic acid. Anatase, brookite, and rutile are the polymorphs of $TiO_2$. There is a strong relation between the ratio of polymorphs and photocatalytic activity. A small amount of rutile (11 wt.%) improved the efficiency of the composite photocatalyst.

The integration of $TiO_2$ into the CNT matrix boosts the conductivity of the $TiO_2$ and assists a faster electron transport by decreasing charge recombination and ensuring high photocatalytic activity. According to the studies of Abega et al. [104], $TiO_2$/CNT NC is an organic /inorganic nanohybrid used for the degradation of MO. 2,4-Dinitrophenol (DNP) is a toxic compound used in agriculture as a pesticide and in the petrochemical industry as a polymerization inhibitor. Wanga et al. [105] reported an MWCNTs/$TiO_2$ composite fabricated by sol-gel technique and tetrabutyl titanate used as a precursor for this synthesis. This MWCNTs/$TiO_2$ nanocatalyst was very effective for the degradation of DNP. Tan and coworkers [106] found that $TiO_2$/MWCNT presented efficient photocatalytic reduction of Cr(VI) to Cr(III), and the photocatalytic reduction is pH-dependent. Cr (VI) reduction increased with decreasing pH. The introduction of $TiO_2$ onto MWCNTs led to a remarkable increase in Cr(VI) elimination through adsorption and photocatalytic reduction under UV irradiation. Modified CNT-$TiO_2$ catalyst showed a higher surface area than pure $TiO_2$. Silva and Faria [107] described the fabrication of CNT-$TiO_2$ NCs via the acid-catalyzed sol-gel method for the degradation of 4-chlorophenol and 4-aminophenol.

Zinc oxide (ZnO) is a semiconductor with photocatalytic potential that has received a lot of attention in recent years due to its excellent electrical and optical properties, low cost, nontoxicity, and stability. At room temperature (RT), it is a semiconductor with an energy gap of 3.37 eV that can only absorb ultraviolet light and hence operate as a photocatalyst. Arsalani et al. [108] fabricated a ZnO/CNT NC for the degradation of malachite green under visible light irradiation. This ZnO/CNT NC was synthesized using a ball milling–hydrothermal technique. Figure 13.2 is a schematic representation of the mechanism of photocatalysis using CNT based catalyst. Prabhu et al. [109] developed an MWCNT/ZnO composite to degrade methylene blue. The activity of MWCNT/ZnO was significantly higher than pure ZnO nanoparticles. Similar to MWCNT, the SWCNTs are also used to improve the activity of ZnO. Sapkota et al. [110] prepared a ZnO–single-walled carbon nanotubes (ZnO-SWCNT) by one-pot-two-chemical recrystallization technique followed by thermal decomposition and this NC was used for the removal of methylene blue dye from wastewater. Zhu and coworkers [111] developed a ZnO/CNTs hierarchical microsphere composites for photocatalytic degradation. The activity of this NC was superior than pure ZnO due to the enhancement of light absorption, surface area and suppression of recombination rate of photo-induced charge carriers.

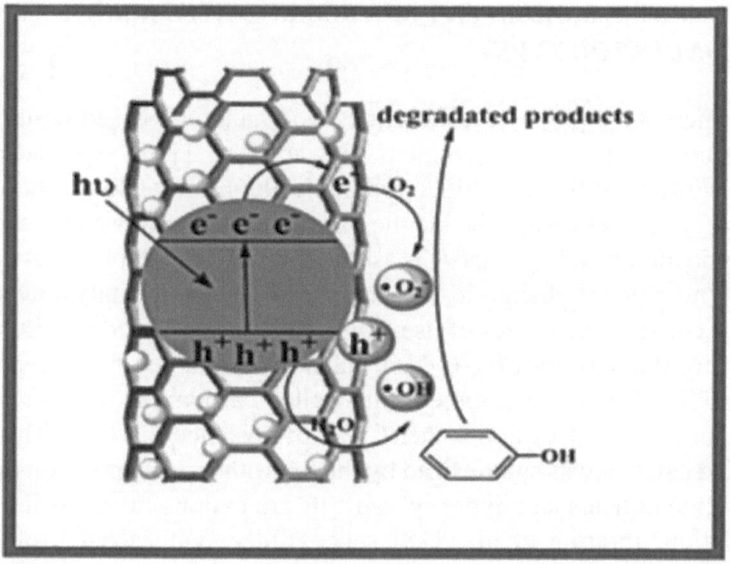


**FIGURE 13.2** Mechanism of photocatalysis using CNT-based binary nanocomposite.
*Source:* Reproduced with permission [112]; Copyright (2012), Elsevier.

For phenol degradation, Xiong et al. [112] created an MWCNT-supported nickel ferrite that is highly photocatalytic and magnetically recyclable. The separation of charged species will be improved by MWCNT. In this system, the photogenerated holes were the main active species. Chatterjee et al. [113] developed a polyaniline/single-walled carbon nanotube (PANI-SWCNT) composite for the removal of hazardous dyes from wastewater. These composites were prepared by polymerization of aniline in the presence of SWCNT with sulfosalicylic acid. Rose Bengal (RB) and MO dyes were degraded with 95.91% and 90.34% efficiency within 10 and 30 minutes. Jiang and coworkers [114] reported that one-pot microwave hydrothermally synthesized CNTs/$CeO_2$ NCs were used for the degradation of acid orange (AO7) dye. $SnO_2$-CNT is another type of carbon nanotube-based binary NC, and this composite was applicable for the effective degradation of methylene blue and MO dyes. $SnO_2$-CNT was developed by Kim and coworkers [115]. Various studies revealed that CNTs-based binary NCs are very effective for wastewater treatment.

### 13.3.2 METAL LOADED CNT NANOCOMPOSITES (METAL/CNT/SEMICONDUCTOR TYPE)

Researchers have prepared different carbon nanotube-based ternary NCs for effective wastewater treatment. Pawar et al. [116] reported a gold nanoparticle-modified graphitic carbon nitride/multi-walled carbon nanotube ($g\text{-}C_3N_4$/CNTs/Au). The synthesis of this photocatalyst proceeded by ultrasonication at RT. $g\text{-}C_3N_4$/CNTs/Au is a very stable photocatalyst, so there is no evident change in the activity of the photocatalyst even after the four consecutive cycles of use. RhB was used as a model pollutant for evaluating the activity of $g\text{-}C_3N_4$/CNTs/Au. Mohamed [117] reported a gold-modified titania nanotube-multi-walled carbon nanotube composite for the oxidation of cyclohexane (CyH) in presence of $H_2O_2$. This gold-modified catalyst was synthesized by the hydrothermal deposition method. Superoxide radicals and hydroxyl radicals are major active species of this degradation. Ibrahim et al. [118] successfully synthesized a platinum-loaded titanium nanotube-multi-walled carbon nanotube hybrid (Pt/TNT-MWCT) fabricated via hydrothermal-deposition methods. The catalytic activity of this nanohybrid was evaluated using methylene blue dye as a model pollutant and compared the efficiency of Pt/TNT-MWCT with Pt/graphene and Pt/TNT photocatalysts. The platinum-loaded titanium nanotube–multi-walled carbon nanotube hybrid showed better catalytic activity due to the formation of the Ti-O-C bond, which promoted electron transportation and charge separation. Zikalalaa et al. [119] developed a nitrogen-doped titania-carbon nanotube nanohybrids for the removal of reactive red 120 (RR120) dye. In this photocatalytic process, 97.4% decolorization and 84% TOC removal were achieved. Different metals were used for improving the activity. Magnesium [120], cadmium [121], cerium [122], strontium [123], yttrium [124], and copper [125] are generally used doping metals for enhancing the catalytic activity of CNTs-based photocatalysts.

### 13.3.3 CNT/SEMICONDUCTOR I/SEMICONDUCTOR II TYPE NANOCOMPOSITES (NC)

CNTs show thermal stability, excellent electron conducting ability, well-defined hollow interior, and high chemical stability. Due to these characteristics, CNTs have broad applications in photocatalysis. Lv et al. [126] developed a ternary NC with the support of MWCNT. In this work, they prepared an excellent photocatalyst, multi-walled carbon nanotube supported CdS-diethylenetriamine (DETA) composite (MWCNT/CdS-DETA) by hydrothermal method. The optimal weight percentage of MWCNT was 6%. Due to the absorption of electrons by MWCNT, the photocorrosion of CdS-DETA was repressed efficiently. Figure 13.3 represents the role of CNTs in ternary systems. Khan and coworkers [127] reported a Z-scheme $Bi_2O_2CO_3$/CNTs/$ZnFe_2O_4$ photocatalyst for the removal of 2,4-dimethyl phenol (DMP) under visible light irradiation. Hydroxyl radicals were major reactive species for the degradation of DMP.

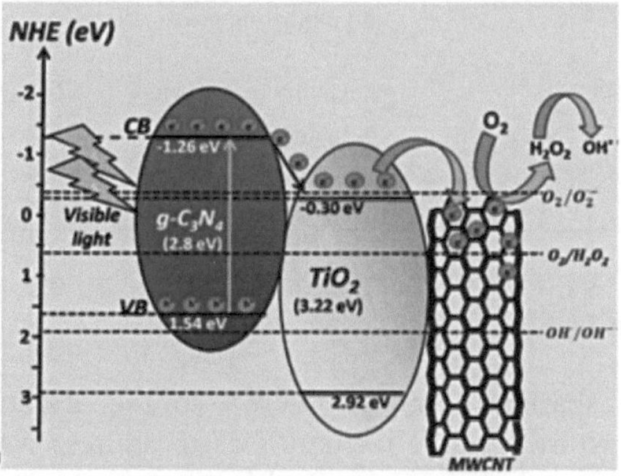


130

**FIGURE 13.3** Schematic representation of the role of CNTs in ternary systems.
*Source:* Reproduced with permission [130]; Copyright (2017), Elsevier.

Shafiee et al. [128] successfully synthesized MWCNT-based ternary NC for the elimination of rhodamine dye from aqueous solution, g-$C_3N_4$ and $Co_3O_4$ are the other two components of this photocatalyst. $Co_3O_4$/g-$C_3N_4$/MWCNT nanophotocatalyst was stabilized in resorcinol formaldehyde hydrogel for the enhancement of photocatalytic activity. Czech and Buda [129] studied the photocatalytic degradation of bisphenol A (BPA) and CBZ using multi-walled carbon nanotubes/$TiO_2$/$SiO_2$ nanocomposites (MWCNT-$TiO_2$-$SiO_2$). This degradation process followed pseudo-first-order kinetics. Chaudhary et al. [130] fabricated g-$C_3N_4$/$TiO_2$/CNT ternary

NC for wastewater treatment. The activity of this ternary NC is five times higher than pure graphitic carbon nitride. Table 13.1 shows different types of ternary NCs based on CNTs.

**TABLE 13.1** Various Types of Carbon Nanotube-based Ternary Nanocomposite

| SL. No. | CNT-based Ternary Systems | Water Pollutant | Light Source | References |
|---|---|---|---|---|
| 1. | $g\text{-}C_3N_4/CNT/Bi_2Fe_4O_9$ | Acid orange 7 (AO7) | UV-Vis light | [131] |
| 2. | $TiO_2/CNTs/PMMA$ | Methyl blue (MB), rhodamine B, sulforhodamine B, and methyl orange (MO) | UV | [132] |
| 3. | $Ag_3PO_4/Bi_2WO_6/MWCNTs$ | Norfloxacin (NOR) | Visible-light | [133] |
| 4. | $GNS/CNT/MnO_2$ | Methylene blue (MB) and malachite green (MG) | UV light | [134] |
| 5. | BiSI/BiOI/CNT | Malachite green (MG) | Visible light | [135] |
| 6. | $Bi_2WO_6/CNT/TiO_2$ | Cephalexin | UV-Vis light | [136] |
| 7. | $Bi_2MoO_6/CNTs/g\text{-}C_3N_4$ | 2,4-Dibromophenol (2,4-DBP) | Visible light | [50] |

## 13.4 CONCLUSION

CNTs have sparked a lot of interest in the scientific community. CNTs can inspire creative technologies to address water scarcity and pollution issues due to their tunable electrical, physical, chemical, and structural capabilities. CNTs serve as a promising adsorbent and catalyst in the water treatment process. Due to the larger adsorption capacity, shorter equilibrium time, superior adsorption selectivity, and easier regeneration, CNTs are very effective adsorbents than other carbonaceous materials like AC. CNTs can easily adsorb a wide range of contaminants in water. Because of their strength, adsorption capacity, and high electrical conductivity, CNTs can be used as catalytic support. Many studies have shown that adding CNTs to semiconducting photocatalysts can significantly increase their activity. The addition of CNT in conventional photocatalysts will aid in improving charge separation and minimize recombination rates that will enhance the photocatalytic activity. Carbon nanotube-based adsorption and photocatalysis are simple, cost-effective, and efficient techniques for wastewater treatment. Yet more study is needed to develop functionalized and purified CNTs and composites based on them for usage in commercial-scale wastewater treatment at reasonable prices and with negligible impact on the environment.

## KEYWORDS

- **adsorption**
- **carbon nanotubes**
- **nanocomposites**
- **photocatalysis**
- **semiconductors**
- **water treatment**

## REFERENCES

1. Reddy, D. H. K., & Lee, S. M., (2012). Water pollution and treatment technologies. *J. Environ. Anal. Toxicol., 2,* 5.
2. Abualnaja, K. M., Alprol, A. E., Ashour, M., & Mansour, A. T., (2021). Influencing multi-walled carbon nanotubes for the removal of ismate violet 2R Dye from wastewater: Isotherm, kinetics, and thermodynamic studies. *Appl. Sci., 11*, 4786.
3. Shawky, H. A., El-Aassar, A. H. M., & Abo-Zeid, D. E., (2011). Chitosan/carbon nanotube composite beads: Preparation, characterization, and cost evaluation for mercury removal from wastewater of some industrial cities in Egypt. *J. Appl. Polym. Sci.,* 125.
4. Anku, W. W., Agorku, E. S., Oppong, S. O. B., & Karikari, A. Y., (2020). MWCNTs attached neodymium doped-ZnO photocatalysts for efficient removal of dyes from wastewater. *SN Appl. Sci., 2*, 5.
5. Gatoo, M. A., Naseem, S., Arfat, M. Y., Mahmood, D. A., Qasim, K., & Zubair, S., (2014). Physicochemical properties of nanomaterials: Implication in associated toxic manifestations. *Biomed Res. Int., 2014*, 1–8.
6. Anjum, M., Miandad, R., Waqas, M., Gehany, F., & Barakat, M. A., (2019). Remediation of wastewater using various nanomaterials. *Arab. J. Chem., 12*, 4897–4919.

7. Camilli, L., Pisani, C., Gautron, E., Scarselli, M., Castrucci, P., D'Orazio, F., Passacantando, M., et al., (2014). A three-dimensional carbon nanotube network for water treatment. *Nanotechnology, 25*, 065701.
8. Das, R., Ali, M. E., Hamid, S. B. A., Ramakrishna, S., & Chowdhury, Z. Z., (2014). Carbon nanotube membranes for water purification: A bright future in water desalination. *Desalination, 336*, 97–109.
9. Crini, G., Lichtfouse, E., Wilson, L. D., & Morin-Crini, N., (2018). Adsorption-oriented processes using conventional and non-conventional adsorbents for wastewater treatment. *Green Adsorbents for Pollutant Removal*, 23–71.
10. Grassi, M., Kaykioglu, G., Belgiorno, V., & Lofrano, G., (2012). Removal of emerging contaminants from water and wastewater by adsorption process. *Emerging Compounds Removal from Wastewater*, 15–37.
11. Sharma, V. K., McDonald, T. J., Kim, H., & Garg, V. K., (2015). Magnetic graphene–carbon nanotube iron nanocomposites as adsorbents and antibacterial agents for water purification. *Adv. Colloid Interface Sci., 225*, 229–240.
12. Jia, K., Ji, Y., He, X., Xie, J., Wang, P., & Liu, X., (2022). One-step fabrication of dual functional $Tb^{3+}$ coordinated polymeric micro/nano-structures for Cr(VI) adsorption and detection. *J. Hazard. Mater., 423*, 127166.
13. Ali, I., & Gupta, V. K., (2006). Advances in water treatment by adsorption technology. *Nat. Protoc., 1*, 2661–2667.
14. Yin, Z., Cui, C., Chen, H., Duoni, X. Y., & Qian, W., (2019). The application of carbon nanotube/graphene-based nanomaterials in wastewater treatment. *Small, 16*, 1902301.
15. Xu, Y., Liu, J., Xie, M., Jing, L., Xu, H., She, X., Li, H., & Xie, J., (2019). Construction of novel CNT/$LaVO_4$ nanostructures for efficient antibiotic photodegradation. *Chem. Eng. J., 357*, 487–497.
16. Gopinath, K. P., Madhav, N. V., Krishnan, A., Malolan, R., & Rangarajan, G., (2020). Present applications of titanium dioxide for the photocatalytic removal of pollutants from water: A review. *J. Environ. Manage., 270,* 110906.
17. Ren, G., Han, H., Wang, Y., Liu, S., Zhao, J., Meng, X., & Li, Z., (2021). Recent advances of photocatalytic application in water treatment: A review. *Nanomaterials, 11*, 1804.
18. Herrmann, J. M., (1999). Water treatment by heterogeneous photocatalysis. *Environmental Catalysis,* 171–194.
19. Shannon, M. A., Bohn, P. W., Elimelech, M., Georgiadis, J. G., Mariñas, B. J., & Mayes, A. M., (2008). Science and technology for water purification in the coming decades. *Nature, 452*, 301–310.
20. Schwarzenbach, R. P., Escher, B. I., Fenner, K., Hofstetter, T. B., Johnson, C. A., Von, G. U., & Wehrli, B., (2006). The challenge of micropollutants in aquatic systems. *Science, 313*, 1072–1077.
21. Wen, Y., Zhang, L., Liu, J., Wen, X., Chen, X., Ma, J., Tang, T., & Mijowska, E., (2019). Hierarchical porous carbon sheets derived on a MgO template for high-performance supercapacitor applications. *Nanotechnology, 30*, 295703.
22. Bhatnagar, A., & Minocha, A. K., (2006). Conventional and non-conventional adsorbents for removal of pollutants from water: A review. *Indian J. Chem. Technol., 13,* 203–217.

23. Vo, M., & Papavassiliou, D., (2016). Effect of sodium dodecyl sulfate adsorption on the behavior of water inside single walled carbon nanotubes with dissipative particle dynamics simulation. *Molecules, 21*, 500.
24. Schnorr, J. M., & Swager, T. M., (2011). Emerging applications of carbon nanotubes. *Chem. Mater, 23*, 645–657.
25. Yang, K., Wang, X., Zhu, L., & Xing, B., (2006). Competitive sorption of pyrene, phenanthrene, and naphthalene on multi-walled carbon nanotubes. *Environ. Sci. Technol., 40*, 5804–5810.
26. Mauter, M. S., & Elimelech, M., (2008). Environmental applications of carbon-based nanomaterials. *Environ. Sci. Technol., 42*, 5843–5859.
27. Yang, K., & Xing, B., (2010). Adsorption of organic compounds by carbon nanomaterials in aqueous phase: Polanyi theory and its application. *Chem. Rev., 110*, 5989–6008.
28. Yu, J. G., Zhao, X. H., Yang, H., Chen, X. H., Yang, Q., Yu, L. Y., Jiang, J. H., & Chen, X. Q., (2014). Aqueous adsorption and removal of organic contaminants by carbon nanotubes. *Sci. Total Environ., 482, 483*, 241–251.
29. Lin, D., & Xing, B., (2008). Adsorption of phenolic compounds by carbon nanotubes: Role of aromaticity and substitution of hydroxyl groups. *Environ. Sci. Technol., 42*, 7254–7259.
30. Ji, L., Chen, W., Duan, L., & Zhu, D., (2009). Mechanisms for strong adsorption of tetracycline to carbon nanotubes: A comparative study using activated carbon and graphite as adsorbents. *Environ. Sci. Technol., 43*, 2322–2327.
31. Jiang, L., Liu, T., He, H., Pham-Huy, L. A., Li, L., Pham-Huy, C., & Xiao, D., (2012). Adsorption behavior of pazufloxacin mesilate on amino-functionalized carbon nanotubes. *J. Nanosci. Nanotechnol., 12*, 7271–7279.
32. Apul, O. G., Shao, T., Zhang, S., & Karanfil, T., (2011). Impact of carbon nanotube morphology on phenanthrene adsorption. *Environ. Toxicol. Chem., 31*, 73–78.
33. Hou, L., Zhu, D., Wang, X., Wang, L., Zhang, C., & Chen, W., (2013). Adsorption of phenanthrene, 2-naphthol, and 1-naphthylamine to colloidal oxidized multi-walled carbon nanotubes: Effects of humic acid and surfactant modification. *Environ. Toxicol. Chem., 32*, 493–500.
34. Yang, K., Jing, Q., Wu, W., Zhu, L., & Xing, B., (2009). Adsorption and conformation of a cationic surfactant on single-walled carbon nanotubes and their influence on naphthalene sorption. *Environ. Sci. Technol., 44*, 681–687.
35. Gotovac, S., Hattori, Y., Noguchi, D., Miyamoto, J., Kanamaru, M., Utsum, S., Kanoh, H., & Kanwko, K., (2006). Phenanthrene adsorption from solution on single wall carbon nanotubes. *J. Phys. Chem. B, 110*, 16219–16224.
36. Zhao, G., Jiang, L., He, Y., Li, J., Dong, H., Wang, X., & Hu, W., (2011). Sulfonated graphene for persistent aromatic pollutant management. *Adv. Mater., 23*, 3959–3963.
37. Chen, W., Duan, L., & Zhu, D., (2007). Adsorption of polar and nonpolar organic chemicals to carbon nanotubes. *Environ. Sci. Technol., 41*, 8295–8300.
38. Pan, B., Lin, D., Mashayekhi, H., & Xing, B., (2008). Adsorption and hysteresis of bisphenol a and 17α-ethinyl estradiol on carbon nanomaterials. *Environ. Sci. Technol., 43*, 548–548.

39. Xu, L., Li, J., & Zhang, M., (2015). Adsorption characteristics of a novel carbon-nanotube-based composite adsorbent toward organic pollutants. *Ind. Eng. Chem. Res., 54*, 2379–2384.
40. Yao, Y., Xu, F., Chen, M., Xu, Z., & Zhu, Z., (2010). Adsorption behavior of methylene blue on carbon nanotubes. *Bioresour. Technol., 101*, 3040–3046.
41. Yunjin, Y., Feifei, X., Ming, C., Zhongxiao, X., & Zhiwen, Z., (2010). Adsorption of cationic methyl violet and methylene blue dyes onto carbon nanotubes. In: *2010 IEEE 5th International Conference on Nano/Micro Engineered and Molecular Systems.*
42. Shahryari, Z., Goharrizi, A. S., & Azadi, M., (2010). Experimental study of methylene blue adsorption from aqueous solutions onto carbon nanotubes. *Int. J. Water Res., 2*, 016–028.
43. Wang, S., Ng, C. W., Wang, W., Li, Q., & Hao, Z., (2012). Synergistic and competitive adsorption of organic dyes on multi-walled carbon nanotubes. *Chem. Eng. J., 197*, 34–40.
44. Zare, K., Sadegh, H., Shahryari-ghoshekandi, R., Maazinejad, B., Ali, V., Tyagi, I., Agarwal, S., & Gupta, V. K., (2015). Enhanced removal of toxic congo red dye using multi walled carbon nanotubes: Kinetic, equilibrium studies and its comparison with other adsorbents. *J. Mol. Liq., 212*, 266–271.
45. Yao, Y., Bing, H., Feifei, X., & Xiaofeng, C., (2011). Equilibrium and kinetic studies of methyl orange adsorption on multi-walled carbon nanotubes. *Chem. Eng. J., 170*, 82–89.
46. Vuono, D., Catizzone, E., Aloise, A., Policicchio, A., Agostino, R. G., Migliori, M., & Giordano, G., (2017). Study of adsorption behavior of multi-walled carbon nanotubes towards dyes applied in textile applications. *Adv. Sci. Lett., 23*, 5851–5854.
47. Alkaim, A. F., Sadik, Z., Mahdi, D. K., Alshrefi, S. M., Al-Sammarraie, A. M., Alamgir, F. M., Singh, P. M., & Aljeboree, A. M., (2015). Preparation, structure and adsorption properties of synthesized multiwall carbon nanotubes for highly effective removal of maxilon blue dye. *Korean J. Chem. Eng., 32*, 2456–2462.
48. Kim, H., Hwang, Y. S., & Sharma, V. K., (2014). Adsorption of antibiotics and iopromide onto single-walled and multi-walled carbon nanotubes. *Chem. Eng. J., 255*, 23–27.
49. Machado, F. M., Bergmann, C. P., Lima, E. C., Royer, B., De Souza, F. E., Jauris, I. M., Calvete, T., & Fagan, S. B., (2012). Adsorption of reactive blue 4 dye from water solutions by carbon nanotubes: Experiment and theory. *Phys. Chem., 14*, 11139.
50. Ma, D., Wu, J., Gao, M., Xin, Y., & Chai, C., (2017). Enhanced debromination and degradation of 2,4-dibromophenol by an Z-scheme $Bi_2MoO_6$/CNTs/g-$C_3N_4$ visible light photocatalyst. *Chem. Eng. J., 316*, 461–470.
51. Machado, F. M., Bergmann, C. P., Fernandes, T. H. M., Lima, E. C., Royer, B., Calvete, T., & Fagan, S. B., (2011). Adsorption of reactive red M-2BE dye from water solutions by multi-walled carbon nanotubes and activated carbon. *J. Hazard. Mater., 192*, 1122–1131.
52. Prola, L. D. T., Machado, F. M., Bergmann, C. P., De Souza, F. E., Gally, C. R., Lima, E. C., Adebayo, M. A., et al., (2013). "Adsorption of direct blue 53 dye from aqueous solutions by multi-walled carbon nanotubes and activated carbon." *Journal of Environmental Management, 130*, 166–175.

53. Ferreira, G. M. D., Ferreira, G. M. D., Hespanhol, M. C., De Paula, R. J., Dos, S. P. A. C., Gurgel, L. V. A., & Da Silva, L. H. M., (2017). Adsorption of red azo dyes on multi-walled carbon nanotubes and activated carbon: A thermodynamic study. *Colloids Surf. A Physicochem. Eng. Asp., 529*, 531–540.
54. Zare, K., Gupta, V. K., Moradi, O., Makhlouf, A. S. H., Sillanpää, M., Nadagouda, M. N., Sadegh, H., et al., (2015). A comparative study on the basis of adsorption capacity between CNTs and activated carbon as adsorbents for removal of noxious synthetic dyes: A review. *J. Nanostructure Chem., 5*, 227–236.
55. Wang, S., Ng, C. W., Wang, W., Li, Q., & Li, L., (2012). A comparative study on the adsorption of acid and reactive dyes on multiwall carbon nanotubes in single and binary dye systems. *J. Chem. Eng. Data, 57*, 1563–1569.
56. Carabineiro, S. A. C., Thavorn-amornsri, T., Pereira, M. F. R., Serp, P., & Figueiredo, J. L., (2012). Comparison between activated carbon, carbon xerogel and carbon nanotubes for the adsorption of the antibiotic ciprofloxacin. *Catal. Today, 186*, 29–34.
57. Zhang, Song, X., Liu, X., Yang, L., Pan, F., & Lv, J., (2011). Studies on the removal of tetracycline by multi-walled carbon nanotubes. *Chem. Eng. J., 178*, 26–33.
58. Zhang, H., & Hu, X., (2015). Adsorption of ceftazidime from aqueous solution by multi-walled carbon nanotubes. *Pol. J. Environ. Stud., 24*, 2285–2293.
59. Mohammadi, A., Kazemipour, M., Ranjbar, H., Walker, R. B., & Ansari, M., (2014). Amoxicillin removal from aqueous media using multi-walled carbon nanotubes. *Fuller. Nanotube. Carbon Nanostructures, 23*, 165–169.
60. Azarpira, H., Mahdavi, Y., Khaleghi, O., & Balarak, D., (2016). Thermodynamic studies on the removal of metronidazole antibiotic by multi-walled carbon nanotubes. *Der Pharm. Lett., 8*, 107–113.
61. Tian, Y., Gao, B., Morales, V. L., Chen, H., Wang, Y., & Li, H., (2013). Removal of sulfamethoxazole and sulfapyridine by carbon nanotubes in fixed-bed columns. *Chemosphere, 90*, 2597–2605.
62. Yu, F., Sun, S., Han, S., Zheng, J., & Ma, J., (2016). Adsorption removal of ciprofloxacin by multi-walled carbon nanotubes with different oxygen contents from aqueous solutions. *Chem. Eng. J., 285*, 588–595.
63. Ghaedi, M., Khajehsharifi, H., Yadkuri, A. H., Roosta, M., & Asghari, A., (2012). Oxidized multi-walled carbon nanotubes as efficient adsorbent for bromothymol blue. *Toxicol. Environ. Chem., 94*, 873–883.
64. Sobhanardakani, S., Zandipak, R., & Sahraei, R., (2013). Removal of Janus green dye from aqueous solutions using oxidized multi-walled carbon nanotubes. *Toxicol. Environ. Chem., 95*, 909–918.
65. Veličković, Z. S., Marinković, A. D., Bajić, Z. J., Marković, J. M., Perić-Grujić, A. A., Uskokovic, P. S., & Ristic, M. D., (2013). Oxidized and ethylenediamine-functionalized multi-walled carbon nanotubes for the separation of low concentration arsenate from water. *Sep. Sci. Technol., 48*, 2047–2058.
66. Al Hamouz, O. C. S., Adelabu, I. O., & Saleh, T. A., (2017). Novel cross-linked melamine based polyamine/CNT composites for lead ions removal. *J. Environ. Manage., 192*, 163–170.
67. Albakri, M. A., Abdelnaby, M. M., Saleh, T. A., & Al Hamouz, O. C. S., (2018). New series of benzene-1,3,5-triamine based cross-linked polyamines and polyamine/CNT composites for lead ion removal from aqueous solutions. *Chem. Eng. J., 333*, 76–84.

68. Alijani, H., & Shariatinia, Z., (2018). Synthesis of high growth rate SWCNTs and their magnetite cobalt sulfide nanohybrid as super-adsorbent for mercury removal. *Chem. Eng. Res. Des., 129*, 132–149.
69. Bandaru, N. M., Reta, N., Dalal, H., Ellis, A. V., Shapter, J., & Voelcker, N. H., (2013). Enhanced adsorption of mercury ions on thiol derivatized single wall carbon nanotubes. *J. Hazard. Mater., 261*, 534–541.
70. Madrakian, T., Afkhami, A., Ahmadi, M., & Bagheri, H., (2011). Removal of some cationic dyes from aqueous solutions using magnetic-modified multi-walled carbon nanotubes. *J. Hazard. Mater., 196*, 109–114.
71. Gong, J. L., Wang, B., Zeng, G. M., Yang, C. P., Niu, C. G., Niu, Q. Y., Zhou, W. J., & Liang, Y., (2009). Removal of cationic dyes from aqueous solution using magnetic multi-wall carbon nanotube nanocomposite as adsorbent. *J. Hazard. Mater., 164*, 1517–1522.
72. Qu, S., Huang, F., Yu, S., Chen, G., & Kong, J., (2008). Magnetic removal of dyes from aqueous solution using multi-walled carbon nanotubes filled with $Fe_2O_3$ particles. *J. Hazard. Mater., 160*, 643–647.
73. Ai, L., Zhang, C., Liao, F., Wang, Y., Li, M., Meng, L., & Jiang, J., (2011). Removal of methylene blue from aqueous solution with magnetite loaded multi-wall carbon nanotube: Kinetic, isotherm and mechanism analysis. *J. Hazard. Mater., 198*, 282–290.
74. Gabal, M. A., Al-Zahrani, N. G., & Al Angari, Y. M., (2018). Beading of Co1-x Znx $Fe_2O_4$ nanoparticles on MWCNTs. Synthesis characterization, effect of Zn-substitution and dye removal capability. *Mater. Chem. Phys., 213*, 220–230.
75. Farghali, A. A., Bahgat, M., El Rouby, W. M. A., & Khedr, M. H., (2012). Decoration of MWCNTs with $CoFe_2O_4$ nanoparticles for methylene blue dye adsorption. *J Solution Chem., 41*, 2209–2225.
76. Fazelirad, H., Ranjbar, M., Taher, M. A., & Sargazi, G., (2015). Preparation of magnetic multi-walled carbon nanotubes for an efficient adsorption and spectrophotometric determination of amoxicillin. *J. Ind. Eng. Chem., 21*, 889–892.
77. Mahmoodian, H., Moradi, O., Shariatzadeha, B., Salehf, T. A., Tyagi, I., Maity, A., Asif, M., & Gupta, V. K., (2015). Enhanced removal of methyl orange from aqueous solutions by poly HEMA–chitosan-MWCNT nanocomposite. *J. Mol. Liq., 202*, 189–198.
78. Sui, K., Li, Y., Liu, R., Zhang, Y., Zhao, X., Liang, H., & Xia, Y., (2012). Biocomposite fiber of calcium alginate/multi-walled carbon nanotubes with enhanced adsorption properties for ionic dyes. *Carbohydr. Polym., 90*, 399–406.
79. Yu, H., & Fugetsu, B., (2010). A novel adsorbent obtained by inserting carbon nanotubes into cavities of diatomite and applications for organic dye elimination from contaminated water. *J. Hazard. Mater., 177*, 138–145.
80. Eskandarian, L., Arami, M., & Pajootan, E., (2014). Evaluation of adsorption characteristics of multi-walled carbon nanotubes modified by a poly(propylene imine) dendrimer in single and multiple dye solutions: Isotherms, kinetics, and thermodynamics. *J. Chem. Eng. Data., 59*, 444–454.
81. Mohammadi, A., & Veisi, P., (2018). High adsorption performance of β-cyclodextrin-functionalized multi-walled carbon nanotubes for the removal of organic dyes from water and industrial wastewater. *J. Environ. Chem. Eng., 6*, 4634–4643.

82. Mallakpour, S., & Rashidimoghadam, S., (2019). Poly(vinyl alcohol)/vitamin C-multi walled carbon nanotubes composites and their applications for removal of methylene blue: Advanced comparison between linear and nonlinear forms of adsorption isotherms and kinetics models. *Polymer, 160,* 115–125.
83. Hosseinzadeh, S., Hosseinzadeh, H., Pashaei, S., & Khodaparast, Z., (2018). Synthesis of magnetic functionalized MWCNT nanocomposite through surface RAFT co-polymerization of acrylic acid and N-isopropyl acrylamide for removal of cationic dyes from aqueous solutions. *Ecotoxicol. Environ. Saf., 161*, 34–44.
84. Chang, P. R., Zheng, P., Liu, B., Anderson, D. P., Yu, J., & Ma, X., (2011). Characterization of magnetic soluble starch-functionalized carbon nanotubes and its application for the adsorption of the dyes. *J. Hazard. Mater., 186*, 2144–2150.
85. Zhu, H. Y., Jiang, R., Xiao, L., & Zeng, G. M., (2010). Preparation, characterization, adsorption kinetics and thermodynamics of novel magnetic chitosan enwrapping nanosized $\gamma$-$Fe_2O_3$ and multi-walled carbon nanotubes with enhanced adsorption properties for methyl orange. *Bioresour. Technol., 101*, 5063–5069.
86. Abbasi, M., (2017). Synthesis and characterization of magnetic nanocomposite of chitosan/$SiO_2$/carbon nanotubes and its application for dyes removal. *J. Clean. Prod., 145*, 105–113.
87. Duman, O., Tunç, S., Polat, T. G., & Bozoğlan, B. K., (2016). Synthesis of magnetic oxidized multi-walled carbon nanotube-κ-carrageenan-$Fe_3O_4$ nanocomposite adsorbent and its application in cationic methylene blue dye adsorption. *Carbohydr. Polym., 147*, 79–88.
88. Yan, L., Chang, P. R., Zheng, P., & Ma, X., (2012). Characterization of magnetic guar gum-grafted carbon nanotubes and the adsorption of the dyes. *Carbohydr. Polym., 87*, 1919–1924.
89. Saber-Samandari, S., Saber-Samandari, S., Joneidi-Yekta, H., & Mohseni, M., (2017). Adsorption of anionic and cationic dyes from aqueous solution using gelatin-based magnetic nanocomposite beads comprising carboxylic acid functionalized carbon nanotube. *Chem. Eng. J. 308*, 1133–1144.
90. Boukhalfa, N., Boutahala, M., Djebri, N., & Idris, A., (2019). Kinetics, thermodynamics, equilibrium isotherms, and reusability studies of cationic dye adsorption by magnetic alginate/oxidized multi-walled carbon nanotubes composites. *Int. J. Biol. Macromol., 123*, 539–548.
91. Tang, Y., Zhang, S., Su, Y., Wu, D., Zhao, Y., & Xie, B., (2021). Removal of microplastics from aqueous solutions by magnetic carbon nanotubes. *Chem. Eng. J., 406*, 126804.
92. Gao, H., Zhao, S., Cheng, X., Wang, X., & Zheng, L., (2013). Removal of anionic azo dyes from aqueous solution using magnetic polymer multi-wall carbon nanotube nanocomposite as adsorbent. *Chem. Eng. J., 223,* 84–90.
93. Hu, D., Wan, X., Li, X., Liu, J., & Zhou, C., (2017). Synthesis of water-dispersible poly- l -lysine-functionalized magnetic $Fe_3O_4$-(GO-MWCNTs) nanocomposite hybrid with a large surface area for high-efficiency removal of tartrazine and Pb(II). *Int. J. Biol. Macromol., 105*, 1611–1621.
94. Qureshi, T., Bakhshpour, M., Çetin, K., Topçu, A. A., & Denizli, A., (2020). *Wastewater Treatment, Photocatalysts in Advanced Oxidation Processes for Wastewater Treatment*, 33–64.

95. Chen, C. S., Liu, T. G., Lin, L. W., Xie, X. D., Chen, X. H., Liu, Q. C., Liang, B., et al., (2012). Multi-walled carbon nanotube-supported metal-doped ZnO nanoparticles and their photocatalytic property. *J. Nanopart. Res.,* 15.
96. Zhou, W., Wang, Y., Wang, M., Qian, B., Li, L., & Gao, B., (2021). The co-effect of ampicillin and multi-walled carbon nanotubes on activated sludge in sequencing batch reactors: Microbial status, microbial community structure and ARGs propagation. *Environmental Science: Nano, 8*, 204–216.
97. Payan, A., Fattahi, M., Jorfi, S., Roozbehani, B., & Payan, S., (2018). Synthesis and characterization of titanate nanotube/single-walled carbon nanotube (TNT/SWCNT) porous nanocomposite and its photocatalytic activity on 4-chlorophenol degradation under UV and solar irradiation. *Appl. Surf. Sci., 434*, 336–350.
98. Murgolo, S., Petronella, F., Ciannarella, R., Comparelli, R., Agostiano, A., Curri, M. L., & Mascolo, G., (2015). UV and solar-based photocatalytic degradation of organic pollutants by nano-sized $TiO_2$ grown on carbon nanotubes. *Catal. Today, 240*, 114–124.
99. Mohammadi, M., & Sabbaghi, S., (2014). Photocatalytic degradation of 2,4-DCP wastewater using MWCNT/$TiO_2$ nanocomposite activated by UV and solar light. *Environ. Nanotechnol. Monit. Manag., 1, 2*, 24–29.
100. Sampaio, M. J., Marques, R. R. N., Tavares, P. B., Faria, J. L., Silva, A. M. T., & Silva, C. G., (2013). Tailoring the properties of immobilized titanium dioxide/carbon nanotube composites for photocatalytic water treatment. *J. Environ. Chem. Eng., 1*, 945–953.
101. Ahmadi, M., Ramezani, M. H., Jaafarzadeh, N., Mostoufi, A., Saeedi, R., Barzegar, G., & Jorfi, S., (2017). Enhanced photocatalytic degradation of tetracycline and real pharmaceutical wastewater using MWCNT/$TiO_2$ nanocomposite. *J. Environ. Manage., 186*, 55–63.
102. Awfa, D., Ateia, M., Fujii, M., & Yoshimura, C., (2019). Novel magnetic carbon nanotube-$TiO_2$ composites for solar light photocatalytic degradation of pharmaceuticals in the presence of natural organic matter. *J. Water Process. Eng., 31*, 100836.
103. Réti, B., Major, Z., Szarka, D., Boldizsár, T., Horváth, E., Magrez, A., Forró, L., Dombi, A., & Hernádi, K., (2016). Influence of $TiO_2$ phase composition on the photocatalytic activity of $TiO_2$/MWCNT composites prepared by combined sol-gel/hydrothermal method. *J. Mol. Catal. A Chem., 414*, 140–147.
104. Abega, A. V., Ngomo, H. M., Nongwe, I., Mukaya, H. E., Kouoh, S. P. M. A., & Yangkou, M. X., (2019). Easy and convenient synthesis of CNT/$TiO_2$ nanohybrid by in-surface oxidation of $Ti^{3+}$ ions and application in the photocatalytic degradation of organic contaminants in water. *Synth. Met., 251*, 1–14.
105. Wang, H., Wang, H. L., Jiang, W. F., & Li, Z. Q., (2009). Photocatalytic degradation of 2,4-dinitrophenol (DNP) by multi-walled carbon nanotubes (MWCNTs)/$TiO_2$ composite in aqueous solution under solar irradiation. *Water Res., 43*, 204–210.
106. Tan, X., Fang, M., & Wang, X., (2008). Preparation of $TiO_2$/multi-walled carbon nanotube composites and their applications in photocatalytic reduction of Cr(VI) study. *J. Nanosci. Nanotechnol., 8*, 5624–5631.

107. Silva, C. G., & Faria, J. L., (2010). Photocatalytic oxidation of phenolic compounds by using a carbon nanotube-titanium dioxide composite catalyst. *ChemSusChem., 3*, 609–618.
108. Arsalani, N., Bazazi, S., Abuali, M., & Jodeyri, S., (2020). A new method for preparing ZnO/CNT nanocomposites with enhanced photocatalytic degradation of malachite green under visible light. *J. Photochem. Photobiol., 389*, 112207.
109. Prabhu, Y. T., Sreedhar, B., & Pal, U., (2019). Achieving enhanced photocatalytic activity of ZnO supported on MWCNTs towards degradation of pollutants under visible light. *Mater. Today: Proc., 8*, 419–426.
110. Sapkota, K. P., Lee, I., Hanif, M. A., Islam, M. A., & Hahn, J. R., (2019). Solar-light-driven efficient ZnO–single-walled carbon nanotube photocatalyst for the degradation of a persistent water pollutant organic dye. *Catalysts, 9*, 498.
111. Zhu, G., Wang, H., Yang, G., Chen, L., Guo, P., & Zhang, L., (2015). A facile synthesis of ZnO/CNT hierarchical microsphere composites with enhanced photocatalytic degradation of methylene blue. *RSC Adv., 5*, 72476–72481.
112. Xiong, P., Fu, Y., Wang, L., & Wang, X., (2012). Multi-walled carbon nanotubes supported nickel ferrite: A magnetically recyclable photocatalyst with high photocatalytic activity on degradation of phenols. *Chem. Eng. J., 195, 196*, 149–157.
113. Chatterjee, M. J., Ghosh, A., Mondal, A., & Banerjee, D., (2017). Polyaniline–single walled carbon nanotube composite – a photocatalyst to degrade rose Bengal and methyl orange dyes under visible-light illumination. *RSC Adv., 7*, 36403–36415.
114. Jiang, H., Ge, S., Zhang, Y., Liu, M., Zhang, J., Lin, J., Dong, M., Wang, J., & Guo, Z., (2020). One-pot microwave-hydrothermally synthesized carbon nanotube-cerium oxide nanocomposites for enhanced visible photodegradation of acid orange 7. *Phys. Chem. Chem. Phys., 22*, 23743–23753.
115. Kim, S. P., Choi, M. Y., & Choi, H. C., (2015). Characterization and photocatalytic performance of $SnO_2$–CNT nanocomposites. *Appl. Surf. Sci., 357*, 302–308.
116. Pawar, R. C., Kang, S., Ahn, S. H., & Lee, C. S., (2015). Gold nanoparticle modified graphitic carbon nitride/multi-walled carbon nanotube (g-$C_3N_4$/CNTs/Au) hybrid photocatalysts for effective water splitting and degradation. *RSC Adv., 5*, 24281–24292.
117. Mohamed, M. M., (2015). Gold loaded titanium dioxide–carbon nanotube composites as active photocatalysts for cyclohexane oxidation at ambient conditions. *RSC Adv., 5*, 46405–46414.
118. Ibrahim, M. M., Ahmed, S. A., Khairou, K. S., & Mokhtar, M., (2014). Carbon nanotube/titanium nanotube composites loaded platinum nanoparticles as high performance photocatalysts. *Appl. Catal. A: Gen., 475*, 90–97.
119. Zikalala, S. A., Kuvarega, A. T., Madhav, H., Mamba, B. B., Mhlanga, S. D., & Nxumalo, E. N., (2020). A comparison of the influence of synthesis methods on the photocatalytic activity of nitrogen doped titania-carbon nanotube nanohybrids. *Appl. Catal. A: Gen., 604*, 117776.
120. Ahmad, I., Akhtar, M. S., Ahmed, E., & Ahmad, M., (2020). Highly efficient visible light driven photocatalytic activity of graphene and CNTs based Mg-doped ZnO photocatalysts: A comparative study. *Sep. Purif. Technol., 245*, 116892.

121. Azqhandi, M. H. A., Vasheghani, F. B., Rajabi, F. H., & Keramati, M., (2017). Synthesis of Cd doped ZnO/CNT nanocomposite by using microwave method: Photocatalytic behavior, adsorption and kinetic study. *Results Phys., 7*, 1106–1114.
122. Elias, M., Amin, M. K., Firoz, S. H., Hossain, M. A., Akter, S., Hossain, M. A., Uddin, M. N., & Siddiquey, I. A., (2017). Microwave-assisted synthesis of Ce-doped ZnO/CNT composite with enhanced photocatalytic activity. *Ceram. Int., 43*, 84–91.
123. Ahmad, I., Shukrullah, S., Naz, M. Y., Rasheed, M. A., Ahmad, M., Ahmed, E., Akhtar, M. S., et al., (2021). Boosted hydrogen evolution activity from Sr doped ZnO/CNTs nanocomposite as visible light driven photocatalyst. *Int. J. Hydrog. Energy, 46*, 26711–26724.
124. Sharma, S. K., Gupta, R., Sharma, G., Vemula, K., Koirala, A. R., Kaushik, N. K., Choi, E. H., et al., (2021). Photocatalytic performance of yttrium-doped CNT-ZnO nanoflowers synthesized from hydrothermal method. *Mater. Today Chem., 20*, 100452.
125. Toloman, D., Popa, A., Stan, M., Stefan, M., Vlad, G., Ulinici, S., Baisan, G., et al., (2021). Visible-light-driven photocatalytic degradation of different organic pollutants using Cu doped ZnO-MWCNT nanocomposites. *J. Alloys Compd., 866*, 159010.
126. Lv, J., Li, D., Dai, K., Liang, C., Jiang, D., Lu, L., & Zhu, G., (2017). Multi-walled carbon nanotube supported CdS-DETA nanocomposite for efficient visible light photocatalysis. *Mater. Chem. Phys., 186*, 372–381.
127. Khan, A. A. P., Singh, P., Raizada, P., & Asiri, A. M., (2020). Synthesis of magnetically separable $Bi_2O_2CO_3$/carbon nanotube/$ZnFe_2O_4$ as Z-scheme heterojunction with enhanced photocatalytic activity for water purification. *J. Sol-Gel Sci. Technol., 95*, 408–422.
128. Shafiee, M. R. M., Parhizkar, J., & Radfar, S., (2019). Removal of rhodamine B by g-$C_3N_4$/$Co_3O_4$/MWCNT composite stabilized in hydrogel via the synergy of adsorption and photocatalysis under visible light. *J. Mater. Sci. Mater., 30*, 12475–12486.
129. Czech, B., & Buda, W., (2015). Photocatalytic treatment of pharmaceutical wastewater using new multiwall-carbon nanotubes/$TiO_2$/$SiO_2$ nanocomposites. *Environ. Res., 137*, 176–184.
130. Chaudhary, D., Vankar, V. D., & Khare, N., (2017). Noble metal-free g-$C_3N_4$/$TiO_2$ /CNT ternary nanocomposite with enhanced photocatalytic performance under visible-light irradiation via multi-step charge transfer process. *J. Sol. Energy, 158*, 132–139.
131. Di, L., Yang, H., Xian, T., & Chen, X., (2018). Construction of Z-scheme g-$C_3N_4$/CNT/$Bi_2Fe_4O_9$ composites with improved simulated-sunlight photocatalytic activity for the dye degradation. *Micromachines, 9*, 613.
132. Li, Y., Wang, Z., Zhao, H., & Yang, M., (2020). Composite of $TiO_2$ nanoparticles and carbon nanotubes loaded on poly(methyl methacrylate) nanofibers: Preparation and photocatalytic performance. *Synth. Met., 269*, 116529.
133. Zhu, Z., Wang, C., Liang, L., Yu, D., Sun, J., Zhang, L., Zhong, S., & Liu, B., (2020). Synthesis of novel ternary photocatalyst $Ag_3PO_4$/$Bi_2WO_6$/Multi-Walled carbon nanotubes and its enhanced visible-light photoactivity for photodegradation of norfloxacin. *J. Nanosci. Nanotechnol., 20*, 2247–2258.

134. Siddeswara, D. M. K., Venkatesh, T., Mahesh, K. R. V., Mylarappa, M., Anantharaju, K. S., Kumara, K. N. S., Raghavendra, N., & Shivakumar, M. S., (2017). One step synthesis of ternary composite of GNS/CNT/$MnO_2$ for the applications of electrochemical and photocatalytic studies. *Mater. Today: Proc., 4*, 11799–11805.
135. Bargozideh, S., Tasviri, M., & Ghabraei, M., (2020). Effect of carbon nanotubes loading on the photocatalytic activity of BiSI/BiOI as a novel photocatalyst. *Environ. Sci. Pollut. Res., 27*, 36754–36764.
136. Rabanimehr, F., Farhadian, M., Nazar, A. R. S., & Moghadam, M., (2021). Fabrication of Z-scheme $Bi_2WO_6$/CNT/$TiO_2$ heterostructure with enhanced cephalexin photodegradation: Optimization and reaction mechanism. *J. Mol. Liq., 339*, 116728.

# Index

## D

## H

## N

## P

## S

## U



## V



## W